T0326282

Compressibility, Turbulence and High Speed Flow

Compressibility, Turbulence and High Speed Flow

Second Edition

Thomas B. Gatski
Institut Pprime CNRS,
Université de Poitiers,
ISAE-ENSMA Poitiers, France

Center for Coastal Physical Oceanography and Ocean,
Earth and Atmospheric Sciences
Old Dominion University, Norfolk, VA, USA

Jean-Paul Bonnet
Institut Pprime CNRS,
Université de Poitiers,
ISAE-ENSMA Poitiers, France

AMSTERDAM • BOSTON • HEIDELBERG • LONDON
NEW YORK • OXFORD • PARIS • SAN DIEGO
SAN FRANCISCO • SINGAPORE • SYDNEY • TOKYO

Academic Press is an Imprint of Elsevier

Academic Press is an imprint of Elsevier
The Boulevard, Langford Lane, Kidlington, Oxford OX5 1GB, UK
Radarweg 29, PO Box 211, 1000 AE Amsterdam, The Netherlands

First edition 2008
Second edition 2013

British Library Cataloguing-in-Publication Data
A catalogue record for this book is available from the British Library

Library of Congress Cataloging-in-Publication Data
A catalog record for this book is available from the Library of Congress

ISBN: 978-0-12-397027-5

Printed and bound in UK

13 14 15 16 17   10 9 8 7 6 5 4 3 2 1

To Rosann and Megan for all their support
A Mary-Christine, Marie, Cécile et Léa

Contents

Preface to the Second Edition

We have been gratified with the response to our first edition, and are pleased to be given the opportunity to further enhance the content of the first edition through both some rewriting and correcting of the original text as well as the addition of some new material including a new chapter dealing with elements of compressible flow control.

Nevertheless, our original goal remains and that is to provide the engineer, teacher, and scientist, involved in experiments, computational and theoretical predictions, or simulations, with a resource where these varied aspects of the compressible and supersonic turbulent flow problem are discussed within a unified framework. This will hopefully contribute to a better overall understanding of the topic and a mutual understanding of the respective challenges faced by those involved in these diverse areas. The volume is not intended to provide the fundamental background to any of the underlying disciplines discussed, such as experimental techniques, numerical methods, or turbulence modeling and theory, but it is intended to provide the reader with some of the additional knowledge necessary to study compressible turbulence and compressible turbulent flows.

Williamsburg, Virginia, USA
Poitiers, France

Thomas B. Gatski
Jean-Paul Bonnet

July 2012

Preface to the First Edition

Our aim is to provide the engineer and scientist who are involved in experiments, computational and theoretical predictions, or simulations, with a resource where these varied aspects of the compressible and supersonic turbulent flow problems are discussed within a unified framework. This will hopefully contribute to a better overall understanding of the topic as well as a mutual understanding of the respective challenges faced by those involved in these different areas. Such a synergism of topical presentation can nurture more active collaborations between those involved in experiments, computations and theory. The volume is not intended, however, to provide the fundamental background to any of the underlying disciplines discussed, such as experimental techniques, numerical methods, or turbulence, it is intended to provide the reader with some of the additional knowledge necessary to study compressible turbulence and compressible turbulent flows. This would significantly lengthen the text and possibly detract from the main focus.

Though intended for the practicing engineer and scientist, this book includes fundamental derivations of conservation equations as well as kinematic and thermodynamic preliminaries necessary in the formulation of such balance equations. It also includes a description of a variety of experimental techniques currently used and the impact compressibility has on their use. These discussions serve to highlight the unique challenges of conducting research in the compressible regime. Where possible, the results from physical and computational experiments, and theoretical analysis are linked to provide a more cohesive overview of the field.

Although not always apparent, thorough analysis and/or solution of practical fluid flow problems require an understanding of the various mechanical forces and thermodynamic energy balances inherent in such systems, and are easily analyzed in dynamically simpler flows. This book provides a fundamental to practical mapping of concepts behind the formulations used in describing compressible fluid flow problems. This mapping is applied to both the instantaneous balance equations as well as to the ensemble mean equations extracted for the study of turbulent flows. Thus, the reader has in one volume, a complete and self-consistent guide for the measurement, analysis, and prediction of compressible, turbulent fluid flows.

The genesis for this book was started while TBG was with the NASA Langley Research Center where he worked on turbulent model development, flow prediction, and simulations. It became apparent as the book developed that its focus was too narrow and needed to be broadened to include physical experiments and more importantly the experimentalist's perspective. J.-P.B. then joined in this effort to produce a book equally appealing to both those involved in physical experiments and those focused on modeling and numerical predictions and simulations. We are fortunate to have been affiliated with supportive organizations that have provided the facilities to pursue our research interests and to have interacted with an excellent array of colleagues. Our thanks go to the support from the Laboratoire d'Étude Aérodynamiques, Centre d'Études Aérodynamiques et Thermiques, the Université de Poitiers, the ENSMA and the Centre National De La Recherche Scientifique (CNRS). TBG would also like to gratefully acknowledge the past support of the National Aeronautics and Space Administration at the Langley Research Center, and the current support of the Center for Coastal Physical Oceanography at Old Dominion University. We have also been fortunate to be associated with colleagues who both challenge and stimulate ideas. Our special thanks go to Chris Rumsey of NASA LaRC for his generous help with our requests, and also to Joe Morrison (NASA LaRC), Stephane Barre, Érwan Collin, Pierre Comte, Laurent Cordier, Joel Delville, and Remi Manceau.

Williamsburg, Virginia, USA   **Thomas B. Gatski**
Poitiers, France   **Jean-Paul Bonnet**

July 2008

Chapter 1

Kinematics, Thermodynamics and Fluid Transport Properties

Compressible fluid flows have long been a topic of study in the fluid dynamics community. Whether in engineering or geophysical flows, there is probably some mass density change in any physical flow. Many flow situations do exist, however, where such changes can be neglected and the flow considered incompressible. In naturally occurring atmospheric and oceanographic flows, that is geophysical flows, mass density changes are neglected in the mass conservation equation and the velocity field is taken as solenoidal, but are accounted for in the momentum and energy balances. In engineering flows, mass density changes are not neglected in the mass balance and, in addition, are kept in the momentum and energy balances. Underlying the discussion of the fluid dynamics of compressible flows is the need to invoke the concepts of thermodynamics and exploit the relations between such quantities as mass density, pressure, and temperature. Such relations, though strictly valid under mechanical and thermal equilibrium conditions, have been found to apply equally well in moving fluids apparently far from the equilibrium state.

Turbulence and turbulent fluid flows have been the focus of research for over 100 years, but complete understanding still eludes engineers and scientists. In the earlier part of the last century, many theoretical and experimental studies were done involving incompressible turbulence and flows; however, the understanding of even the incompressible problem was incomplete, and the analyses constrained by the techniques available at the time. The study of compressible turbulence and compressible turbulent flows thus merge together two topical areas of fluid dynamics that have been thoroughly investigated but yet remain elusive to complete prediction and control.

Since an important focus of the material will be on engineering aerodynamic flows, it is useful to provide some historical background on the research

Compressibility, Turbulence and High Speed Flow. http://dx.doi.org/10.1016/B978-0-12-397027-5.00001-0

trends to date. In the 1930s research groups became focused on the study of compressible, high speed boundary layer flows. This appears to have been motivated by the extensive experimentation with rockets in the late 1920s and 1930s in Europe, Russia, and the United States. While such experimentation had been going for about a decade prior to this, much of the fundamental theoretical work was primarily interested in incompressible flows. Prandtl's lecture notes of the time (see Prandtl & Tietjens, 1934) include a brief section on compressibility, but predominantly focused on minimizing its effects in practical instances. By the late 1930s and throughout the 1940s extensive work was done on compressible turbulent flows (Frankl & Voishel, 1943; Frössel, 1938). In the 1950s, research in compressible flows was an active topic from both an experimental and theoretical standpoint, and the fundamental work of this period is of relevance in today's analysis of turbulent flows.

With the development of high speed aircraft and the development of manned and unmanned space programs in the second half of the last century, compressible, turbulent flows became a topic for study with wide ranging international significance. An illustrative example of the complexity of the flow fields around manned space vehicles is shown in Figure 1.1. The shadowgraph figure of an Apollo-like capsule in a Mach 2.2 flow, and Crew Exploration Vehicle-like (CEV-like) capsule in a Mach 1.2 contains the essential ingredients of these flows including: presence of multiple shock waves, separated zones, turbulent boundary layers and wakes, and large scale structures.

Although aeronautics and space may now be the primary areas where compressible, turbulent flows are relevant, there exists a diverse range of several

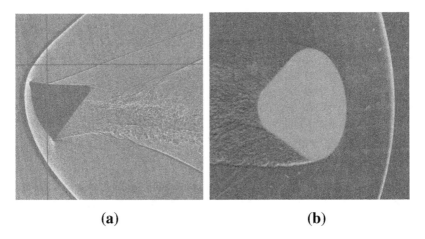

(a)   **(b)**

FIGURE 1.1 Shadowgraph of supersonic flow around space crew modules: (a) Mach 2.2 flow around an Apollo-like capsule at 25° angle-of-attack (sideview, from Kruse, 1968; Schneider, 2006); (b) Mach 1.2 flow around a CEV-like capsule at −33° angle-of-attack (top view—heat shield pitched down to flow, from Brown et al., 2008, with permission).

industrial applications where supersonic flows can be encountered that are not related to aerospace or aeronautics. These applications follow in the same spirit as the first application of the supersonic nozzle to steam turbines by Gustaf de Laval in the 1890s. As an example, in a recovery boiler where heat is used to produce high pressure stream, such as in the conversion of wood into wood pulp, sootblowers are used to remove fireside deposits from tube surfaces by blasting the deposits with high-pressure steam jets (Jameel, Cormack, Tran, & Moskal, 1994; Tran, Tandra, & Jones, 2007). Since the steam flow through the sootblower nozzle is compressible and supersonic (see Figure 1.2a), the nozzle characteristics can be optimized to increase the penetration depth (potential core length) of the jet flow while minimizing the use of high pressure steam in the process. Another example is in the metal processing industry such as steel production. Supersonic jets inject oxygen gas into the molten bath of electric arc furnaces (EAF) in order to optimize the carbon oxidation. Figure 1.2b shows the oxygen injector of an EAF protruding from the wall of the furnace. Once again optimized penetration depth (increase of potential core length) of the jet is desired. The previous two examples involved jets with an embedded shock structure. A final example involves an airblast nozzle that ejects a liquid at high speed into a reactor such as those used in the syngas industry. As Figure 1.3a shows, a liquid to be injected into a reactor is embedded within a multiple-layered coaxial stream of gas and liquid. Clearly the shearing interfaces between

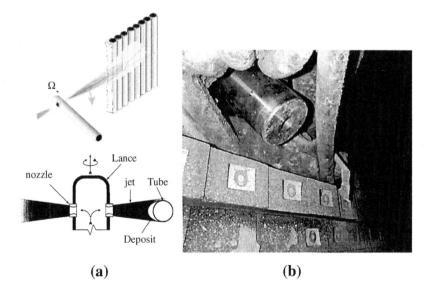

(a)   **(b)**

FIGURE 1.2 Diverse utilization of high speed jets: (a) schematic use of supersonic jets in sootblower cleaning process (Tran, Univ. of Toronto, private communication, Jameel et al., 1994, with permission); (b) nozzle exit of supersonic oxygen jet installed in electric arc furnace (from Allemand et al., 2001, with permission).

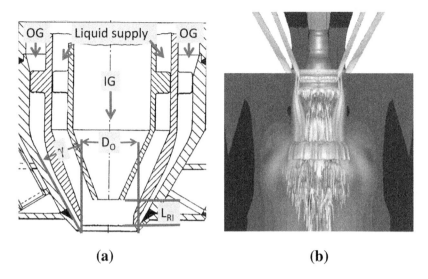

(a)   (b)

FIGURE 1.3 Three-stream coaxial airblast injector: (a) schematic of injector geometry: IG, OG, inner and outer gas supply (from Strasser, 2011 with permission); (b) CFD solution of turbulent pulsatile flow field (Strasser, Eastman Chemical Company, private communication, with permission).

the multiphase fluid streams interact and can have a significant influence on performance depending on exit velocity and angle of ejection. Figure 1.3b shows an instantaneous snapshot of the complex turbulent self-exciting pulsatile flow field that can arise.

While these examples attempt to exploit the underlying physics associated with high speed jet and nozzle optimization, their introduction into the industrial process is far less intricate than applications associated with aircraft engine designs. While crude in appearance these devices have an important effect on the manufacturing process and are of no less value than the operational characteristics of more sophisticated applications. Nevertheless, fundamental knowledge and exploitation of compressible fluid dynamics can significantly improve the operating efficiency of the processes and reduce the associated costs.

1.1 KINEMATIC PRELIMINARIES

It is necessary at the outset to go through some mathematical preliminaries that will prove useful in the development of the governing equations for compressible flows as well as in their analysis. Of course, such kinematic preliminaries can be found in innumerable fluid mechanics resources. The discussion and presentation here will be kept as general as possible, and may, at times, reflect more of a continuum mechanics slant. Such a bias is intentional and seeks to emphasize that fluid mechanics is a direct subset of the broader continuum mechanics field that also includes solid mechanics. Of course, the exclusions are fluids and flows where the continuum hypothesis no longer applies, such as in rarefied gas flows.

1.1.1 Motion of Material Elements

With the focus on compressible fluid motions, consider the motion of a material element of fluid undergoing an arbitrary deformation. Let the material or Lagrangian coordinates of a particle within the element at some reference state be represented by X_α, and the spatial (Eulerian) coordinates of the element at some later time t be represented by ξ_i. A continuous deformation, from some reference time t_0, of this material element to a state at time t is assumed. This mapping can be expressed as[1]

$$\xi_i = \chi_i(X_\alpha,t), \quad \boldsymbol{\xi} = \boldsymbol{\chi}(\mathbf{X},t) \quad (i = 1,2,3), \quad (1.1a)$$

or

$$X_\alpha = \chi_\alpha^{-1}(\xi_i,t), \quad \mathbf{X} = \boldsymbol{\chi}^{-1}(\boldsymbol{\xi},t) \quad (\alpha = 1,2,3), \quad (1.1b)$$

where $\boldsymbol{\chi}$ is the deformation function. In the present context, a continuous deformation implies that the transformations in Eqs. (1.1a) and (1.1b) possess continuous partial derivatives with respect to their arguments. The corresponding velocity of any particle within the material element is then

$$\frac{d}{dt}\xi_i(X_\alpha,t) = \left.\frac{\partial}{\partial t}\xi_i(X_\alpha,t)\right|_{X_\alpha} = u_i(\xi_k,t) \quad (1.2)$$

where Eq. (1.1b) has been used.

Although these equations are often the starting point for deriving several important kinematic relationships, caution is necessary in performing any subsequent calculations. The complication arises because the deformation is being described in two different frames at the same time, that is, a material frame based in the element described by X_α and a spatial (Eulerian) frame described by ξ_i. Thus, it is more convenient to use the spatial coordinate frame at time t as the reference configuration and measure changes in the same spatial coordinates at a later time $t' \geqslant t$. If the material point $\mathbf{X}$ at the later time t' is described by the spatial coordinates $\mathbf{x}$, then

$$x_i = \chi_i(X_\beta,t') = \chi_{(t)i}(\xi_k,t'), \quad \mathbf{x} = \boldsymbol{\chi}(\mathbf{X},t') = \boldsymbol{\chi}_{(t)}(\boldsymbol{\xi},t'), \quad (1.3a)$$

---

[1] In this section both index and boldface notation will be used in the mathematical description and formulation. In subsequent chapters, the index notation will prevail for the most part, since the algebraic manipulations of the equations are easier. However, the boldface notation will be retained where compactness of the mathematical expression is paramount.

where $\chi_{(t)}$ is called the relative deformation function. Since the deformations of the fluid elements are continuous, the mapping given in Eq. (1.3a) is invertible, so that

$$\xi_i = \chi_{(t)i}^{-1}(x_k,t), \qquad \boldsymbol{\xi} = \boldsymbol{\chi}_{(t)}^{-1}(\mathbf{x},t). \tag{1.3b}$$

The velocity of any particle at t' is then given by

$$\frac{dx_i}{dt'} = \frac{d}{dt'}\chi_i(X_\beta,t') = \left.\frac{\partial}{\partial t'}\xi_i(X_\beta,t')\right|_{X_\beta} = u_i(x_k,t'). \tag{1.4}$$

At $t' = t$, the deformations are equivalent and Eqs. (1.4) and (1.2) are the same. The corresponding acceleration can be written as

$$\left.\frac{d}{dt'}u_i(x_k,t')\right|_{t'=t} = \frac{\partial u_i}{\partial t} + u_j(x_k,t)\frac{\partial u_i}{\partial x_j}, \tag{1.5}$$

where the repeated indices implies summation over the index range (Einstein summation convention). It is worth emphasizing here that the kinematic developments within this chapter are based on basis representations in Cartesian coordinate frames. As such, spatial gradients simply revert to partial differentiations with the respective coordinate. Consideration of other non-Cartesian frames would require a more general definition of spatial gradients accounting for variations of the basis vectors. The interested reader is referred to texts on tensor analysis (e.g. Aris, 1962) where the generalization to co- and contra-variant differentiation is discussed. With these relations, it is now possible to develop a description of the motion of the material elements based on a single coordinate system $\mathbf{x}$ rather than the two systems given by $\mathbf{X}$ and $\boldsymbol{\xi}$.

1.1.2 Deformation

With the particle motion described by Eq. (1.1a), the evolution of a material line element dX_α, composed of several fluid particles, from its reference state to its state at some time t can be determined from

$$d\xi_i = \frac{\partial \chi_i(\mathbf{X},t)}{\partial X_\alpha}dX_\alpha = \mathcal{F}_{i\alpha}(t)dX_\alpha, \tag{1.6}$$

where the second-order tensor $\mathcal{F}_{i\alpha}(t)(\,=\partial\xi_i/\partial X_\alpha)$ is the deformation gradient tensor. Correspondingly, a line element $d\boldsymbol{\xi}^{(s)}$ within the initial material volume at time t evolves so that at time t' it is given by $d\mathbf{x}^{(s)}$. These two line elements are then related by

$$dx_i^{(s)} = \frac{\partial x_i}{\partial \xi_k}d\xi_k^{(s)} = \mathcal{F}_{(t)ik}d\xi_k^{(s)}, \qquad d\mathbf{x}^{(s)} = \boldsymbol{\mathcal{F}}_{(t)}d\boldsymbol{\xi}^{(s)}, \tag{1.7}$$

where the second-order tensor $\mathcal{F}_{(t)ik}$ ($\boldsymbol{\mathcal{F}}_{(t)}$) is the relative deformation gradient. The extension to both surface area and volume follows directly (see Figure 1.4),

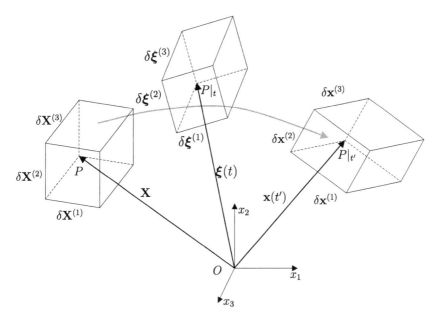

FIGURE 1.4 Evolution of an undeformed body at point P with sides $\delta\mathbf{X}^{(k)}$ to deformed states at point $P|_t$ at time t and at point $P|_{t'}$ at time t'.

so that at time t', the volume of the corresponding infinitesimal material element with edges $d\mathbf{x}^{(1)}, d\mathbf{x}^{(2)}$, and $d\mathbf{x}^{(3)}$ is

$$d\mathcal{V}(t') = \underbrace{\left(e_{ijk}dx_j^{(1)}dx_k^{(2)} \right)}_{n_i\,d\mathcal{S}} dx_i^{(3)}, \quad d\mathcal{V} = \left(d\mathbf{x}^{(1)} \times d\mathbf{x}^{(2)} \right) \cdot d\mathbf{x}^{(3)}, \quad (1.8)$$

where n_i is the unit surface normal with element surface area $d\mathcal{S}$, and e_{ijk} is the Levi–Civita permutation tensor. This can be related to the initial (reference) volume $dV_r (= e_{\alpha\beta\gamma}dX_\beta^{(1)}dX_\gamma^{(2)}dX_\alpha^{(3)})$ using Eq. (1.7),

$$d\mathcal{V}(t') = e_{ijk}\left[\frac{\partial x_j}{\partial \xi_q}\frac{\partial x_k}{\partial \xi_r}\frac{\partial x_i}{\partial \xi_p} \right] d\xi_q^{(1)}d\xi_r^{(2)}d\xi_p^{(3)}$$

$$= e_{ijk}\frac{\partial x_j}{\partial X_\beta}\frac{\partial x_k}{\partial X_\gamma}\frac{\partial x_i}{\partial X_\alpha}dX_\beta^{(1)}dX_\gamma^{(2)}dX_\alpha^{(3)}$$

$$= \det \mathcal{F}e_{\alpha\beta\gamma}dX_\beta^{(1)}dX_\gamma^{(2)}dX_\alpha^{(3)} \quad (1.9a)$$

or

$$d\mathcal{V}(t') = \mathcal{F}_{(t)}(t')e_{ijk}d\xi_q^{(1)}d\xi_r^{(2)}d\xi_p^{(3)}$$

$$= \mathcal{F}_{(t)}(t')\mathcal{F}(t)e_{ijk}dX_\beta^{(1)}dX_\gamma^{(2)}dX_\alpha^{(3)}$$

$$= \det \mathcal{F}(t')e_{\alpha\beta\gamma}dX_\beta^{(1)}dX_\gamma^{(2)}dX_\alpha^{(3)} \quad (1.9b)$$

where $\det \mathcal{F}(t')$ is the determinant of the deformation gradient at t', and the identity

$$e_{\alpha\beta\gamma}\det \mathcal{F} = e_{ijk}\frac{\partial x_j}{\partial X_\beta^{(1)}}\frac{\partial x_k}{\partial X_\gamma^{(2)}}\frac{\partial x_i}{\partial X_\alpha^{(3)}},\tag{1.10}$$

has been used. Since $\det \mathcal{F}$ is simply the Jacobian of the transformation J at t', Equation (1.9b) becomes

$$d\mathcal{V}(t') = J dV_r.\tag{1.11}$$

The deformation gradient is a second-order tensor so its determinant, $\det \mathcal{F}$, is simply the third invariant $\mathrm{III}_{\mathcal{F}}$ of it. From the Cayley-Hamilton theorem, $\mathrm{III}_{\mathcal{F}}$ can be written as a function of $\mathcal{F}$, that is

$$\mathrm{III}_{\mathcal{F}} = \frac{1}{3}\left(\{\mathcal{F}^3\} - I_{\mathcal{F}}\{\mathcal{F}^2\} + \mathrm{II}_{\mathcal{F}}\{\mathcal{F}\}\right)\tag{1.12}$$

where $\{\cdot\}$ represents the trace. The second invariant $\mathrm{II}_{\mathcal{F}}$ is defined as

$$\mathrm{II}_{\mathcal{F}} = \frac{1}{2}\left[\{\mathcal{F}\}^2 - \{\mathcal{F}^2\}\right],\tag{1.13}$$

so that

$$\mathrm{III}_{\mathcal{F}} = \frac{1}{6}\left(\{\mathcal{F}\}^3 - 3\{\mathcal{F}\}\{\mathcal{F}^2\} + 2\{\mathcal{F}^3\}\right).\tag{1.14}$$

Thus, the deformation gradient provides information on the transformation of line, surface, and volume elements. It should be noted that the second and third invariants just defined for second-order tensors can play an important role in analyzing the turbulent statistical moments such as the turbulent Reynolds stress tensor. The applicability of these such concepts to compressible flows is discussed in detail in Chapter 5.

In the study of fluid flows, the focus is on the rate of change of physical variables. As such, the interest is not necessarily on the deformation gradient itself or the material volume, but rather on the rate of change of these quantities. For example, useful results can also be derived that relate the relative deformation gradient to the velocity field. Consider the material derivative of the Jacobian,

$$\begin{aligned}
\frac{dJ}{dt} &= \frac{d\mathrm{III}_{\mathcal{F}}}{dt} = \frac{\partial \mathrm{III}_{\mathcal{F}}}{\partial \mathcal{F}_{ij}}\frac{d\mathcal{F}_{ij}}{dt} = \frac{d\mathcal{F}}{dt}\cdot\frac{\partial \mathrm{III}_{\mathcal{F}}}{\partial \mathcal{F}^T} \\
&= J(\mathcal{F}_{ji})^{-1}\frac{d\mathcal{F}_{ij}}{dt} = J\frac{d\mathcal{F}}{dt}\mathcal{F}^{-1} \\
&= J(\mathcal{F}_{ji})^{-1}\frac{\partial u_i}{\partial x_k}\mathcal{F}_{kj} = J\left(\mathrm{grad}\,\mathbf{u}\right)\mathcal{F}\mathcal{F}^{-1} \\
&= \delta_{ki}J\frac{\partial u_i}{\partial x_k} = J\frac{\partial u_i}{\partial x_i} = J\,\mathrm{div}\,\mathbf{u},
\end{aligned}\tag{1.15}$$

where $\mathcal{F}^{-1}|||_{\mathcal{F}} = \mathcal{F}^2 - |_{\mathcal{F}}\mathcal{F} + ||_{\mathcal{F}}\mathbf{I}$ has been used[2] from the Cayley-Hamilton theorem, and the velocity u_i is dx_i/dt. With the material volume relationship given in Eq. (1.11), Eq. (1.15) can be rewritten as

$$\frac{1}{\mathcal{V}}\frac{d\mathcal{V}}{dt} = \frac{\partial u_i}{\partial x_i} = \operatorname{div} \mathbf{u}. \tag{1.16}$$

Equation (1.16) shows that the divergence of the velocity is a measure of the rate of change of the dilatation J. When the condition $\operatorname{div} \mathbf{u} = 0$ holds the fluid can be considered as incompressible. As will be shown in Chapter 2, the mass conservation is a direct consequence of this relation when the total mass of the element is kept constant.

Though, at first, these kinematic relationships may seem distantly related to the dynamics associated with compressible fluid motion, it is the deformation gradient that is the foundation for the specification of the fluid stress constitutive equation. Within the continuum mechanics framework, the stress field $\mathbf{\Sigma}$ associated with a fluid element is inherently constrained to have a functional dependency solely dependent on the deformation gradient (see, for example, Huilgol & Phan-Thien, 1997; Truesdell & Rajagopal, 2000), that is

$$\mathbf{\Sigma}(\mathbf{X},t) = \mathcal{S}(\mathcal{F}(t - s); \mathbf{X},t), \qquad s \geqslant 0, \tag{1.17}$$

where t' has been replaced by $t - s$ to emphasize the fact that the material element history enters into the stress field specification. However, it was shown in Eq. (1.9b) that the deformation gradient $\mathcal{F}(t - s)$ is related to the relative deformation gradient $\mathcal{F}_{(t)}(t - s)$ through the relation

$$\mathcal{F}(t - s) = \mathcal{F}_{(t)}(t - s)\mathcal{F}(t), \tag{1.18}$$

so that the stress constitutive relationship in Eq. (1.17) can be written as

$$\mathbf{\Sigma}(\mathbf{X},t) = \mathcal{S}(\mathcal{F}_{(t)}(t - s), \mathcal{F}(t); \mathbf{X},t), \qquad s \geqslant 0. \tag{1.19}$$

Within the continuum mechanics framework, the development of stress constitutive equations involves the imposition of constraints on the functionals

---

[2] The tensorial differentiation $\partial|||_{\mathcal{F}}/\partial\mathcal{F}$ can be expanded so that

$$\frac{\partial|||_{\mathcal{F}}}{\partial\mathcal{F}_{ij}} = \frac{\partial}{\partial\mathcal{F}_{ij}}\left[\frac{1}{3}\left(\{\mathcal{F}^3\} - \{\mathcal{F}\}^3\right) + ||_{\mathcal{F}}\{\mathcal{F}\}\right],$$

where, for example, the differentiation operation follows

$$\frac{\partial\{\mathcal{F}^3\}}{\partial\mathcal{F}_{ij}} = \frac{\partial}{\partial\mathcal{F}_{ij}}\left(\mathcal{F}_{kl}\mathcal{F}_{lm}\mathcal{F}_{mk}\right)$$
$$= \delta_{ik}\delta_{jl}\mathcal{F}_{lm}\mathcal{F}_{mk} + \delta_{il}\delta_{jm}\mathcal{F}_{kl}\mathcal{F}_{mk} + \delta_{im}\delta_{jk}\mathcal{F}_{kl}\mathcal{F}_{lm}$$
$$= 3\mathcal{F}_{jm}\mathcal{F}_{mi} = 3\mathcal{F}_{ji}^2 = 3\left(\mathcal{F}^2\right)^T.$$

$\mathfrak{S}$ (for example, Huilgol & Phan-Thien, 1997). Although such a discussion is outside the scope of this text, suffice it to note that the imposition of these constraints yields an isotropic representation for $\mathcal{F}(t)$ that is proportional to $\det \mathcal{F}$, that is $\mathcal{F}(t) = (\det \mathcal{F})\mathbf{I}$. This reduces the functional relationship for the stress field in Eq. (1.19) to

$$\mathbf{\Sigma}(\mathbf{X},t) = \mathfrak{S}(\mathcal{F}_{(t)}(t-s),\rho(t);\mathbf{X},t), \qquad s \geqslant 0, \qquad (1.20)$$

where the dependency on $\det \mathcal{F}$ (the Jacobian J) has been replaced by the mass density which follows directly from Eq. (1.16) and mass conservation equation. (see Eq. (2.3) in Chapter 2).

1.1.3 Reynolds Transport Theorem

A final kinematic result that is a prerequisite for obtaining the various conservation equations in the next chapter can be extracted using the result in Eq. (1.16). Of interest in the study of any fluid flow is the evolution of physical variables. Within a volume element $\mathcal{V}$ moving with the fluid, the evolution of these physical variables can be obtained from a knowledge of a corresponding density function of space and time, $\mathcal{F}(\mathbf{x},t)$, and would be given by

$$\int_{\mathcal{V}(t)} \mathcal{F}(\mathbf{x},t)\mathrm{d}\mathcal{V}. \qquad (1.21)$$

The term density function is used in a broad context. Up to this point, the mass density (mass per unit volumes) has only been considered. In Chapter 2 additional (thermodynamically extensive) variables such as the momentum density and energy density will also be considered. As noted previously, the study of fluid flows in general is focused on the rate of change of variables, so the interest is actually in the rate of change of this integral, that is, its material derivative $\mathrm{d}/\mathrm{d}t$. Unfortunately, in its present form the integral is not readily amenable to differentiation since the volume $\mathcal{V}$ varies with time and the differentiation cannot be taken through the integral sign. However, recall from Eq. (1.9b) that this changing volume $\mathrm{d}\mathcal{V}$ can be related to a volume in $\boldsymbol{\xi}$-space, and that $\mathrm{d}/\mathrm{d}t$ is differentiation with respect to time with $\boldsymbol{\xi}$ constant. With these relations,

$$\frac{\mathrm{d}}{\mathrm{d}t} \int_{\mathcal{V}(t)} \mathcal{F}(\mathbf{x},t)\mathrm{d}\mathcal{V} = \frac{\mathrm{d}}{\mathrm{d}t} \int_{V} \mathcal{F}[\boldsymbol{\chi}_{(t)}(\boldsymbol{\xi},t'),t]J\mathrm{d}V$$

$$= \int_{V} \left(\frac{\mathrm{d}\mathcal{F}}{\mathrm{d}t}J + \mathcal{F}\frac{\mathrm{d}J}{\mathrm{d}t} \right) \mathrm{d}V$$

$$= \int_{V} \left(\frac{\mathrm{d}\mathcal{F}}{\mathrm{d}t} + \mathcal{F}\frac{\partial u_i}{\partial x_i} \right) \mathrm{d}\mathcal{V}. \qquad (1.22)$$

The material derivative d/dt can be re-expressed in an Eulerian frame by the relation

$$\frac{d}{dt}\bigg|_{\xi} = \frac{D}{Dt} = \frac{\partial}{\partial t}\bigg|_{\xi} + \frac{\partial}{\partial x_j}\frac{\partial x_j}{\partial t}\bigg|_{\xi} = \frac{\partial}{\partial t} + u_j\frac{\partial}{\partial x_j}, \tag{1.23}$$

so that Eq. (1.22) can be written as

$$\frac{d}{dt}\int_{\mathcal{V}(t)} \mathcal{F}(\mathbf{x},t)d\mathcal{V} = \int_{\mathcal{V}}\left[\frac{\partial\mathcal{F}}{\partial t} + \frac{\partial}{\partial x_j}(\mathcal{F}u_j)\right]d\mathcal{V}. \tag{1.24}$$

A clearer physical interpretation of this relation is obtained by rewriting the last term on the right-side using Green's theorem. The resulting expression,

$$\frac{d}{dt}\int_{\mathcal{V}(t)} \mathcal{F}(\mathbf{x},t)d\mathcal{V} = \int_{\mathcal{V}}\frac{\partial\mathcal{F}}{\partial t}d\mathcal{V} + \int_{\mathcal{S}}\mathcal{F}u_j n_j d\mathcal{S}, \tag{1.25}$$

where $\mathcal{S}(t)$ is the surface of $\mathcal{V}(t)$, and n_i is the unit normal to the surface, shows that the rate of change of the integral of $\mathcal{F}$ within the moving volume $\mathcal{V}$ is the rate of change at a point plus the net flow of $\mathcal{F}$ over the surface of $\mathcal{V}$.

1.2 EQUILIBRIUM THERMODYNAMICS

The thermodynamic equilibrium state of any fluid element can be uniquely characterized by two state parameters, say the density ρ and the pressure p.[3] Thus, any other (scalar) quantity can be extracted from a knowledge of these two. Of course, the temperature of the element then is a function of these two quantities so that

$$T = T(\rho, p). \tag{1.26}$$

Other, equivalent permutations of this relation would be equally valid. (The discussion throughout will focus almost exclusively on engineering flows. As such, a fluid such as sea water, which is a chemical solution, is not considered. In such cases, a third state parameter, such as the salinity would need to be added.) In the absence of fluid motion, where the elements are in thermal equilibrium (according to the zeroth law), the first law of thermodynamics simply states that the energy of the system is conserved,

$$\Delta E = Q - W, \tag{1.27}$$

where ΔE is the energy change of the system, Q is the net heat input and W is the net work output.

---

[3] For notational convenience, we will simply designate this equilibrium pressure as p. In a moving fluid, a distinction between the equilibrium and mechanical pressure will be made.

The thermodynamic system of interest is the fluid element which is considered a closed system since the mass is fixed but the associated boundary and volume are not. If both the effects of friction and body forces are neglected, and our system undergoes a series of changes so slowly that at each step the system is in (local) equilibrium, then the process is reversible. Such reversible work on a fluid element is best exemplified by the expansion or contraction of its boundaries. In such instances work is done on the element through pressure forces which, per unit mass, can be written as $-p\,dv$ (where $v = 1/\rho$ is the specific volume or volume per unit mass and p is the static or equilibrium pressure). The first law can then be written as

$$de = dq - p\,dv. \tag{1.28}$$

Correspondingly from the second law, in a reversible transition from one equilibrium state to another involving no gain of heat, adiabatic lines dq can be parameterized by a state variable, the entropy s defined as

$$ds = \frac{dq}{T} = \frac{1}{T}\left(de + p\,dv\right), \tag{1.29}$$

with $ds = 0$ for a reversible process and $ds > 0$ for an irreversible one (Clausius inequality).

If it is assumed that s is a function of p and T then

$$ds = \left(\frac{\partial s}{\partial T}\right)_p dT + \left(\frac{\partial s}{\partial p}\right)_T dp, \tag{1.30}$$

where the right hand side can be written in terms of the specific heat at constant pressure, c_p

$$c_p \equiv \left(\frac{dq}{dT}\right)_{dp=0} = T\left(\frac{\partial s}{\partial T}\right)_p = \left(\frac{\partial e}{\partial T}\right)_p + p\left(\frac{\partial v}{\partial T}\right)_p, \tag{1.31a}$$

and the coefficient of thermal expansion, β,

$$\beta \equiv -\frac{1}{\rho}\left(\frac{\partial \rho}{\partial T}\right)_p = \frac{1}{v}\left(\frac{\partial v}{\partial T}\right)_p = -\rho\left(\frac{\partial s}{\partial p}\right)_T. \tag{1.31b}$$

The equality on the right in Eq. (1.31b) is one of Maxwell's identities.[4] Equation (1.30) can then be rewritten in terms of c_p and β as

$$T\,ds = de + p\,dv = de - \frac{p}{\rho^2}d\rho \tag{1.32a}$$

$$= c_p\,dT - \frac{\beta T}{\rho}dp. \tag{1.32b}$$

---

[4] The remaining three relations are: $\left(\frac{\partial p}{\partial s}\right)_v = \left(\frac{\partial T}{\partial v}\right)_s$, $\left(\frac{\partial v}{\partial s}\right)_p = \left(\frac{\partial T}{\partial p}\right)_s$, and $\left(\frac{\partial p}{\partial T}\right)_v = \left(\frac{\partial s}{\partial v}\right)_T$.

Alternately, if it is assumed that s is a function of p and ρ then

$$ds = \left(\frac{\partial s}{\partial \rho}\right)_p d\rho + \left(\frac{\partial s}{\partial p}\right)_\rho dp, \tag{1.33}$$

where

$$\left(\frac{\partial s}{\partial \rho}\right)_p = \frac{(\partial s/\partial T)_p}{(\partial \rho/\partial T)_p} = -\frac{c_p}{\rho\beta T} \tag{1.34a}$$

$$\left(\frac{\partial s}{\partial p}\right)_\rho = -\left(\frac{\partial s}{\partial \rho}\right)_p \left(\frac{\partial \rho}{\partial p}\right)_s$$

$$= -\frac{(\partial s/\partial T)_p}{(\partial \rho/\partial T)_p}\left(\frac{1}{c^2}\right)$$

$$= \left(\frac{c_p}{\rho\beta T}\right)\left(\frac{1}{c^2}\right) \tag{1.34b}$$

where Eq. (1.31a) has been used, and c is the sound speed with $c^2 = (\partial p/\partial \rho)_s$. Equation (1.33) can then be rewritten as

$$\beta T ds = -\frac{c_p}{\rho}\left(d\rho - \frac{dp}{c^2}\right). \tag{1.35}$$

Analogous to the definition of specific heat at constant pressure in Eq. (1.31a), the definition of the specific heat at constant volume c_v can be written as

$$c_v \equiv \left(\frac{dq}{dT}\right)_{dv=0} = T\left(\frac{\partial s}{\partial T}\right)_v = \left(\frac{\partial e}{\partial T}\right)_v. \tag{1.36}$$

A condition that is often assumed in the study of compressible flows is that of an isentropic change of state. The consequences of such a condition are readily seen by considering Eq. (1.32b), and writing it in the form

$$T\left(\frac{\partial s}{\partial \rho}\right)_T = \left(\frac{\partial e}{\partial \rho}\right)_T - \frac{p}{\rho^2} \tag{1.37a}$$

$$T\left(\frac{\partial p}{\partial T}\right)_\rho = p - \rho^2\left(\frac{\partial e}{\partial \rho}\right)_T, \tag{1.37b}$$

where one of the Maxwell's identities has been used. For a perfect gas, $p = \rho\mathcal{R}T$ (where $\mathcal{R}$ is the specific gas constant), the internal energy is solely a function of temperature $e = e(T)$, so that Eq. (1.36) yields $e = c_v T$ and Eq. (1.37b) simply reduces to

$$\left(\frac{\partial p}{\partial T}\right)_\rho = \frac{p}{T}. \tag{1.38}$$

Correspondingly, another quantity which reduces to a form analogous to the internal energy is the enthalpy. It is an energy per unit mass and can be written as

$$h = e + pv = e + \frac{P}{\rho}.$$ (1.39)

For a perfect gas, this reduces to the simple form

$$h = c_v T + \frac{P}{\rho} = (c_v + \mathcal{R})T = c_p T,$$ (1.40)

where the relation between the gas constant $\mathcal{R}$ and the specific heats $\mathcal{R} = c_p - c_v$ is used. A final simple thermodynamic expression worth noting comes from the first law, Eq. (1.28), which when integrated for an isentropic change of state yields,

$$\int \left(\frac{\mathcal{R}T}{\rho'} \right) d\rho' = \int de$$

$$\int_{\rho_0}^{\rho} \frac{d\rho'}{\rho} = \frac{1}{(\gamma - 1)} \int_{T_0}^{T} \frac{dT'}{T}$$

$$\left(\frac{\rho}{\rho_0} \right) = \left(\frac{T}{T_0} \right)^{\frac{1}{(\gamma - 1)}}$$ (1.41a)

where the ratio of specific heats $\gamma \, (= c_p/c_v)$ has been used. Alternate expressions for pressure–temperature and pressure–density can be obtained by varied application of the perfect gas law, and written as

$$\left(\frac{p}{p_0} \right) = \left(\frac{T}{T_0} \right)^{\frac{\gamma}{(\gamma - 1)}} \quad \text{and} \quad \left(\frac{p}{p_0} \right) = \left(\frac{\rho}{\rho_0} \right)^{\gamma}.$$ (1.41b)

1.3 COMPRESSIBLE SUBSONIC AND SUPERSONIC FLOWS

Many flows of practical and/or fundamental interest can, in general, have regions where the flow is supersonic and other regions where the flow is subsonic and even incompressible. This is the case in any wall-bounded flow, but also for many supersonic free shear flows including the near-wakes, jets and (subsonic/supersonic) mixing layers. Very complex flows can be encountered in supersonic flight (such as military aircraft and civil aircraft—until recently such as Concorde) or for most aircraft the transonic regime. In both cases, the flows exhibit several characteristic features that are affected by compressibility. For example, in maneuvering a civil aircraft, airbrakes (spoilers) are sometimes deployed. In Figure 1.5, a numerical calculation of the flow over the aircraft at $M = 0.8$ highlights the wall-bounded and free shear flow regions that compressibility influences. Additionally, in cruise conditions shock waves can interact with the boundary layers having a parasitic effect and causing unstable separation.

In such flows, the mathematical character of the equations governing the motion changes as well as exhibiting an elliptic character in the subsonic regions and a hyperbolic character in the supersonic region. Any perturbations existing in the flow propagate with a velocity relative to the speed of sound c in the fluid medium. In laminar or turbulent flows where mixed subsonic/supersonic regions exist, perturbations initiating in and/or introduced into the flow can be easily amplified through, for example, feedback effects. The screech tone effect, or transonic buffeting are illustrations of this phenomenon. The transonic buffeting over an airfoil is illustrated by a typical CFD calculation shown in Figure 1.6. Both subsonic (aft portion) and supersonic (cone shaped forward portion) regimes exist in the same flow which induces an unsteadiness of the shock system. The movement of the shock is depicted in Figure 1.6 where the maximum difference in shock movement is obtained from the difference between the (labeled) shock locations in the figures on the left and right $(\Delta x = 0.08c)$.

In engineering flows, density changes in a fluid medium can be initiated in several ways. Often, compressibility is associated with high speed flow where the fluid speed has the same order of magnitude as the sound speed. However, compressibility does play a significant role in flows with wave propagation within the fluid and body force driven convection caused, for example, by thermal expansion.

Even with the somewhat limited mathematical infrastructure that has been presented so far, it is possible to delve more deeply into the nature of compressible flows, and deduce some velocity field criteria that will enable us to determine when density changes can be neglected. In the next chapter,

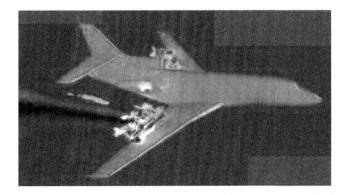

FIGURE 1.5 Numerical calculation over a Falcon airplane at $M = 0.8$. High pressure regions are shown as darker shades. The turbulent structures downstream of the spoilers are also shown. From Dassault Aviation, Aether code, courtesy J.C. Courty.

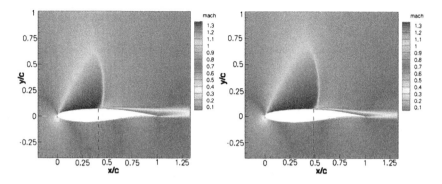

FIGURE 1.6 Typical CFD computation of transonic (external Mach number 0.73) buffeting on a ONERA OAT15A airfoil at 4.5° angle-of-attack: mean Mach number contours for two extreme positions of the shock during a cycle shown. G. Browaeys, private communication, with permission.

the various conservation equations associated with the motion of the fluid will be considered. As a prelude to this discussion, consider the equation for the evolution of the material volume given in Eq. (1.16). The underlying assumption for the mass balance equation to be discussed is that, in the absence of any sources or sinks, the mass of the system is unchanged. As such, the (mass) density only varies as a result of the change of material volume and is also described by Eq. (1.16). However, from the equation of state given in Eq. (1.26), it is apparent that any condition on the density field necessarily influences the pressure and temperature fields. Neglecting changes in density then requires a divergence free or solenoidal velocity field.

In steady flow, pressure variations associated with flow velocities that approach the speed of sound produce density changes that require compressibility effects to be taken into account. The quantitative measure of such effects is the Mach number U/c. Although this is the most common source of density variations in fluid flows, other sources induced by pressure variations need to be considered. If the flow is unsteady, due to the passage of some wave field, pressure differences can occur and their magnitude is proportional to the frequency of the wave. In geophysical flows, body forces arising from the effects of gravity can induce pressure differences requiring compressibility effects be accounted for. Such a situation occurs in the atmosphere if the static pressure difference at two different elevations is comparable with the absolute pressure. These three examples show how variations in pressure can have an effect on the density field that requires that such compressibility effects be taken into account. A second contributor to density variations in the flow field are changes due to temperature. In the absence of combustion or external heat sources, these changes can be due to internal dissipative heating or molecular conduction. Such effects are rarely strong enough to cause density variations significant enough to be explicitly accounted for.

In addition to the fundamental physical characteristics just discussed, replication of practical engineering flow conditions in the laboratory are complicated by the often conflicting requirements needed to simultaneously satisfy the Reynolds number and Mach number scaling parameters. The limited spatial dimensions of the model configurations that have to be used in the relatively small size wind tunnels necessitates high velocities to replicate the requisite value of the Reynolds number. It then becomes difficult, if not impossible, to maintain the corresponding Mach number value. For example, in a configuration in subsonic flow, the Reynolds number similarity for small dimensional models can impose a requirement for high velocities leading to a higher Mach number value in the laboratory experiment that is inconsistent with the real flow to be replicated. In contrast, if the Mach number is imposed to provide representative compressibility effects, this fixes the velocity of the flow, and then the Reynolds number cannot be adjusted to the required value since the size of the model is delimited by the geometry of the wind tunnel. It is possible to overcome this limitation by modifying (increasing) the density in two ways. The first increases the density by pressurization of the wind tunnel, and the second acts on the temperature such as in cryogenic tunnels (in which the settling chamber temperature can be as low as 77 K (Goodyer, 1997)). These cryogenic tunnels are much more complex and expensive when compared with academic tunnels and are generally in use in national or international research centers. However, the Mach number and Reynolds number can be coupled to form the Knudsen number from which the rarefaction effects on the flow can be assessed at a macroscopic scale. As will be discussed in Chapter 4, the Knudsen number is a useful parameter for analyzing the physics of sensing probes and tracer particles for intrusive and non-intrusive experimental methods. From a theoretical standpoint, it can be useful in assessing the continuum nature of the compressible turbulence at the smallest Kolmogorov scales, and from a numerical standpoint it is a central parameter in the development of Lattice Boltzmann Methods (LBM) for compressible flows (see Chapter 5). The physical constraints imposed by a laboratory experiment do not always carry over to numerical experiments; however, it is not always straightforward to correctly replicate the physical experiments either due to an incomplete knowledge of the boundary conditions of the laboratory experiment or inaccuracies in the physical "measurements." Correspondingly, the computational constraints imposed by numerical experiments do not always carry over to physical experiments. For example, numerical errors associated with algorithms and closure models for governing equations are rather unique to the numerical experiment, although limitations imposed by computer hardware on memory and speed are applicable to both, albeit within somewhat different contexts. Nevertheless, it is often possible to perform complementary physical and numerical experiments that can confirm important design and performance characteristics. Such is the case shown in Figure 1.7 where results from an unstructured grid numerical solver, FUN3D (see FUN3D Product Manual,

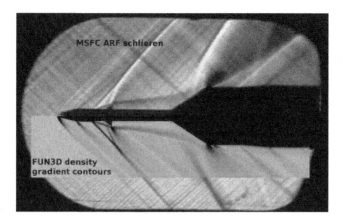

FIGURE 1.7 Comparison of experimental and CFD schlieren visualizations of the NASA ARES I crew launch vehicle (CLV) in a supersonic flow. (Schlieren courtesy of A. Frost, see http://fun3d.larc.nasa.gov/.)

http://fun3d.larc.nasa.gov/), are compared with laboratory results for a next generation NASA launch vehicle in supersonic turbulent flow. Qualitatively, the results show an excellent reproduction of the general shock structure of the flow using relatively low-order eddy viscosity turbulence models but, more importantly, while the experimental data consists of such schlieren visualizations and total forces and moments, the computational study includes distributed forces along the vehicle that are important for structural analysis. These distributed forces are then calibrated by the measured forces. Of course, caution needs to be exercised in making comparisons, both qualitative and quantitative, between practical engineering flows, and "corresponding" physical and numerical experiments.

1.4 TURBULENT FLOWS AND COMPRESSIBLE TURBULENCE

The description of turbulent flow fields by ensemble averaged correlations has long been the prediction measure of choice. Van Driest (van Driest, 1951) appears to have been the first to publish the derivation of the differential equations (applicable to thin boundary layer flows) governing the motion of compressible, turbulent flows. Prior to this, the analyses relied on using von Kármán's similarity theory for turbulent flows and were thus constrained to a mixing layer type analysis (Frankl & Voishel, 1943). Over the last three decades there have been significant strides in the modeling of these correlations and in associating their behavior with characterizing features of the turbulence. This process has been aided by several texts (Durbin & Pettersson-Reif, 2001;

Heinz, 2003; Jovanovic, 2003; Pope, 2000) that have clearly explained the mathematical derivation and rationale behind the theory.

In the last two decades, a series of direct numerical simulations (DNS) coupled with experimental validation have shed considerable light on the flow dynamics. The detailed results from these simulations allow for improved model development for flow prediction. Since it is still not possible to accurately compute complex, engineering compressible turbulent flow fields directly by numerically solving the conservation equations for mass, momentum and energy (DNS), it is necessary to use the available alternatives that include large eddy simulations (LES), Reynolds-averaged Navier–Stokes (RANS) formulations, and variants of these. As an engineering tool, the RANS approach is currently the most popular methodology.

It is worth emphasizing that while many fluid flows may require compressibility effects be taken into account in describing the (ensemble) mean motion, this does not necessarily imply that the turbulent fluctuations exhibit the analogous dependence on the density variations. A particular illustration of this point is the strong Reynolds analogy (SRA, see Section 3.5), and its variants relating the density (or temperature) fluctuations to the velocity fluctuations. Forms of this analogy have been verified numerically (e.g. Huang, Coleman, and Bradshaw, 1995) and experimentally (Gaviglio, 1987; Panda & Seasholtz, 2006; Smith & Smits, 1993) from analyses of the relevant statistical correlations. A more stringent constraint, that assumes the relationship between density (or temperature) fluctuations and velocity fluctuations holds instantaneously, has not been verified experimentally nor investigated in numerical simulations. From this, one can conclude that a necessary condition for the statistical correlations to exhibit compressibility effects is for the instantaneous turbulent fluctuations to be compressible; however, it is not sufficient that the statistical correlations display compressibility effects for the instantaneous field to be compressible. Such behavior might be expected since instantaneous time series comparisons between velocity and temperature fluctuations in a supersonic boundary layer (e.g. Smith and Smits, 1993) have obvious different frequency characteristics, and spanwise spectra comparison between velocity and density in both a supersonic boundary layer (Pirozzoli, Grasso, & Gatski, 2004) and channel flow (Coleman, Kim, & Moser, 1995) display different spectral characteristics over a large range of wavenumbers. A consequence of these results is that one can use statistical characteristics of the velocity field to infer the statistical characteristics of the temperature, or density, field. However, a knowledge of the structural character of the fluctuating velocity field does not necessarily imply a structural knowledge of the temperature or density field. This relationship between the statistical characteristics simplifies the problem of turbulence model development. For example, models developed for correlations involving the turbulent velocity field can be adapted in an analogous manner to the temperature and density fields. In addition, a further consequence (see Section 3.5.2) is the ability to relate the statistical correlations for the fluctuating velocity field to their incompressible counterparts through simple variable density extensions. In both cases, models must inherently satisfy some consistency constraints which can be used as guidance in developing accurate closure models for such flows.

One of the measures for determining whether the mean flow is compressible is the mean flow Mach number. One of the first complicating factors of such a measure is the differing effects of compressibility on free shear flows and wall-bounded flows. Another companion measure of compressibility is often needed and this can be the deformation or gradient Mach number. In this case, instead of a characteristic mean flow velocity being used, a composite scale obtained from the product of a characteristic mean shear rate and turbulent length scale is used in ratio with the sound speed ($M_g = Sl/c$). This measure has been found to quantify the effects of mean compression and shear on the turbulence. Both the mean flow Mach number and the deformation, or gradient, Mach number M_g, can then be used to assess the potential impact of compressibility effects on the mean flow.

For the fluctuating turbulence itself, one such measure of compressibility is the turbulent Mach number M_t where the characteristic velocity is associated with the turbulence itself. For example, the turbulent velocity field is often characterized by the square root of the turbulent kinetic energy. Thus, the turbulent Mach number is a statistically averaged quantity whose value is a root-mean-square (*rms*) measure of the phenomena. This, of course, is distinct from a corresponding instantaneous value which could span a wide range of values. For a quantity such as the turbulent Mach number, the distinction between a statistically averaged measure and an instantaneous measure is important due to the particular physical meaning of the parameter and, most importantly, its sonic limit. As an example, consider a turbulent flow with a turbulent Mach number $M_t = 0.4$ (i.e. a flow with a 10% turbulent intensity and a mean flow Mach number of 4). At first, the $M_t = 0.4$ value can be assumed small enough to consider that the turbulent field behaves in an incompressible manner. However, the instantaneous values within the fluctuating velocity field can reach levels as much as 3 times the *rms* values (e.g. for a Gaussian process the probability of this occurring is 3×10^{-3}), so that the turbulent velocity field can include sonic or supersonic regions that can lead to the presence of shocklets. Additionally, the flow field itself can reach instantaneous supersonic values for lower turbulent fluctuation levels. For a local mean velocity Mach number of 0.8, for example, with 10% fluctuations ($M_t = 0.08$), the local Mach number can reach a value of 1.04. This is a particularly complex situation that has been experimentally observed by some authors and at times been associated with important compressibility effects. While such situations may, at first, seem surprising they highlight the subtlety and complexity of analyses associated with compressible flows.

The presence of shocklets has been identified in a variety of high speed turbulent flows—from very moderate Mach numbers to hypersonic speeds—but are more easily identified in the higher Mach number flows due to their increased strength. These shocklets are, in general, associated with eddies that are coherent and have sufficiently large (observable) spatial and temporal scales. Figure 1.8 gives a typical example of an isolated shocklet occurring around an eddy observed in a supersonic–supersonic counterflow. In addition, it has also been shown at low supersonic Mach numbers that the effect of the Reynolds number cannot be neglected (Johnson, Zhang, & Johnson, 1988). The visualization illustrates the fact that shocklets are generally more complex to characterize due to the three-dimensional character of the eddies. A similar situation arises in numerical simulations where in two-dimensional simulations the shocklets are easily found; whereas in three-dimensional simulations (in the same regime) the shocklets are hardly in evidence.

A more practical manifestation of high speed flow and the influence of a shock arises in flow over a wing (or airfoil at very slight angle of attack) in the form of a substantial increase of drag, that is, the appearance of wave drag. Below freestream Mach numbers $M \approx 0.5$ there is an insensitivity of drag to changes in Mach number. However, as the Mach number increases, there are regions of the airfoil upper surface where the flow has accelerated to near sonic conditions. A critical freestream Mach number exists ($M \approx 0.75$) that produces local sonic flow on the airfoil surface. This, fortunately, has minimal effect until the freestream Mach number is increased such that the sonic line on the airfoil has moved sufficiently far downstream and occurs at the airfoil crest (for wings the occurrence is along the crest-line). A further small increase in Mach number causes the drag to begin to increase significantly with Mach number. The speed

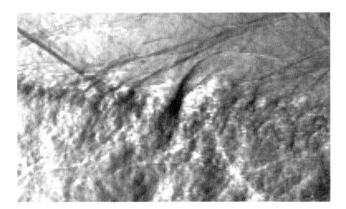

FIGURE 1.8 Enlarged shocklet view in shear layer. From Papamoschou (1995) with permission.

at which this occurs is the force divergence Mach number M_{dDiv}. This is the parameter range where flow instabilities termed buffet occur with ensuing shock oscillations. Such oscillations induce structural vibrations, buffeting, that can lead to fatigue and possibly structural failure. A direct consequence of this is a reduction in cruising speed as well as a reduction in maximum lift coefficient. Another consequence is, of course, an increase in drag. Control of such parasitic phenomena has been a topic of interest for over a half-century, first, as a necessary step in achieving supersonic speeds and now as a means of improving vehicle efficiency.

Chapter 2

Compressible Flow Dynamics

It is useful at the outset to discuss the conservation equations applicable to the description of compressible turbulent flows since these equations form the basis for the analysis to follow in subsequent chapters. Both the modeled statistical transport equations (Reynolds-averaged Navier–Stokes, RANS) and the filtered transport equations (large eddy simulation, LES) used in numerical simulation of compressible flows have their origin in these equations.

The starting point in the development of a mathematical description of compressible flows is the mass, momentum, and energy conservation equations. The derivation of these equations can be found in almost all fluid dynamics texts so it will not be necessary here to go into their detailed formulation. However, each will be presented to introduce the reader to the notational convention used as well as highlight the various assumptions used in deriving the commonly used forms.

The mathematical basis for these balance equations lies in the Reynolds transport theorem which simply equates the time rate of change of an arbitrary moving material element, characterized by some physical property (e.g. mass density, momentum density, etc.), to the sum of the time change of the physical property within the volume, the rate of change of the surface of the element, and the cumulative effect of (body) forces on the element.

2.1 MASS CONSERVATION

Consider a body of fluid whose mass density is $\rho(\mathbf{x},t)$. The total mass of the body is given by (cf. Eq. (1.21))

$$m(t) = \int_{\mathcal{V}(t)} \rho(\mathbf{x},t)\mathrm{d}\mathcal{V}. \tag{2.1}$$

If it is assumed that during the motion the total mass is unchanged within the volume (there are no sources or sinks within the material volume), the rate of

Compressibility, Turbulence and High Speed Flow. http://dx.doi.org/10.1016/B978-0-12-397027-5.00002-2

change of mass is then

$$\frac{dm}{dt} = \int_{\mathcal{V}} \left[\frac{\partial \rho}{\partial t} + \frac{\partial}{\partial x_j}(\rho u_j) \right] d\mathcal{V} = 0, \tag{2.2}$$

where Eq. (1.24) has been used with $\rho(\mathbf{x},t) = \mathcal{F}(\mathbf{x},t)$. Since $\mathcal{V}$ is an arbitrary volume, the integrand must vanish everywhere, so that the mass conservation equation is

$$\frac{\partial \rho}{\partial t} + \frac{\partial}{\partial x_j}(\rho u_j) = 0 \quad \text{or} \quad \frac{D\rho}{Dt} = -\rho \frac{\partial u_j}{\partial x_j}. \tag{2.3}$$

Equation (2.3) is also known as the continuity equation, and it shows that if the fluid volume is density preserving $D\rho/Dt = 0$, then the velocity field is solenoidal (source free) $\partial u_j/x_j = 0$ and the fluid is incompressible. As was pointed out previously, Eq. (1.16) shows that the solenoidal velocity field implies a volume conserving (or isobaric) motion as well. Thus the terms, density preserving motions and volume preserving motions can be used interchangeably. In terms of the specific (per unit mass) volume $\upsilon = 1/\rho$, the mass conservation equation (2.3) can be written as (cf. Eq. (1.16))

$$\frac{1}{\upsilon}\frac{D\upsilon}{Dt} = \frac{\partial u_j}{\partial x_j}. \tag{2.4}$$

Many important flow fields can be assumed incompressible and as such the mathematical description of the flow problem is simplified. The conditions under which such a field exists are discussed in Section 2.4.

2.2 MOMENTUM CONSERVATION

The conservation equation for the (linear) momentum of a body can be stated as a balance between the time rate of change of the momentum of the body and the resultant force acting on the body,

$$\int_{\mathcal{V}} \left[\frac{\partial \rho u_i}{\partial t} + \frac{\partial}{\partial x_j}(\rho u_i u_j) \right] d\mathcal{V} = \mathcal{F}_i, \tag{2.5}$$

where Eq. (1.24) has been used with $\mathcal{F}(\mathbf{x},t)$ replaced with the momentum density ρu_i. The vector total force $\mathcal{F}_i$ consists of the sum of all the forces acting on the body. This total force can be partitioned into a surface force and a body force,

$$\mathcal{F}_i(\mathbf{x},t) = \oint_{\mathcal{S}} \Sigma_{ij} n_j \, d\mathcal{S} + \int_{\mathcal{V}} \rho f_i \, d\mathcal{V}, \tag{2.6}$$

where $\Sigma_{ij} n_j \, d\mathcal{S}$ is the surface force exerted across an element of area $d\mathcal{S}$, and f_i is the body force per unit mass. The surface force is proportional to the amount of surface area acted upon; whereas, the body force is assumed to act uniformly on all elements within the fluid volume.

In integral form over the material boundary, the balance equation given in Eq. (2.5) can be written as

$$\int_{\mathcal{V}} \left[\frac{\partial \rho u_i}{\partial t} + \frac{\partial}{\partial x_j}(\rho u_i u_j) - \frac{\partial}{\partial x_j} \Sigma_{ij} - \rho f_i \right] d\mathcal{V} = 0, \qquad (2.7)$$

where the divergence theorem has been used on the integral of the surfaces forces. Since Eq. (2.7) holds for all choices of the material volume $\mathcal{V}$, and the integrands are continuous functions in space, then the corresponding differential form can be written as

$$\rho \frac{Du_i}{Dt} = \frac{\partial(\rho u_i)}{\partial t} + \frac{\partial(u_i \rho u_j)}{\partial x_j} = \frac{\partial}{\partial x_j} \Sigma_{ij} + \rho f_i. \qquad (2.8)$$

An alternate form of this equation, that can be useful in the subsequent analysis of compressible flows, can be obtained. It is easily extracted by introducing the vorticity vector $\boldsymbol{\omega}$ $(= \nabla \times \mathbf{u})$ or ω_i $(e_{ijk}(\partial u_k / \partial x_j))$ into the formulation using the vector identity

$$\boldsymbol{\omega} \times \mathbf{u} = \mathbf{u} \cdot \nabla \mathbf{u} - \nabla \left(\frac{|u|^2}{2} \right) \qquad (2.9a)$$

or, in Cartesian tensor notation

$$e_{ijk}\omega_j u_k = e_{ijk}e_{jlm} \left(\frac{\partial u_m}{\partial x_l} \right) u_k = u_j \frac{\partial u_i}{\partial x_j} - \frac{\partial}{\partial x_i} \left(\frac{u_j u_j}{2} \right), \qquad (2.9b)$$

where the (two-dimensional) tensor identity $e_{ijk}e_{jlm} = \delta_{kl}\delta_{im} - \delta_{km}\delta_{il}$ is also used. The alternate form for the momentum conservation equation can then be written as

$$\rho \frac{\partial u_i}{\partial t} + \rho e_{ijk}\omega_j u_k = -\rho \frac{\partial}{\partial x_i} \left(\frac{u_j u_j}{2} \right) + \frac{\partial}{\partial x_j} \Sigma_{ij} + \rho f_i. \qquad (2.10)$$

It is also straightforward to derive the corresponding compressible vorticity equation from Eq. (2.8), by simply taking the curl, in order to obtain

$$\rho \frac{D\omega_i}{Dt} = \rho \frac{\partial u_i}{\partial x_k}\omega_k - \rho \frac{\partial u_k}{\partial x_k}\omega_i - \frac{1}{\rho} \left(e_{ijk} \frac{\partial \rho}{\partial x_j} \frac{\partial \Sigma_{kl}}{\partial x_l} \right)$$
$$+ e_{ijk} \frac{\partial}{\partial x_j} \left(\frac{\partial \Sigma_{kl}}{\partial x_l} \right) + \rho e_{ijk} \frac{\partial f_k}{\partial x_j}. \qquad (2.11)$$

It remains to identify the exact forms of both the surface and body forces acting on a material element. This requires the specification of the local stress field Σ_{ij} and the applicable body force (per unit volume) ρf_i.

2.2.1 Surface Forces: The Stress Tensor

It is not necessary here in the discussion of the stress tensor to revert back to the elementary balance of forces presented in introductory texts. The starting point here will simply highlight important aspects with relation to compressible flows. In this context, it is assumed that no applied torques are imposed on the fluid and that the conservation of angular momentum simply shows the symmetry of the stress tensor Σ_{ij} ($= \Sigma_{ji}$).

In a fluid at rest, the only non-zero stresses are the normal stresses and these are independent of the surface normals on which they act. Thus, the stress tensor Σ_{ij} can be written in the isotropic form

$$\Sigma_{ij} = -p\delta_{ij}, \tag{2.12}$$

where p is the static-fluid or equilibrium pressure and is a thermodynamic variable related to ρ and T by an equation of state. For a fluid in motion, the stress not only consists of an isotropic part but also a non-isotropic or deviatoric part σ_{ij} to account for the fluid motion. The total stress tensor can then be written as

$$\Sigma_{ij} = -P\delta_{ij} + \sigma_{ij}, \tag{2.13}$$

where it should be noted that now the pressure P is different than the static-fluid pressure shown in Eq. (2.12). The mechanical pressure P is the pressure at a point in the moving fluid and is proportional to the mean normal stress acting on an element of fluid. Equation (2.13) then requires the specification of both the mechanical pressure as well as a constitutive equation for the stress tensor σ_{ij}.

A defining representation for the viscous stress tensor σ_{ij} for a Newtonian fluid can be formally deduced from a continuation of the discussion in Section 1.1 wherein Eq. (1.20) described the functional dependency required for the specification of the stress field. The dependency was based on an isotropic dependence on the density and a second-order tensorial dependency on the relative deformation gradient. Although beyond the scope of the current discussion, it is a straightforward exercise to show that the relative deformation gradient can be related to the Rivlin–Ericksen tensors (see Deville and Gatski, 2012; Huilgol and Phan-Thien, 1997). It is the Fundamental Theorem of Fluids first put forth by Noll and discussed, for example, in Truesdell (1991). The most general linear tensor representation for σ_{ij} in terms of the mean velocity gradient can be written as

$$\sigma_{ij} = \alpha_{ijkl} \frac{\partial u_k}{\partial x_l} = \alpha_{ijkl} \left(S_{kl} + W_{kl} \right), \tag{2.14}$$

where the mean velocity gradient has been partitioned into its symmetric and skew-symmetric parts, S_{kl} and W_{kl}, respectively, and are given by

$$S_{kl} = \frac{1}{2} \left(\frac{\partial u_k}{\partial x_l} + \frac{\partial u_l}{\partial x_k} \right), \quad W_{kl} = \frac{1}{2} \left(\frac{\partial u_k}{\partial x_l} - \frac{\partial u_l}{\partial x_k} \right). \tag{2.15}$$

Since the molecular structure of the fluid is statistically isotropic, α_{ijkl} is a fourth order isotropic tensor function (81 components) and as such can be written as the sum of the products of the Kronecker delta δ_{ij}, so that

$$\alpha_{ijkl} = \mu \left(\delta_{ik}\delta_{jl} + \delta_{il}\delta_{jk} \right) + \mu_1 \delta_{ij}\delta_{kl} + \mu_2 \left(\delta_{ik}\delta_{jl} - \delta_{il}\delta_{jk} \right), \quad (2.16)$$

where μ, μ_1, and μ_2 are scalar coefficients that depend on the local thermodynamic state. Since the stress tensor is symmetric, it immediately follows from Eq. (2.16), that $\mu_2 = 0$. Substituting this expansion into Eq. (2.14) yields

$$\sigma_{ij} = 2\mu S_{ij} + \mu_1 S_{kk}\delta_{ij}. \quad (2.17)$$

Recall that at the outset σ_{ij} was taken as the traceless (deviatoric) part of the stress tensor Σ_{ij}. Imposing this condition on Eq. (2.17) yields $\mu_1 = -2\mu/3$, and results in the Newtonian constitutive equation for the deviatoric part of the viscous stress tensor

$$\sigma_{ij} = 2\mu \left(S_{ij} - \frac{\delta_{ij}}{3} S_{kk} \right), \quad (2.18)$$

with μ the shear viscosity. It should be recognized that $-P\delta_{ij} + \sigma_{ij}$ represents the total stress acting on a moving fluid element with $-P\delta_{ij}$ the isotropic part and σ_{ij} the deviatoric part.

Also in a fluid at rest, the pressure P is now an equilibrium pressure p that can be extracted from a thermodynamic equation of state. However, when there is fluid motion, it is necessary to determine P in order to completely specify the surface stress. An estimate can be obtained for the difference between the mechanical and thermodynamic pressures $P - p$ by assuming that this difference is only a function of the mean velocity gradient. Following the same rationale that led to the functional relation for the stress tensor in Eq. (2.17) using Eq. (2.16), but now for a second-order isotropic tensor function yields

$$P - p = \mu\delta_{kl} \left(S_{kl} + W_{kl} \right)$$
$$= -\mu_v S_{kk}, \quad (2.19)$$

where $\mu_v (> 0)$ is the bulk or expansion viscosity. This latter terminology becomes apparent when Eq. (1.16) is used

$$P = p - \mu_v \left(\frac{1}{V} \frac{dV}{dt} \right). \quad (2.20)$$

With this, the constitutive equation for the viscous stress tensor can be written as

$$\Sigma_{ij} = - \left(p - \mu_v S_{kk} \right) \delta_{ij} + 2\mu \left(S_{ij} - \frac{\delta_{ij}}{3} S_{kk} \right). \quad (2.21)$$

For the most part, Stokes's hypothesis is usually invoked so that $\mu_v = 0$ is assumed. Unfortunately, this assumption is not borne out by experimental measurements, though commonly invoked. Nevertheless, unless otherwise noted, Stokes hypothesis will be used throughout so that,

$$\Sigma_{ij} = -p\delta_{ij} + 2\mu \left(S_{ij} - \frac{\delta_{ij}}{3} S_{kk} \right) = -p\delta_{ij} + \sigma_{ij}, \tag{2.22}$$

where $P = p$ will be taken for the remainder of the book.

2.2.2 Body Forces

As noted previously, body forces act uniformly on all elements within the fluid volume. Their action is a result of the fluid (volume) being embedded in a force field that acts directly on the fluid elements within the fluid. Common examples are gravitational, electrostatic and (electro)magnetic effects. Of these only the former gravitational force will be considered further here. A convenient feature of the gravitational force is that it can be written in terms of a force potential, that is,

$$\rho f_i = -\rho g_i = -\rho \frac{\partial \mathcal{F}_p}{\partial x_i}, \tag{2.23}$$

where $\mathcal{F}_p = g_k x_k$ is the force potential or potential energy per unit mass (the negative sign accounts for proper orientation of the gravitational force). Of course in geophysical applications where such effects are often encountered, it is also necessary to transform to a rotating frame of reference corresponding to the Earth's rotation.

2.3 ENERGY CONSERVATION

The conservation equation for the total energy is composed of contributions from both the kinetic energy and the internal energy of the body. It is a balance between the time rate of change of this total energy and the rate at which energy is transferred to the body through work and heat. Using Reynolds transport theorem, this balance statement can be written as

$$\int_{\mathcal{V}} \left[\frac{\partial (\rho E)}{\partial t} + \frac{\partial}{\partial x_j} (\rho E u_j) \right] d\mathcal{V} = \oint_{\mathcal{S}} u_i \Sigma_{ij} n_j \, d\mathcal{S} + \int_{\mathcal{V}} \rho u_i f_i \, d\mathcal{V} - \oint_{\mathcal{S}} q_j n_j \, d\mathcal{S}, \tag{2.24}$$

where the total energy ρE is

$$\rho E = \rho \left(e + \frac{u_i u_i}{2} \right), \tag{2.25}$$

with ρe the internal energy and $\rho u_i u_i / 2$ the kinetic energy. The right-hand side is necessarily composed of the same contributions attributed to the change of momentum of a fluid element, $\mathcal{F}_i(\mathbf{x}, t)$, as well as a contribution representing any addition (subtraction) of heat to the body. This heat change is quantified through the use of a heat flux vector q_i. (Note that the projection of the heat flux vector onto the surface unit normal $q_j n_j$ is defined as a positive quantity; however, a positive heat flux represents a loss of energy to the body and necessitates the minus sign in this term.) In final integral form over the volume, the balance equation for the total energy is

$$\int_{\mathcal{V}} \left[\frac{\partial(\rho E)}{\partial t} + \frac{\partial(u_j \rho E)}{\partial x_j} - \frac{\partial(u_i \Sigma_{ij})}{\partial x_j} - \rho u_i f_i + \frac{\partial q_j}{\partial x_j} \right] d\mathcal{V} = 0, \quad (2.26)$$

where the divergence theorem is used once again. With Eq. (2.26) holding for all choices of the material volume $\mathcal{V}$, the corresponding differential form can be written as

$$\frac{\partial(\rho E)}{\partial t} + \frac{\partial(u_j \rho E)}{\partial x_j} = \frac{\partial}{\partial x_j} \left[u_i \left(-p\delta_{ij} + \sigma_{ij} \right) \right]$$

$$+ \rho u_i f_i - \frac{\partial q_j}{\partial x_j}, \quad (2.27a)$$

$$\frac{\partial(\rho E)}{\partial t} + \frac{\partial(u_j \rho H)}{\partial x_j} = \frac{\partial(u_i \sigma_{ij})}{\partial x_j} + \rho u_i f_i - \frac{\partial q_j}{\partial x_j}, \quad (2.27b)$$

where Eq. (2.22) for Σ_{ij} has been used, and the total, or stagnation, enthalpy H (cf. Eq. (1.39)) has been introduced,

$$\rho H = \rho c_p T_t = \rho \left(E + \frac{p}{\rho} \right) = \rho \left(e + \frac{p}{\rho} + \frac{u_i u_i}{2} \right), \quad (2.28)$$

where T_t is the total, or stagnation, temperature. The conservation of total energy within the system is composed of both the kinetic and internal energy, and an equation for each contribution can be obtained.

An equation for the kinetic energy $\rho u_i u_i / 2$ can be derived from Eq. (2.8) by simply forming the scalar product, that is, multiplying by u_i. This yields

$$\rho \frac{D}{Dt} \left(\frac{u_i u_i}{2} \right) = \frac{\partial}{\partial t} \left(\rho \frac{u_i u_i}{2} \right) + \frac{\partial}{\partial x_j} \left(u_j \rho \frac{u_i u_i}{2} \right)$$

$$= u_i \frac{\partial}{\partial x_j} \Sigma_{ij} + \rho u_i f_i$$

$$= \frac{\partial}{\partial x_j} \left[u_i \left(-p\delta_{ij} + \sigma_{ij} \right) \right] - \left(-p\delta_{ij} + \sigma_{ij} \right) S_{ij} + \rho u_i f_i.$$

$$(2.29)$$

The internal energy contribution is obtained from Eq. (2.27) by subtracting out the kinetic energy contribution given in Eq. (2.29). The internal energy contribution can be written as

$$\rho \frac{De}{Dt} = \frac{\partial(\rho e)}{\partial t} + \frac{\partial(u_j \rho e)}{\partial x_j} = \left(-p\delta_{ij} + \sigma_{ij} \right) S_{ji} - \frac{\partial q_j}{\partial x_j}. \quad (2.30)$$

The internal energy of the system can be related to the first law of thermodynamics so that the first term on the right, $\Sigma_{ij} S_{ji}$, represents the rate at which surface forces do work on a fluid element and the second term represents the heat transferred (q_j the heat flux vector) to a fluid element. Many texts include the pressure term in the definition of σ_{ij}; however, since the pressure plays a central role in the analysis of compressible flows it is advantageous to separate it from the viscous stress tensor. For an ideal gas, for example, the internal energy is obtained from $e = \int c_v(T)\,dT$ and for a perfect gas, the internal energy is obtained from $e = c_v T$, where c_v is the specific heat at constant volume. A linear conduction model is usually assumed for the heat flux vector q_j yielding the relation (Fourier conduction law) $q_j = -k_T\, \partial T/\partial x_j$ (k_T is the thermal conductivity).

It is a straightforward extension of Eq. (2.30) to formulate the transport equation corresponding to the second law of thermodynamics. This relationship for the entropy s is given by (cf. Eq. (1.29))

$$T\frac{Ds}{Dt} = \frac{De}{Dt} + p\frac{D\upsilon}{Dt} = \frac{\sigma_{ij}}{\rho}S_{ji} - \frac{1}{\rho}\frac{\partial q_j}{\partial x_j}. \tag{2.31}$$

Alternative forms can be obtained by extending the equilibrium relations found in Eqs. (1.32) and (1.35) to a moving fluid. This gives

$$T\frac{Ds}{Dt} = c_p\frac{DT}{Dt} - \frac{\beta T}{\rho}\frac{Dp}{Dt} \tag{2.32a}$$

$$= -\frac{c_p}{\rho\beta}\left(\frac{D\rho}{Dt} - \frac{1}{c^2}\frac{Dp}{Dt}\right). \tag{2.32b}$$

If the flow is isentropic $Ds/Dt = 0$, then the temperature, pressure, and density variations can be related by

$$c_p\frac{DT}{Dt} = \frac{\beta T}{\rho}\frac{Dp}{Dt} = \frac{\beta T c^2}{\rho}\frac{D\rho}{Dt}, \tag{2.33a}$$

which for a perfect gas $\beta = 1/T$, reduces to

$$c_p\frac{DT}{Dt} = \frac{1}{\rho}\frac{Dp}{Dt} = \frac{c^2}{\rho}\frac{D\rho}{Dt}, \tag{2.33b}$$

and which shows that the temperature has no direct role on the fluid motion. It is a passive quantity changing in response to the pressure (density) field.

These equations will form the mathematical basis for the analysis and discussion to follow in subsequent chapters. Both the modeled statistical transport equations and the filtered transport equations used in numerical simulation of compressible flows have their origin in these equations.

There are alternate forms to the energy equation that help analyze the dynamics of compressible flows. The Bernoulli equation can be obtained by rewriting the total energy conservation, Eq. (2.27a). First, the body force term can be rewritten using a material derivative since it has a steady force potential, $g_k x_k$ so that

$$\rho u_i f_i = -\rho u_i \frac{\partial \mathcal{F}_p}{\partial x_i} = -\rho \frac{D \mathcal{F}_p}{Dt}. \tag{2.34}$$

Second, the pressure term that appears on the right side of Eq. (2.27a) can be written in the form

$$-\frac{\partial}{\partial x_j}\left(u_i p \delta_{ij}\right) = \frac{p}{\rho}\frac{D\rho}{Dt} - u_i \frac{\partial p}{\partial x_i}$$

$$= -\rho \frac{D}{Dt}\left(\frac{p}{\rho}\right) + \frac{\partial p}{\partial t}. \tag{2.35}$$

Using Eqs. (2.34) and (2.35), the total energy equation (2.27) can then be written as

$$\frac{D}{Dt}\left[\rho\left(E + \frac{p}{\rho} + \mathcal{F}_p\right)\right] = \frac{\partial p}{\partial t} + \frac{\partial}{\partial x_j}\left(u_i \sigma_{ij}\right) - \frac{\partial q_j}{\partial x_j}. \tag{2.36}$$

An obvious consequence of this equation is that for a steady pressure field in a flow where viscous and molecular heat conduction effects can be neglected, the quantity

$$\mathcal{B} = \rho\left(E + \frac{p}{\rho} + \mathcal{F}_p\right) = \rho\left(H + \mathcal{F}_p\right) \tag{2.37}$$

is constant at all points along the path (streamline) of the fluid volume. This result is known as Bernoulli's theorem—a result first established almost three centuries ago, and $\mathcal{B}$ is Bernoulli's constant. In addition, it can be seen from Eq. (2.31) that under the conditions when viscous and molecular heat conduction effects can be neglected, the flow can also be assumed to be isentropic. If the force potential can also be neglected relative to the other terms, $\mathcal{B}$ is equivalent to the stagnation enthalpy H. For a perfect gas $\mathcal{B}$ can be simplified further using the definition of the enthalpy given in Eq. (1.40). Equation (2.37) can then be written as

$$\mathcal{B} = \rho\left(\frac{u_i u_i}{2} + h + \mathcal{F}_p\right)$$

$$= \rho\left(\frac{u_i u_i}{2} + c_p T + \mathcal{F}_p\right). \tag{2.38}$$

The connection between the Bernoulli constant and the entropy can be further explored in simple steady flow situations. This can be done by considering the spatial variation of each quantity, $\mathcal{B}$ and s, within the flow. For the entropy, Eq. (1.29) can be generalized and written as,

$$T \frac{\partial s}{\partial x_i} = \frac{\partial e}{\partial x_i} + p \frac{\partial v}{\partial x_i}. \tag{2.39}$$

The corresponding equation for $\mathcal{B}$ can be easily extracted from Eq. (2.37),

$$\frac{\partial \mathcal{B}}{\partial x_i} = T \frac{\partial s}{\partial x_i} + \frac{\partial}{\partial x_i}\left(\frac{u_i u_i}{2} + \mathcal{F}_p\right) + \frac{1}{\rho}\frac{\partial p}{\partial x_i}, \tag{2.40}$$

where Eq. (2.25) has been used. If the alternate form of the momentum conservation equation is used along with the forms established for Σ_{ij} and ρf_i in Eqs. (2.22) and (2.23), respectively, then the variation of $\mathcal{B}$ across the flow can be written as

$$\frac{\partial \mathcal{B}}{\partial x_i} = T \frac{\partial s}{\partial x_i} + e_{ijk} u_j \omega_k, \tag{2.41}$$

which is a statement of Crocco's theorem. Note that this relation shows that the Bernoulli constant $\mathcal{B}$ and the entropy s have similar variations only when the flow is irrotational. Alternatively, if the entropy is uniform over the flow, then $\partial s / \partial x_i = 0$ (homentropic) and $\mathcal{B}$ is constant along vortex lines. If the flow is both irrotational and homentropic then $\mathcal{B}$ does not vary across the flow.

2.4 SOLENOIDAL VELOCITY FIELDS AND DENSITY CHANGES

With the general form of the dynamic equations governing the motion of a fluid flow just given, it is now possible to examine under what conditions the flow may not need to be considered compressible and when such effects can be neglected. When the density variations in the flow can be neglected; the mass conservation equation (2.3), simply reduces to the requirement that the velocity field be solenoidal, that is,

$$\left|\frac{1}{\rho}\frac{D\rho}{Dt}\right| < \left|\frac{\partial u_i}{\partial x_i}\right|. \tag{2.42}$$

Flow fields in which Eq. (2.42) hold are incompressible, and it is possible to show under what circumstances the approximation in Eq. (2.42) is valid. To show this, consider a form of the equilibrium thermodynamic relation given in Eq. (1.33) generalized for a fluid in motion,

$$\begin{aligned}\frac{1}{\rho}\frac{D\rho}{Dt} &= -\frac{\beta T}{c_p}\frac{Ds}{Dt} - \frac{1}{\gamma p}\frac{Dp}{Dt} \\ &= -\frac{\beta}{\rho c_p}\left[\sigma_{ij} S_{ji} - \frac{\partial q_j}{\partial x_j}\right] - \frac{1}{\gamma p}\frac{Dp}{Dt},\end{aligned} \tag{2.43}$$

where γ ratio of specific heats (c_p/c_v), and Eq. (2.31) is used. Each of the terms in Eq. (2.43) can be assessed relative to the compressibility condition given in Eq. (2.42), by a dimensional scaling analysis of the terms relative to the solenoidal velocity condition. Consider length, temperature, and velocity scales of the motion being given by L, θ, and U, respectively.

For the first term associated with the change of entropy Ds/Dt due to internal dissipative heating and molecular diffusion, this relationship can be written as

$$\underbrace{\left|\frac{\beta}{\rho c_p}(\sigma_{ij}S_{ji})\right|}_{\dfrac{\beta}{\rho c_p}\dfrac{\mu U^2}{L^2}} < \underbrace{\left|\frac{\partial u_i}{\partial x_i}\right|}_{\dfrac{U}{L}} \quad \text{and} \quad \underbrace{\left|\frac{\beta}{\rho c_p}\frac{\partial q_j}{\partial x_j}\right|}_{\dfrac{\beta}{\rho c_p}\dfrac{k_T \theta}{L^2}} < \underbrace{\left|\frac{\partial u_i}{\partial x_i}\right|}_{\dfrac{U}{L}} \tag{2.44}$$

or

$$\left(\frac{\mu\beta}{\rho c_p}\right)\frac{U}{L} < 1 \quad \text{and} \quad \frac{\beta\theta}{Pr\,Re} < 1, \tag{2.45}$$

where the Prandtl number Pr is defined as $\mu c_p/k_T$, and the Reynolds number Re as $\rho U L/\mu$. It is easy to show that these inequalities are satisfied and that changes in the dissipative heating and molecular diffusion are not sufficient to require accounting for compressibility effects. Consider air at 10 km, an altitude at which many commercial aircraft fly. (The fluid properties of air at this height are approximately given by: $\rho = 1.58$ kg/m^3, $\mu = 1.46 \times 10^{-5}$ kg/m–s, $c_p = 1.01 \times 10^3$ J/kg–K, $\beta = 4.48 \times 10^{-3}$ K^{-1}, and $Pr = 0.73$.) At this altitude, the temperature θ varies about 2°C for every 300 m (L), with a corresponding variation of wind speed (U) of 3 m/s. The result is a Reynolds number Re of 9.7×10^7. At these conditions, the values in Eq. (2.45) associated with dissipative heating and molecular diffusion are 4.1×10^{-13} and 1.3×10^{-10}, respectively. This shows that the changes in density due to changes in the entropy s through dissipative heating or molecular diffusion are not significant. Thus, changes in the entropy s, that is Ds/Dt, can be considered small and the rather common assumption of isentropic flow is justified. If the flow is isentropic, then it behaves thermodynamically as a reversible or adiabatic process.

In order to complete the assessment when compressibility effects become important, consider now the second term in Eq. (2.43) associated with changes in pressure Dp/Dt. Since the previous analysis of the first term shows that effects of viscosity and conductivity can be neglected, the inequality in Eq. (2.42) can be rewritten using the inviscid form of Eqs. (2.8), (2.12), and (2.23) as

$$\begin{aligned}
\frac{1}{\gamma p}\frac{Dp}{Dt} &= \frac{1}{\rho c^2}\frac{Dp}{Dt} \\
&= \frac{1}{\rho c^2}\frac{\partial p}{\partial t} - \frac{1}{2c^2}\frac{Du_i^2}{Dt} - \frac{u_i g_i}{c^2},
\end{aligned} \tag{2.46}$$

where a perfect gas is assumed, and

$$c^2 \equiv \left(\frac{\partial p}{\partial \rho}\right)_s = \frac{\gamma p}{\rho} = \gamma \mathcal{R} T \tag{2.47}$$

is the speed of sound. Once again each term can be examined relative to the velocity dilatation,

$$\underbrace{\left|\frac{1}{\rho c^2}\frac{\partial p}{\partial t}\right| < \left|\frac{\partial u_i}{\partial x_i}\right|}_{(i)}, \quad \underbrace{\left|\frac{1}{2c^2}\frac{Du_i^2}{Dt}\right| < \left|\frac{\partial u_i}{\partial x_i}\right|}_{(ii)} \quad \text{and} \quad \underbrace{\left|\frac{u_i g_i}{c^2}\right| < \left|\frac{\partial u_i}{\partial x_i}\right|}_{(iii)}. \quad (2.48)$$

In contrast to the assessment of the entropic term which involved properties of the fluid, the individual term assessment in Eq. (2.48) is primarily related to the behavior of the flow, and is a distinct dynamic effect that needs to be considered separately.

The first term (i) in Eq. (2.48) is directly associated with flow unsteadiness. If the unsteady pressure field is characterized by the frequency f (and associated time scale f^{-1}), then an estimate for the magnitude of the pressure wave is extracted from an estimate of the momentum (per unit area) acting over this time scale, that is, ρULf. The inequality given in Eq. (2.48) (i) can then be written as

$$\underbrace{\left|\frac{1}{\rho c^2}\frac{\partial p}{\partial t}\right|}_{\dfrac{ULf^2}{c^2}} < \underbrace{\left|\frac{\partial u_i}{\partial x_i}\right|}_{\dfrac{U}{L}},$$

$$\frac{L^2 f^2}{c^2} < 1, \quad (2.49)$$

which is in the form of a condition on the Strouhal number (St) squared. For an acoustic wave propagating through the fluid, L is the wavelength and Lf/c is unity. Thus Eq. (2.49) does not hold and the full mass conservation equation needs to considered. Alternately, if the frequency f of the pressure wave is represented by the characteristic length L and velocity U scales (that is U/L), then Eq. (2.49) is equivalent to a condition on the Mach number (squared), that is $M^2 = U^2/c^2 < 1$. Then in cases where $f \gg U/L$, the condition given in Eq. (2.49) may once again be violated.

The second term (ii) in Eq. (2.48) is associated with the kinetic energy transport and is easily scaled as

$$\underbrace{\left|\frac{1}{2c^2}\frac{Du_i^2}{Dt}\right|}_{\dfrac{U^3}{c^2 L}} < \underbrace{\left|\frac{\partial u_i}{\partial x_i}\right|}_{\dfrac{U}{L}},$$

$$\frac{U^2}{c^2} = M^2 < 1. \quad (2.50)$$

This Mach number condition is the one most familiar to readers and the one most closely associated with the need to account for compressibility effects.

Since the only body force contribution that will be considered is due to gravity, the remaining term (iii) in Eq. (2.48) is associated with the condition

$$\underbrace{\left|\frac{u_i g_i}{c^2}\right|}_{\dfrac{Ug}{c^2}} < \underbrace{\left|\frac{\partial u_i}{\partial x_i}\right|}_{\dfrac{U}{L}},$$

$$\frac{gL}{c^2} = \frac{\rho g L}{\gamma p} < 1, \tag{2.51}$$

where the isentropic relation for the speed of sound in a perfect gas equation (2.47) was used. Such a condition has a direct aeronautical relevance, since at commercial aircraft cruise altitudes the fluid is compressible. In an atmospheric context, the inequality in Eq. (2.51) states that if the static-fluid pressure between two points a (vertical) distance L apart is less than the absolute pressure, then the velocity field can be considered solenoidal. Another interpretation, based on atmospheric scale height $p/\rho g$ (e.g. Thompson, 1988), is more illustrative. (The atmospheric scale height for Earth under standard conditions is approximately 8.4 km.) In this case Eq. (2.51) reduces to $(L/\gamma) < (p/\rho g)$, and the velocity field in the lower atmosphere can be considered solenoidal over layers of several hundred meters. In comparison, for a planet such as Mars with its surface gravity only about one-third that of Earth, the atmospheric scale height is approximately 18.1 km and compressibility effects need to be considered over a much larger range of heights.

The inequalities in Eqs. (2.45), (2.49)–(2.51) have provided important insights into the practical conditions under which density variations, and hence compressibility effects, need to be considered. Equation (2.45) showed that under most situations the changes in entropy due to dissipative heating and molecular diffusion effects can be neglected relative to dilatation effects. From the inequality in Eq. (2.49) it was shown that certain characteristic frequencies associated with the passage of pressure waves (such as acoustic waves) can require that compressibility conditions be taken into account. The inequality extracted from the transport of kinetic energy equation (2.50), is the most familiar to those involved in aeronautical and aerospace applications. It will be shown later, that for turbulent flows, different characteristic velocities used in defining a Mach number can have significantly different effects on the flow analysis.

The final inequality just discussed involved a condition associated with a body force on the fluid due to gravity. For completeness, it is worth considering this condition further, since invoking a solenoidal condition on the velocity field simplifies the mass conservation equation, density variations through the Boussinesq approximation can still be allowed and accounted for. Even though this is a topic of relevance in geophysical flows, it is worth including here in the discussion.

If the velocity field is solenoidal, then the momentum equation (2.8) reduces to

$$\rho \frac{Du_i}{Dt} = -\frac{\partial p}{\partial x_i} - \rho g_i + \mu \frac{\partial^2 u_i}{\partial x_j^2}. \qquad (2.52)$$

In oceanographic flows, for example, the fluid motion is customarily described about some static reference state

$$p = p_s + \overline{p} \quad \text{and} \quad \rho = \rho_0 + \rho', \qquad (2.53)$$

where the reference state pressure is written as $p_s = \rho_0 g_i x_i$, and $\overline{p}$ and ρ' are changes in pressure and density, respectively, from these reference states. The momentum equation becomes

$$\left(1 + \frac{\rho'}{\rho_0}\right) \frac{Du_i}{Dt} = -\frac{1}{\rho_0} \frac{\partial \overline{p}}{\partial x_i} - \frac{\rho'}{\rho_0} g_i + v \frac{\partial^2 u_i}{\partial x_j^2}, \qquad (2.54)$$

where v is a kinematic viscosity $v = \mu/\rho_0$. With the assumption of small variations in density $\rho' < \rho_0$, Eq. (2.54) becomes

$$\rho_0 \frac{Du_i}{Dt} = -\frac{\partial \overline{p}}{\partial x_i} - (\rho - \rho_0) g_i + \mu \frac{\partial^2 u_i}{\partial x_j^2}. \qquad (2.55)$$

The buoyancy term $(\rho - \rho_0) g_i$ in the momentum equation is important and cannot be neglected. Interestingly, Lu (2001) has suggested to reformulate some Boussinesq circulation models into terms of density-weighted variables (see Section 3.2). Such a modification, if proven applicable, will more closely bind the discussion here to these geophysical flows.

Since the primary topical focus of this book is on engineering compressible flows rather than geophysical flows, the inclusion of the buoyancy term in either the momentum or total energy equation will not be considered further. Suffice it to say, however, that the averaging and filtering procedures discussed in the next chapter can be applied to the different conservation equations under the Boussinesq approximation in geophysical applications.

2.5 TWO-DIMENSIONAL FLOW AND A REYNOLDS ANALOGY

Some insights into the underlying dynamics of compressible flows can be easily extracted by considering the case of two-dimensional steady wall bounded flow. For simplicity of notation, the flow is along a flat plate in the $(x_1, x_2) = (x, y)$ plane with x the streamwise direction, and y the direction of the mean shear (normal to plate), and with the respective velocities (u_1, u_2) given by (u, v). The flow is assumed to be slowly varying in the streamwise direction $(v \ll u)$, and the equation of state for a perfect gas is used. The applicable mass, momentum and energy conservation equations (2.3), (2.8), and (2.27b), respectively, in this boundary layer flow can be approximated as

$$\frac{\partial(\rho u)}{\partial x} + \frac{\partial(\rho v)}{\partial y} = 0 \tag{2.56a}$$

$$\frac{\partial(u\rho u)}{\partial x} + \frac{\partial(v\rho u)}{\partial y} = -\frac{\partial p}{\partial x} + \frac{\partial \sigma_{xy}}{\partial y} \tag{2.56b}$$

$$\frac{\partial(u\rho H)}{\partial x} + \frac{\partial(v\rho H)}{\partial y} = \frac{\partial(u\sigma_{xy})}{\partial y} - \frac{\partial q_y}{\partial y}, \tag{2.56c}$$

where body forces are neglected, and the total or stagnation enthalpy is approximated as

$$H \approx c_p T + \frac{u^2}{2}, \tag{2.57}$$

which is equivalent to the Bernoulli constant in Eq. (2.38). Consistently, the viscous stress tensor can be approximated as

$$\sigma_{xy} = \sigma_{12} \approx \mu \frac{\partial u}{\partial y}, \tag{2.58}$$

and the heat flux vector q_y, given by the Fourier conduction law, can be written as

$$q_y = q_2 \approx -k_T \frac{\partial T}{\partial y} = -\left(\frac{k_T}{c_p}\right) \frac{\partial}{\partial y} \left(H - \frac{u^2}{2}\right). \tag{2.59}$$

In the flow field under consideration here, both the thermal conductivity k_T and viscosity μ are assumed constant across the layer. Equations (2.56b) and (2.56c) can then be rewritten as

$$\frac{\partial(u\rho u)}{\partial x} + \frac{\partial(v\rho u)}{\partial y} = -\frac{\partial p}{\partial x} + \frac{\partial}{\partial y}\left(\mu \frac{\partial u}{\partial y}\right) \tag{2.60}$$

$$\frac{\partial(u\rho H)}{\partial x} + \frac{\partial(v\rho H)}{\partial y} = \frac{\partial}{\partial y}\left[\left(\frac{k_T}{c_p}\right) \frac{\partial H}{\partial y}\right.$$
$$\left. + \mu\left(1 - \frac{1}{Pr}\right) \frac{\partial}{\partial y}\left(\frac{u^2}{2}\right)\right]. \tag{2.61}$$

If the assumption that $Pr = 1$ is made, which implies a relative balance between viscous dissipation and thermal transport, Eq. (2.56c) has the particular solution $H = $ const. This solution is the first of two Crocco–Busemann relations (e.g. White, 1991) and, as a consequence of the solution of constant total enthalpy (temperature), implies adiabatic conditions at the wall. If, additionally, the assumption of zero pressure gradient is imposed, it is clear from Eqs. (2.60) and (2.61) that $H \propto u$, and

$$H = a_0 u + a_1, \tag{2.62}$$

where a_1 is the total enthalpy at the wall H_w, and a_0 is related to both the adiabatic wall and wall temperatures $(T_{aw} - T_w)/u_e$ (with u_e the streamwise edge velocity) so that (cf. White, 1991)

$$H - H_w = \frac{c_p(T_{aw} - T_w)}{u_e} u. \tag{2.63}$$

If this relation, which is the second Crocco–Busemann relation, is used in the heat flux equation (2.59), and evaluated at the wall, it follows that

$$q_y\big|_w \approx -k_T \frac{(T_{aw} - T_w)}{u_e} \left(\frac{\sigma_{xy}}{\mu} \right)\bigg|_w, \tag{2.64}$$

where Eq. (2.58) has been used. The Crocco–Busemann relation in Eq. (2.63) then becomes,

$$H - H_w = Pr_w \left(\frac{q_y}{\sigma_{xy}} \right)\bigg|_w u, \tag{2.65}$$

with $Pr_w = \mu_w c_p/k_T$, and is a form of the Reynolds analogy.

While such manipulations are straightforward and necessarily involve some simplifying assumptions, they provide a general framework from which many useful relations between the velocity and temperature fields can be obtained. Additionally, these introductory ideas are the basis from which many useful relations between temperature and velocity correlations in turbulent flows can be deduced.

Chapter 3

Compressible Turbulent Flow

In the previous chapter, the underlying equations used to describe the motion of a compressible fluid flow have been derived. The solution of these equations with appropriate boundary and initial conditions will yield a description of a laminar or turbulent compressible flow. For practical flows of engineering interest, numerical solutions of these equations are required. In the case of turbulent flows, such direct simulations require large computational meshes in order to capture enough scales of motion to accurately reproduce the dynamics of the flow. Even with current computational capabilities that allow for simulations with computational meshes of $\mathcal{O}(10^{10})$, Reynolds numbers are still limited to $\mathcal{O}(10^4)$ which is generally well below relevant engineering Reynolds numbers. As in the case of incompressible flows, alternate formulations for the prediction of turbulent flows are used. In general terms, these alternatives require the solution of the conservation equations, but now in terms of ensemble-averaged or filtered independent variables. In this chapter, these averaging or filtering procedures will be presented and the equations describing the motion in terms of these variables will be presented.

3.1 AVERAGED AND FILTERED VARIABLES

Currently and for the foreseeable future, it will be necessary for the numerical simulation of practical engineering turbulent flow fields to solve a set of equations for flow variables that represent the motion of a limited spectral range of scales. This description holds true for methods such as Reynolds-averaged Navier-Stokes (RANS), large eddy simulation (LES), and any of the newly developed hybrid or composite methodologies. As such, the equations describing the averaged or filtered motions in any of these formulations are form-invariant, that is, they contain terms in any coordinate frame of reference that represent transport, production, redistribution and diffusion. They are not, of course, frame-invariant since in non-inertial frames or under Euclidean

Compressibility, Turbulence and High Speed Flow. http://dx.doi.org/10.1016/B978-0-12-397027-5.00003-4

transformations additional terms in the general tensor forms of the equations appear, and depending on the filtering process, they obviously differ with respect to the flow field motions being described. This difference in filtering dictates how the higher-order correlations are parameterized.

As in the incompressible case, the statistical analysis of compressible turbulent flows begins with an averaging or filtering process that effectively partitions the dependent variables, such as density, pressure, temperature, and velocity. This yields in all formulations, the flow variable f being decomposed into an averaged or filtered part, $\overline{f}$, and a fluctuating or subfiltered, unresolved part, f', given as

$$f(\mathbf{x},t) = \overline{f}(\mathbf{x},t) + f'(\mathbf{x},t). \tag{3.1}$$

The filtering process can be defined as a subset of the general operation (see Sagaut, 2006)

$$\overline{f}(\mathbf{x},t) = \mathcal{G} * f = \int \mathcal{G}(\mathbf{x}-\mathbf{x}',t-t')f(\mathbf{x}',t')d\mathbf{x}'dt', \tag{3.2}$$

where different forms for the convolution kernel can be associated with the various solution methodologies.

3.1.1 Reynolds Average

For the RANS formulation, the partitioning is usually expressed in terms of an ensemble mean and fluctuating part called a Reynolds decomposition. If the turbulence field is statistically steady or stationary, ergodicity is assumed and a (long) time average over many turbulent time scales corresponds to an ensemble average. Similarly, if the turbulence is statistically homogeneous a spatial average over many turbulent length scales corresponds to an ensemble average.

Most practical flow fields are inhomogeneous flows so that the Reynolds average is often formulated in terms of time averages. Nevertheless, there are flow fields where assumptions of homogeneity can be applied in at least one coordinate direction. In these directions, spatial averaging is used and combined with a time average. This type of situation often arises in direct numerical simulations of developing two-dimensional boundary layers, where in the spanwise coordinate direction the condition of homogeneity is applied, or in fully developed channel flow where the condition of homogeneity is applied in the streamwise direction as well as in the spanwise direction. In these cases, the combination of space and time averaging results in converged statistics much faster than if a single time average is employed.

For the Reynolds average, the filter function (Gatski, Rumsey, & Manceau, 2007) in Eq. (3.2) is then given by

$$\mathcal{G}(\mathbf{x} - \mathbf{x}', t - t') \equiv \mathcal{G}_T(\mathbf{x} - \mathbf{x}', t - t') = G(\mathbf{x} - \mathbf{x}')G_T(t - t'; T)$$

$$= \delta(\mathbf{x} - \mathbf{x}')\mathcal{H}(t - t')\frac{\mathcal{H}(T - t + t')}{T}, \qquad (3.3)$$

so that

$$\overline{f}_T(\mathbf{x}, t; T) = \mathcal{G}_T * f = \frac{1}{T}\int_{t-T}^{t} f(\mathbf{x}, t')dt', \qquad (3.4)$$

with $\mathcal{H}$ representing the Heaviside function. In such flows, the entire spectral range of scales is modeled so the subfiltered part f' is a fluctuating quantity whose average is zero $\overline{f'} = 0$, and the filtered, or mean quantity $\overline{f}$ can be extracted as

$$\overline{f}(\mathbf{x}) = \lim_{T \to \infty} \overline{f}_T(\mathbf{x}; T) = \lim_{T \to \infty} \frac{1}{T}\int_0^T f(\mathbf{x}, t)dt. \qquad (3.5)$$

Note that if the Reynolds average is expressed as

$$\overline{f}(\mathbf{x}) = \left(\lim_{T \to \infty} \mathcal{G}_T\right) * f, \qquad (3.6)$$

then it cannot be interpreted as a convolution filter since $\lim_{T \to \infty} \mathcal{G}_T(t) = 0$. However, as shown by Eq. (3.5) and recognized very early by Kampé de Fériet and Betchov (1951), the Reynolds average can be expressed as a series of "truncated" functions obtained by convolution. As a practical matter, Reynolds averaging is evaluated as long-time averaging, where "long-time" implies long compared to the turbulence time scale. Reynolds averaging can then be considered as the convolution filter $\mathcal{G}_T$ with a sufficiently large T. Once established, the important condition that the Reynolds-average is idempotent $(\overline{\overline{f}} = \overline{f})$ holds, and fluctuations are centered about the mean value, $\overline{f'} = 0$.

3.1.2 Average Over Fixed Phase

For a flow statistically periodic or cyclic in time, a filtering procedure associated with the various phases of the underlying low wavenumber/frequency unsteadiness can be formulated. In this case, an average over fixed phase is formalized with a corresponding filter function given by

$$\mathcal{G}(\mathbf{x} - \mathbf{x}', t - t') \equiv \mathcal{G}_{T_N}(\mathbf{x} - \mathbf{x}', t - t') = G(\mathbf{x} - \mathbf{x}')G_{T_N}(t - t'; T_p)$$

$$= \delta(\mathbf{x} - \mathbf{x}')\left[\sum_{n=0}^{N}\frac{\delta(T_n - t + t')}{N + 1}\right], \qquad (3.7)$$

where $T_n = nT_p + \phi$ for given phase ϕ and period T_p. The corresponding average over a particular phase is then given by

$$\langle f(\mathbf{x},t; T_p) \rangle_N = \mathcal{G}_{T_N} * f = \frac{1}{N+1} \sum_{n=0}^{N} f(\mathbf{x}, t - T_n) \qquad (3.8a)$$

$$\langle f(\mathbf{x},t) \rangle = \lim_{N \to \infty} \langle f(\mathbf{x},t; T_p) \rangle_N = \lim_{N \to \infty} \frac{1}{N+1} \sum_{n=0}^{N} f(\mathbf{x}, t - T_n), \qquad (3.8b)$$

and the phase average of the subfiltered part f' is zero, $\langle f' \rangle = \overline{f'} = 0$. As Eq. (3.8b) shows, this filtering procedure produces a time varying filtered field. If coupled with the Reynolds averaging procedure derived in Section 3.1.1, a time-dependent representation of the resolved scale (filtered) motions can be extracted from $\langle f(\mathbf{x},t) \rangle - \overline{f}(\mathbf{x})$. This procedure is the basis of the so-called triple-decomposition formalism (Hussain & Reynolds, 1970; Reynolds & Hussain, 1972) since it further partitions the filtered field in Eq. (3.1) into time-independent and time-dependent parts. These ideas can be generalized, although less rigorously, to conditional averages where the time-averaged dependency is cyclic rather than periodic.

3.1.3 Temporal LES Filters

Although currently not a common formalism for the large eddy simulation methodology, the Eulerian temporal filtering of the governing equations is a straightforward extension of the filtering concepts under discussion. This relatively new procedure has received little attention in the past although some work (Pruett, 2000; Pruett, Gatski, Grosch, & Thacker, 2003) suggests it might be an effective alternative to the common spatial filtering process (see Section 3.1.4).

Causal time domain filters are used for the temporal LES since they are applicable to real time processes and only depend on past and present inputs. The associated filter function in Eq. (3.2) is then given by (cf. Eq. (3.3))

$$\mathcal{G}(\mathbf{x},t) \equiv \mathcal{G}_\Delta(\mathbf{x},t; \Delta) = G(\mathbf{x})G_\Delta(t; \Delta) = \delta(\mathbf{x}) \frac{\mathcal{H}(\Delta + t)}{\Delta}, \qquad (3.9)$$

where Δ is here a temporal filter width. Unlike the earlier Reynolds and phase-averaged approaches, only a portion of the spectral range of scales is modeled through the subfiltered part f'. The filtered quantity $\overline{f}$ can then be constructed in an analogous manner from Eq. (3.4),

$$\overline{f}_\Delta(\mathbf{x},t; \Delta) = \mathcal{G}_\Delta * f = \frac{1}{\Delta} \int_{t-\Delta}^{t} f(\mathbf{x},t')dt'. \qquad (3.10)$$

Within the realm of temporal filtering, it is possible to develop a more rigorous linkage between the large eddy and Reynolds-averaged approaches (Pruett et al., 2003), since the Reynolds-averaged function $\overline{f}(\mathbf{x})$ is the limit of the temporally-filtered function $\overline{f}_\Delta$ when the temporal filter width Δ goes to infinity. In the context of spatial filtering to be discussed next, such a linkage can only be rigorously established in homogeneous flows. Nevertheless, it should be recognized that an inherent linkage exists between any temporal filtering process and a corresponding spatial one.

3.1.4 Spatial LES Filters

For the most part, the large eddy simulation methodology has been based on spatial filtering and was originally formulated for geophysical (atmospheric) applications (Deardorff, 1970; Lilly, 1966; Smagorinsky, 1963). The formalism behind the requisite spatial filters is well established and documented (e.g. Sagaut, 2006), and a discussion analogous to that for the time domain filters can also be presented for spatial filters in homogeneous flows. For the most part, the top-hat, Gaussian, and spectral filters are the three convolution filters most commonly employed for the scale partitioning. Of these, the top-hat filter most directly connects with the Reynolds and phase-averaged filters just discussed. However, the spectral or sharp cutoff filters have the idempotent property which, as will be shown in Section 3.2, is central in relating some Reynolds-averaged variable and density-weighted variables. This may prove important in future developments where alternative approaches are considered, such as in spectral element or discontinuous Galerkin methods.

Consider the top-hat filter with a spatial filter cut-off length of $\mathbf{\Delta}$, and a corresponding (centered) filter function written as

$$\mathcal{G}(\mathbf{x},t) \equiv \mathcal{G}_\mathbf{\Delta}(\mathbf{x},t; \mathbf{\Delta}) = G_\mathbf{\Delta}(\mathbf{x}; \mathbf{\Delta})G(t) = \frac{\mathcal{H}\left(\frac{\Delta_i}{2} - |x_i|\right)}{\Delta}\delta(t). \quad (3.11)$$

The filter width $\mathbf{\Delta}$ is written here as a vector to emphasize that the spatial filtering process should, in general, be inhomogeneous. This leads to several additional considerations that are beyond the intent and scope of the discussion here. The intent here is to introduce the reader to a common filtering framework that links many of the current methodologies together. Thus, for the remainder of the discussion in this section, only processes associated with filters that are homogeneous and isotropic will be considered. Within these limits, a filtered quantity $\overline{f}$ can then be constructed from Eq. (3.2) as

$$\overline{f}_\Delta(\mathbf{x},t; \Delta) = \mathcal{G}_\Delta * f = \frac{1}{\Delta} \int_{x-\frac{\Delta}{2}}^{x+\frac{\Delta}{2}} f(\mathbf{x}',t)d\mathbf{x}'. \quad (3.12)$$

A comparison of the filtered variable in Eq. (3.12), for flows where homogeneous and isotropic filters apply, with the corresponding form for the temporally filtered variable in Eq. (3.10), shows that in the limit of "infinite" filter sizes both processes yield filtered variables consistent with a Reynolds-averaged mean variable. Although the concept of a large eddy simulation methodology for engineering flows is over 30 years old (Reynolds, 1976), several issues associated with procedure remain unresolved. The most prominent among these are the computational requirements associated with wall-bounded flows. This has led to the development of zonal or global hybrid models to help alleviate the computational overhead. Other issues are the inherent limitations associated with grid anisotropies for both structured and unstructured computational meshes. Many of these issues, for example, are thoroughly discussed in the book by Sagaut (2006) (the interested reader is also referred to the books by Geurts (2004) and Lesieur, Métais, and Comte (2005) for additional references and discussion).

It should be recognized, that even though a common linkage exists between the various filters, both the spatial and temporal LES filters only replicate the Reynolds-averaged filter in a large filter-width limit. In practice, these LES filters act as low-pass filters in space and time of the Fourier components of any flow variable. As such the LES filters presented here do not replicate the Reynolds filter so that $\overline{\overline{f}} \neq \overline{f}$ and $\overline{f'} \neq 0$, that is, they are not idempotent in physical (or spectral) space. Fortunately, this behavior does not impact the form of filtered conservation equations derived for each of the processes just described.

3.2 DENSITY-WEIGHTED VARIABLES

From an historical perspective, it is worth noting that in Osborne Reynolds's original paper (Reynolds, 1895) on "...incompressible viscous fluids ...," the (spatial) average was defined relative to density-weighted variables. Extending the ideas developed for the study of atmospheric flows, Favre (1965a, 1965b) (see also Favre, 1969) investigated the statistical equations of turbulent compressible flows. This study was used to identify an optimal partitioning method for the flow field variables to be decomposed into macroscopic and fluctuating parts. In this method, originally termed in French "Méthode B," three primary factors were taken into consideration in order to provide a simpler form of the equations. These factors were: mathematical treatment, physical interpretation, and measurement. The results of this study suggested that a decomposition of the flow field variables in terms of density-weighted or, as often now termed, Favre variables, was the best choice. Throughout, the terminology mass-weighted will often be interchanged with density-weighted and *vice versa*. The two terms are both commonly used but as the derivations show, the term density-weighted provides the more accurate description.

For a dependent variable f, the density-weighted average is defined as

$$\tilde{f} = \frac{\overline{\rho f}}{\overline{\rho}}, \qquad (3.13)$$

where any of the filtering processes discussed in Section 3.1 can be applied. The associated partitioning into filtered and subfiltered or unresolved part is usually now written as

$$f(\mathbf{x},t) = \tilde{f}(\mathbf{x}, t) + f''(\mathbf{x}, t), \qquad (3.14)$$

where $\tilde{f}(\mathbf{x}, t)$ is the averaged or filtered part and $f''(\mathbf{x}, t)$ the fluctuating or unresolved part. Before examining how this partitioning method relates to the mathematical treatment, physical interpretation, and measurement factors mentioned above, it is worth highlighting some inherent consequences between methods based on averaging and ones based on filtering.

One of the most fundamental is easily obtained by equating the two variable decompositions and applying an averaging or filtering procedure. For an averaging procedure

$$\overline{f} = \overline{\overline{f}} + \overline{f'} = \overline{\tilde{f}} + \overline{f''} = \tilde{f} + \overline{f''}, \qquad (3.15a)$$

which yields

$$\overline{f''} = \overline{f} - \tilde{f} = f'' - f'. \qquad (3.15b)$$

However, for a filtering procedure,

$$\begin{aligned} \overline{f} &= \overline{\overline{f}} + \overline{f'} = \overline{\tilde{f}} + \overline{f''} \\ &= \overline{\frac{\overline{\rho f}}{\overline{\rho}}} + \overline{f''} \\ &\neq \tilde{f} + \overline{f''}, \end{aligned} \qquad (3.16)$$

since the filtering procedure is not idempotent as noted in Section 3.1.4, and

$$\overline{f''} \neq \overline{f} - \tilde{f} \qquad (3.17a)$$
$$\neq f'' - f'. \qquad (3.17b)$$

Unfortunately, the fact that this relation (and others to be derived later in this section and in Section 4.4), is invalid because of the filtering operation has a direct impact on the ability to compare numerical results from filtered equations with experimental results. As will be discussed further in Chapter 4, Reynolds fluctuating quantities, either f' or $(\rho f)'$, are measured experimentally. In order to compare to computational results, which are formulated in terms of density-weighted variables, relations analogous to Eq. (3.15b) are required. In addition, validation of results obtained by numerical solutions of filtered transport equations against some well-established theoretical relations, such as the strong Reynolds analogy (SRA) discussed in Section 3.5, is precluded. It is possible, however, to obtain an *a priori* assessment of such numerical solutions by applying the appropriate filters to direct numerical simulation data. Nevertheless, questions associated with the ability of such methods to be quantitatively evaluated, or used as a quantitative measure of flow dynamics, naturally arise.

Another useful relation can be obtained directly from Eqs. (3.13) and (3.14). First the density field ρ, for either a Reynolds or a Favre decomposition, is partitioned such that the fluctuations are centered, that is $\overline{\rho'} = 0$ and $\rho = \overline{\rho} + \rho'$. Then, using the decomposition given in Eq. (3.14), the averaging operation applied to ρf can be written as

$$\overline{\rho f} = \overline{\rho \widetilde{f}} + \overline{\rho f''}$$
$$= \overline{\rho} \widetilde{f} + \overline{\rho f''}, \tag{3.18}$$

but with $\overline{\rho f''} = \overline{\rho} \widetilde{f''}$, Eq. (3.18) only holds if

$$\widetilde{f''} = 0. \tag{3.19}$$

This shows that the density-weighted average of the Favre fluctuation is centered about the corresponding mean. In Eq. (3.15b), it was shown that the Reynolds average of the Favre fluctuating quantity was non-zero and a measure of the difference between the two mean and the two fluctuating values. Alternatively, an analogous relation can be obtained from the definition of the Favre average, that is

$$\widetilde{f} = \frac{\overline{\rho f}}{\overline{\rho}} = \overline{f} + \frac{\overline{\rho f'}}{\overline{\rho}} = \widetilde{f} + \frac{\overline{\rho f''}}{\overline{\rho}}, \tag{3.20a}$$

or, after rearranging,

$$\frac{\overline{\rho' f'}}{\overline{\rho}} = \frac{\overline{\rho f'}}{\overline{\rho}} = \widetilde{f'} = \widetilde{f} - \overline{f}. \tag{3.20b}$$

Equation (3.15b) provides a measure based on the Reynolds average of the Favre fluctuating velocity (which is nonzero), and Eq. (3.20b) provides a measure based on the correlation between the density (fluctuations) and the fluctuations associated with the variable f. While Eq. (3.15b) shows that in general $\overline{f''} \neq 0$, Eq. (3.19) shows that the Favre average of the Favre fluctuations $\widetilde{f''} = 0$. This is in direct contrast with the averaging of the Reynolds fluctuations where $\overline{f'} = 0$ and $\widetilde{f'} \neq 0$. Equations (3.15b), (3.19), and (3.20b) can be combined to show that Reynolds-averaged Favre fluctuations can be related to the fluctuating density correlations by

$$\overline{f''} = -\frac{\overline{\rho' f'}}{\overline{\rho}} = -\frac{\overline{\rho' f''}}{\overline{\rho}} = -\widetilde{f'}. \tag{3.21}$$

These relations between the two decompositions form the basis for evaluating the three factors that were taken into account in formulating the equations governing the compressible, turbulent motion. The factor associated with the mathematical treatment relates directly to the issue of a simpler form of the equations. As Favre (1965a, 1965b) showed, the form of the mass and momentum conservation equations are simpler with the density-weighted variables. However, when constitutive equations are introduced, such as for the viscous stress, the Reynolds-variable partitioning of these terms yield the simpler form (Rubesin & Rose, 1973). For this reason, the transport equations for the various correlations that arise in compressible turbulent flows often have a mixed notation where some terms appear with density-weighted variables and others with Reynolds-averaged variables.

There are two aspects associated with the physical interpretation factor. One is related to the general view of turbulent flows, and the other is related to the closure problem. In general, the choice of a density-weighted variable imposes on the averaged form of the state equation and the mass conservation equation a strict form-invariance relative to the incompressible formulation. This means that the surfaces delimited by the mean streamlines are tangent to the density-weighted average velocity vector, and that these surfaces are on average sealed so that there is no mass flux through these surfaces (see Figure 3.1). Locally, the streamlines associated with the mass flux $\overline{\rho}\,\widetilde{\mathbf{u}}$ make an angle in the coordinate frame of $\tan^{-1}(\widetilde{u}_2/\widetilde{u}_1)$. If the Reynolds variables are used, such an interpretation is not possible since there is a fluctuating mass flux correlation contribution, $\overline{\rho'\mathbf{u}'}/\overline{\rho}$ as seen from Eq. (3.20a). Locally, the streamlines associated with the mass flux $\overline{\rho}\,\overline{\mathbf{u}}$ make an angle in the coordinate frame of $\tan^{-1}(\overline{u}_2/\overline{u}_1)$. The difference between the angle associated with the Reynolds variables and that associated with the Favre variables, a_{dif}, is easily

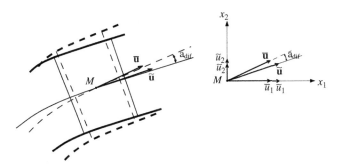

FIGURE 3.1 Sketch of streamline bounded mass-elements of fluid associated with Reynolds variables and density-weighted variables.

extracted from the trigonometric identity and written as

$$a_{\text{dif}} = \tan^{-1}\left[\frac{\overline{u}_2\widetilde{u}_1 - \widetilde{u}_2\overline{u}_1}{\overline{u}_1\widetilde{u}_1 + \overline{u}_2\widetilde{u}_2}\right]. \tag{3.22}$$

Clearly the angle orientation of the streamline given by the Favre-variables ($\arctan \widetilde{u}_2/\widetilde{u}_1$) and by the Reynolds-variables ($\arctan \overline{u}_2/\overline{u}_1$) differ. The difference being related to the respective mass flux terms given in Eq. (3.22) (cf. Gaviglio, 1976). In boundary layer flows (without shocks), the streamwise fluctuating mass flux $\overline{\rho'u_1'}/\overline{\rho}$ is much smaller than $\widetilde{u}_1$ and the cross-stream fluctuating mass flux $\overline{\rho'u_2'}/\overline{\rho}$ is of the same order as $\widetilde{u}_2$ but of opposite sign.

For the closure problem aspect, at the lowest level using the mixing-length hypothesis, all the underlying physical arguments have to be re-visited within the context of density-weighted averages, and at the highest level using the statistical moment transport equations, where the degree of empiricism is reduced, the effect of the density-weighted averages on the higher-order unknown correlations is less well known. These issues will be discussed in Chapter 5 where the various closure models will be introduced. It can be pointed out that if the turbulence is homogeneous (spatial gradients of statistical quantities vanish) then the mass flux $\overline{\rho'u_i'}$ itself also vanishes (Blaisdell, Mansour, & Reynolds, 1993; Feiereisen, Reynolds, & Ferziger, 1981), so that the density-weighted mean and fluctuating velocity are equal to the Reynolds-averaged mean and fluctuating velocity. In the absence of any directional bias introduced by (or independent of) the initial conditions, the directionality of the flow is dictated by the mean motion, specifically the mean velocity gradient $\partial \overline{u}_i/\partial x_j$. Consider now a simple rotational transformation such that the new coordinate frame $x_i^* = -x_i$ and similarly in the new frame $u_i^* = -u_i$. For the tensor transformation of the velocity gradient, $\partial \overline{u}_i^*/\partial x_j^* = \partial \overline{u}_i/\partial x_j$ so there is no change. This means the turbulent statistical field should be the same and as such $\overline{\rho'u_i'^*} = \overline{\rho'u_i'}$; however, this would contradict the coordinate and velocity transformation originally proposed unless $\overline{\rho'u_i'} = 0$.

Nevertheless, one can anticipate that it will be desirable to relate the turbulent correlations involving Favre variables with correlations involving Reynolds variables. Such relationships will become necessary when relating experimental results and numerical predictions, for example. The results just obtained for the mean and fluctuating quantities can be used to obtain such relations. A Favre fluctuating variable is related to the Reynolds fluctuation by

$$\begin{aligned} f'' &= f' + \overline{f} - \widetilde{f} \\ &= f' - \frac{\overline{\rho'f'}}{\overline{\rho}}, \end{aligned} \tag{3.23}$$

where Eq. (3.20b) has been used. Then a correlation between the two variables f'' and g'' given by $\overline{\rho} \widetilde{f'' g''} = \overline{\rho f'' g''}$ can be written as

$$
\begin{aligned}
\overline{\rho} \widetilde{f'' g''} = \overline{\rho f'' g''} &= \overline{\rho \left(f' - \frac{\overline{\rho' f'}}{\overline{\rho}} \right) \left(g' - \frac{\overline{\rho' g'}}{\overline{\rho}} \right)} \\
&= \overline{\rho f' g'} - \frac{\left(\overline{\rho' f'} \right) \left(\overline{\rho' g'} \right)}{\overline{\rho}} \\
&= \overline{\rho}\ \overline{f' g'} + \overline{\rho' f' g'} - \frac{\left(\overline{\rho' f'} \right) \left(\overline{\rho' g'} \right)}{\overline{\rho}}.
\end{aligned}
\tag{3.24}
$$

Once again, the difference between the Favre- and Reynolds-averaged quantities is related to the respective mass flux terms associated with each variable.

As will become apparent in the derivation of the mean flow equations, it is often beneficial to cast the higher-order correlations appearing in the density-weighted mean equations in terms of Reynolds variables. This is often possible since a correlation between a Reynolds fluctuating variable and Favre-fluctuating variable, that is the correlation $\overline{f' g''}$

$$
\begin{aligned}
\overline{f' g''} &= \overline{f' \overline{\overline{g}}} - \overline{f' \widetilde{g}} + \overline{f' g'} \\
&= \overline{f' g'},
\end{aligned}
\tag{3.25}
$$

reduces to a correlation between fluctuating Reynolds variables. It is worth reiterating that all these results are a direct consequence of the properties associated with the Reynolds average operator, and are necessarily kinematic relations. In addition, as Eq. (3.24) clearly shows, if a filtering operation were employed, it would not be possible to form such a relationship. This means that in solution methods where filtered variables are the dependent variables, it is not possible to use relations such as those in Eq. (3.25). In those methods, the filtered variables need to be averaged and the fluctuating component obtained. The averaged and fluctuating components of the filtered quantities will then satisfy the relations such as those in Eq. (3.25). In the large eddy simulation methods, these filtered variables are the resolved scale motions. If the flow is stationary, these resolved scale motions can then be time averaged, and the corresponding fluctuating part extracted as the difference between the instantaneous resolved motion and the averaged mean.

The last primary factor that was considered in the method selection was applicability to measurements. At the time, the hot-wire anemometer (HWA) was the only instrument capable of detailed turbulence measurement. The output of the HWA is a function of the mass flux and total temperature fluctuations. By varying the temperature of the wire, it is possible to separate the two variables, either instantaneously or on average; although the latter method is less difficult and is the procedure generally followed. In Section 4.4, relations between the mass flux and the fluctuating velocities for both Reynolds- and Favre-averaged decompositions are derived corresponding to negligible total temperature fluctuations. Of course, in the intervening four decades other more sophisticated techniques have been developed such as laser Doppler velocimetry (LDV), particle image velocimetry (PIV), and collective light scattering (CLS). These alternative methods are discussed more fully in Chapter 4.

As a final remark to complete the discussion of the density-weighted averages, it is also possible to express them in terms of a Favre probability density function $\widetilde{P}(f)$ (Bilger, 1975). The Favre *pdf*, $\widetilde{P}(f)$, can be written in terms of a conventional joint *pdf* function $P(\rho, f)$ as

$$\widetilde{P}(f) = \frac{1}{\overline{\rho}} \int_0^\infty \rho P(\rho, f) \mathrm{d}\rho. \tag{3.26}$$

Then, the Favre mean, statistical moments and general average can be written as

$$\widetilde{f} = \frac{\overline{\rho f}}{\overline{\rho}} = \int_f f \widetilde{P}(f) \mathrm{d}f, \tag{3.27}$$

and

$$\widetilde{f''^2} = \frac{\overline{\rho f''^2}}{\overline{\rho}} = \int_f (f - \widetilde{f})^2 \widetilde{P}(f) \mathrm{d}f \tag{3.28a}$$

$$\widetilde{f''^3} = \frac{\overline{\rho f''^3}}{\overline{\rho}} = \int_f (f - \widetilde{f})^3 \widetilde{P}(f) \mathrm{d}f \tag{3.28b}$$

$$\vdots$$

$$\widetilde{\phi(f)} = \frac{\overline{\rho \phi(f)}}{\overline{\rho}} = \int_f \phi(f) \widetilde{P}(f) \mathrm{d}f. \tag{3.28c}$$

The utilization of the *pdf* expressions for the Favre variables is quite popular within the combustion community, and the interested reader is referred to the book by Bray and Libby (1994).

While the utilization of Favre variables is by far the most common decomposition currently used in the development of the governing equations for turbulent compressible flows, it is not the only approach that has been proposed. Based on an earlier decomposition proposed by Bauer, Zumwalt, and Fila (1968), Ha Minh, Launder, and MacInnes (1981) (see also Chassaing, 2001; Cousteix & Aupoix, 1990) proposed a mixed-weighted procedure where the mean momentum is represented by a density-weighted variable, but the Reynolds stress is represented by Reynolds variables. While the mean variable equations remain form-invariant with respect to the Favre mean equations, for example, the requisite turbulent velocity second-moment equations are more complex. The complexity is due to the fact that the fluctuating variables are now centered in the mean (in contrast to the Favre decomposition), but the second-order stress tensors are no longer symmetric. Although the form-invariance of the mean equations made this mixed-weighted method easily adaptable to existing RANS-type numerical solvers, the added complexity of now asymmetric stress correlations made both the closure and numerical solution of these equations problematic.

In another alternative formulation (Chassaing, 1985 see also Aupoix, 2000; Chassaing, 2001; Cousteix & Aupoix, 1990), the Reynolds decomposition is retained so that the kinematic interpretation of the strain rate and rotation rate tensors for both the instantaneous and mean fields is the same (a constraint not followed in the density-weighted formulation), but the concept of fluid elements being enclosed by a stream tube is violated since the mass conservation equation now has a source term contribution related to the mass flux term. The motivation for such a formulation lies in the fact that the influence of the density variations is directly observable in the governing equations, and therefore sensitizes any closure schemes more closely to the effect of the density variations. Although both the alternative formulations (Chassaing, 1985; Minh et al., 1981) have merit, their adaptability to existing RANS-type numerical solvers with inherent incompressible or density-weighted forms of the equations is not straightforward. Thus, their utilization and development is limited and the necessary validations difficult to perform.

As has been noted, many of the useful relations between Reynolds-averaged variables and density-weighted variables do not carry over to other filtering procedures. Fortunately, it is still possible to derive a set of filtered conservation equations, but the relationships between the Reynolds variables and Favre variables need to be assessed carefully. The subgrid or subfiltered scale motions will not apparently satisfy some of the relations between density-weighted and Reynolds variables derived in this section. However, as has been discussed, an averaging operator can be applied to the resolved scale motions and the various correlations resulting from the fluctuating resolved motions can satisfy the various relationships derived here.

3.3 TRANSPORT EQUATIONS FOR THE MEAN/RESOLVED FIELD

In Chapter 2, the equations governing the compressible flow of a Newtonian fluid were presented. As noted earlier in this chapter, such equations provide the mathematical description for the fluid motion and if numerically solved with sufficient accuracy would provide the complete description of the compressible motion. Unfortunately, practical limitations on resources and current technology prevent this in almost all engineering relevant flows. This necessitates the need for filtered sets of equations that can describe the motion of as broad a range of resolved scales as possible and be coupled with closure models for the unresolved scales.

In this section, the set of governing equations for the resolved or mean motions of a turbulent flow are formulated. No distinction at this point is necessary on whether the filtering process employed is based on a Reynolds average, average over fixed phase, or temporal or spatial LES filters. Since these equations are applicable to both Reynolds-averaged approaches as well as filtered approaches, the time derivative appearing in the equations will be retained, and not systematically dropped as in the traditional stationary RANS approach. The intent is to emphasize, that at the resolved scale level, the equations are form-invariant with respect to the filtering procedure; however, it is important to recognize that closure in these equations is fundamentally different.

In developing the equations for the resolved or mean motions for compressible turbulent flows, the procedure followed is analogous to that in the incompressible formulation, and is first applied to the conservation equations (2.3), (2.8), and (2.27). The flow variables can then be partitioned into resolved and fluctuating parts consistent with the decompositions given in Eq. (3.1) or (3.14). As has been suggested in Section 3.2, it has been found that a rewriting of the turbulent compressible equations into mass-weighted, or Favre (Favre, 1965a), variables is more desirable since the equations take a more compact form relative to the Reynolds-averaged variables and the terms in the equations can be shown to have analogous counterparts in the incompressible formulation.

Consider first the mass conservation equation given in Eq. (2.3). The averaging/filtering procedure applied to this equation yields

$$\frac{\partial \overline{\rho}}{\partial t} + \frac{\partial}{\partial x_j}(\overline{\rho u_j}) = 0. \tag{3.29}$$

If the decomposition in terms of density-weighted variables Eq. (3.14) is used, Eq. (3.29) can be written as

$$\frac{\partial \overline{\rho}}{\partial t} + \frac{\partial}{\partial x_j}(\overline{\rho}\widetilde{u}_j) = 0, \tag{3.30a}$$

or in terms of the material derivative D/Dt,[1]

$$\frac{D\overline{\rho}}{Dt} = \frac{\partial \overline{\rho}}{\partial t} + \widetilde{u}_j \frac{\partial \overline{\rho}}{\partial x_j} = -\overline{\rho}\frac{\partial \widetilde{u}_j}{\partial x_j}. \tag{3.30b}$$

In general, issues associated with the commutativity of the various filters exist; however, for the purposes here, the commutativity property of the operators is assumed to hold unless otherwise noted.

---

[1] The convention is adopted here that application of the material derivative to filtered quantities requires a filtered convection velocity. For compressible flows this will be the Favre-averaged velocity $\widetilde{u}_j$.

The averaged/filtered momentum conservation equation is extracted from Eq. (2.8), and given by

$$\frac{\partial \overline{\rho u}_i}{\partial t} + \frac{\partial}{\partial x_j}\left(\overline{u_i \rho u}_j\right) = -\frac{\partial \overline{p}}{\partial x_i} + \frac{\partial \overline{\sigma}_{ij}}{\partial x_j}. \tag{3.31}$$

In terms of the Favre-variable partitioning, it can be written as

$$\overline{\rho}\frac{D\tilde{u}_i}{Dt} = \frac{\partial(\overline{\rho}\tilde{u}_i)}{\partial t} + \frac{\partial}{\partial x_j}(\tilde{u}_i\overline{\rho}\tilde{u}_j) = -\frac{\partial \overline{p}}{\partial x_i} + \frac{\partial \overline{\sigma}_{ij}}{\partial x_j} - \frac{\partial \overline{R}_{ij}}{\partial x_j}, \tag{3.32}$$

where the averaged/filtered viscous stress tensor $\overline{\sigma}_{ij}$ and turbulent stress tensor are written as

$$\overline{\sigma}_{ij} = \overline{2\mu\left(S_{ij} - \frac{1}{3}S_{kk}\delta_{ij}\right)} \tag{3.33}$$

$$\overline{R}_{ij} = \overline{\rho u_i u_j} - \overline{\rho}\tilde{u}_i\tilde{u}_j = \overline{\rho}\left(\widetilde{u_i u_j} - \tilde{u}_i\tilde{u}_j\right), \tag{3.34}$$

respectively. As was pointed out in Section 3.2, using density-weighted variables in the partitioning of the viscous stress tensor Eq. (3.33) results in a more complex form than using Reynolds variables. In addition, as will be discussed in Section 5.3, in developing closures for the higher-order correlations appearing in the density-weighted formulation, it is useful to write some correlations in terms of Reynolds variables in order to provide linkage with the corresponding incompressible terms. Using Reynolds variables, Eq. (3.33) can be written as

$$\begin{aligned}\overline{\sigma}_{ij} &= \overline{2\mu\left(S_{ij} - \frac{1}{3}S_{kk}\delta_{ij}\right)} \\ &= 2\overline{\mu}\left(\overline{S}_{ij} - \frac{1}{3}\overline{S}_{kk}\delta_{ij}\right) + \overline{2\mu'\left(s'_{ij} - \frac{1}{3}s'_{kk}\delta_{ij}\right)}, \end{aligned} \tag{3.35}$$

where the Reynolds fluctuating velocity gradient $\partial u'_i/\partial x_j$ is partitioned as,

$$\frac{\partial u'_i}{\partial x_j} = s'_{ij} + w'_{ij}, \tag{3.36}$$

with the fluctuating strain rate s'_{ij} and rotation rate w'_{ij} tensors given by

$$s'_{ij} = \frac{1}{2}\left(\frac{\partial u'_i}{\partial x_j} + \frac{\partial u'_j}{\partial x_i}\right) \quad \text{and} \quad w'_{ij} = \frac{1}{2}\left(\frac{\partial u'_i}{\partial x_j} - \frac{\partial u'_j}{\partial x_i}\right). \tag{3.37}$$

The mean strain rate $\overline{S}_j$ is formed from velocity gradients using Reynolds-averaged variables. Even if fluctuations in the viscosity μ' can be neglected, $\mu \approx \overline{\mu}$, the remaining contribution to the viscous stress is still expressed in terms of the Reynolds-averaged mean velocity rather than the Favre mean velocity. Of course, for any manipulations of the equations, computational or analytical, the dependent variable is the Favre mean velocity. An additional approximation is required, and this involves using Eq. (3.20a) for the velocity field, that is

$$\frac{\tilde{u}_i}{\overline{u}_i} = 1 + \frac{\overline{\rho' u'_i}}{\overline{\rho}\,\overline{u}_i}. \tag{3.38}$$

Then, the final approximations to the viscous stress tensor are

$$\overline{\sigma}_{ij} \approx 2\overline{\mu}\left(\overline{S}_{ij} - \frac{1}{3}\overline{S}_{kk}\delta_{ij}\right) \approx 2\overline{\mu}\left(\tilde{S}_{ij} - \frac{1}{3}\tilde{S}_{kk}\delta_{ij}\right)$$

$$\approx 2\tilde{\mu}\left(\tilde{S}_{ij} - \frac{1}{3}\tilde{S}_{kk}\delta_{ij}\right). \tag{3.39}$$

The approximation between the average velocities implies that the normalized mass flux $\overline{\rho' u'_i}/\overline{\rho}\,\overline{u}_i$ is small, and the introduction of $\tilde{\mu}$ simply reflects the fact that the viscosity can be considered as an extensive variable (Favre, 1965b, see p. 406) and that viscosity laws, such as Sutherland's law, necessarily will use the density-weighted temperature. It is difficult to provide an *a priori* estimate of this ratio directly from experiments because the value of $\overline{\rho' u'_i}$ is not known from usual methods; however, as will be discussed in Section 3.5, this value can be estimated indirectly if relations from the SRA are applied. In contrast, this ratio can be obtained directly without any assumptions if direct numerical simulation results are available. In Figure 3.2, the normalized mass flux variation, $\overline{\rho' u'_1}/\overline{\rho}\,\overline{u}_1$, obtained from a direct numerical simulation

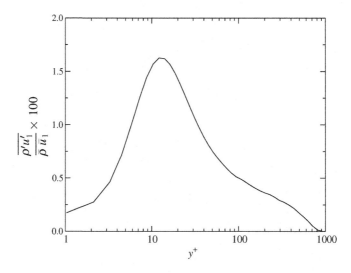

FIGURE 3.2 Scaled mass flux variation across a supersonic boundary layer.

(DNS) of a $M_\infty = 2.25$ supersonic boundary layer flow with adiabatic walls (Pirozzoli, Grasso, & Gatski, 2004), is shown. As can be seen, the maximum variation for the streamwise mass flux component occurs in the buffer-layer region and is less than 2%. Thus, the assumptions used in forming the mean conservation equations appear valid for this flow. Similar results were obtained from simulations of a cold-walled channel flow (Huang, Coleman, & Bradshaw, 1995). In those simulations the maximum normalized mass flux ranged from 1% to 3% with the maximum increasing with an increase in the ratio between centerline and wall temperature. For high Reynolds number engineering flows (without shocks), these approximations of the viscous stress tensor may not be critical since $\overline{\sigma}_{ij}$ represents molecular transport effects that are generally less important than turbulence effects except in the vicinity of solid boundaries.

In contrast to the mass conservation equation, which did not introduce any higher-order correlations that required closure, the momentum equation (3.32) introduced a correlation involving the velocity second-moments. For Reynolds and fixed-phase averages, these lead to the familiar Reynolds stresses, and for the other filtering processes are the subgrid scale or sub-filter scale stresses. In either case, substituting a decomposition for velocity given by Eq. (3.14) into Eq. (3.34) yields

$$\overline{R}_{ij} = \overline{\rho \tilde{u}_i \tilde{u}_j} - \overline{\rho} \tilde{u}_i \tilde{u}_j + \overline{\rho \tilde{u}_i u_j''} + \overline{\rho u_i'' \tilde{u}_j} + \overline{\rho u_i'' u_j''}. \tag{3.40}$$

In this form, the turbulent stress tensor would correspond to a form consistent with a Leonard decomposition used in incompressible LES subgrid scale modeling. For the RANS approach, the averaging procedure is idempotent with $\overline{\rho u_i''} = 0$, so Eq. (3.40) simply reduces to the form

$$\overline{R}_{ij} = \overline{\rho} \tau_{ij} = \overline{\rho u_i'' u_j''} = \overline{\rho} \widetilde{u_i'' u_j''}. \tag{3.41}$$

Nevertheless, whether an averaging or filtering procedure is used, models or parameterizations for these second-velocity moments are required to close the equations.

The remaining conservation equation to be considered is the total energy equation given in Eq. (2.27b). Applying the averaging/filtering procedure yields

$$\frac{\partial(\overline{\rho E})}{\partial t} + \frac{\partial(\overline{u_j \rho H})}{\partial x_j} = \frac{\partial(\overline{u_i \sigma_{ij}})}{\partial x_j} - \frac{\partial \overline{q}_j}{\partial x_j}, \tag{3.42}$$

where the averaged/filtered total energy field, $\overline{\rho E}$, can be written in Favre-variables as

$$\overline{\rho E} = \overline{\rho} \tilde{E} = \overline{\rho} \left(\tilde{e} + \frac{\widetilde{u_i u_i}}{2} \right). \tag{3.43}$$

The total energy equation can then be written as

$$\frac{\partial(\overline{\rho} \tilde{E})}{\partial t} + \frac{\partial(\tilde{u}_j \overline{\rho} \tilde{H})}{\partial x_j} = \frac{\partial(\overline{u_i \sigma_{ij}})}{\partial x_j} - \frac{\partial \overline{q}_j}{\partial x_j} - \frac{\partial \overline{Q}_j}{\partial x_j}, \tag{3.44}$$

with

$$\bar{\rho}\tilde{E} = \bar{\rho}\tilde{e} + \frac{\overline{\rho\tilde{u}_i\tilde{u}_i}}{2} + \frac{\overline{R}_{ii}}{2} \tag{3.45a}$$

$$\bar{\rho}\tilde{H} = \bar{\rho}\left(\tilde{E} + \frac{\bar{p}}{\bar{\rho}}\right) \tag{3.45b}$$

$$\bar{q}_j = -k_T\frac{\overline{\partial T}}{\partial x_j} \tag{3.45c}$$

$$\overline{Q}_j = \overline{\rho u_j H} - \bar{\rho}\tilde{u}_j\tilde{H} = \bar{\rho}\left(\widetilde{u_j H} - \tilde{u}_j\tilde{H}\right). \tag{3.45d}$$

If one assumes a perfect gas equation of state,

$$\bar{p} = \bar{\rho}\mathcal{R}\tilde{T}, \tag{3.46a}$$

the specific heats c_p and c_v are constant so that $\tilde{e} = c_v\tilde{T}$, and

$$\bar{p} = (\gamma - 1)\left[\bar{\rho}\tilde{E} - \frac{\overline{\rho\tilde{u}_i\tilde{u}_i}}{2} - \frac{\overline{R}_{ii}}{2}\right]. \tag{3.46b}$$

The perfect gas assumption also allows Eq. (3.45) to be expanded and written as

$$\bar{\rho}\tilde{E} = \bar{\rho}c_v\tilde{T} + \frac{\overline{\rho\tilde{u}_i\tilde{u}_i}}{2} + \frac{\overline{R}_{ii}}{2} \tag{3.47a}$$

$$\bar{\rho}\tilde{H} = \bar{\rho}c_p\tilde{T} + \frac{\overline{\rho\tilde{u}_i\tilde{u}_i}}{2} + \frac{\overline{R}_{ii}}{2} = \bar{\rho}c_p\tilde{T}_t \tag{3.47b}$$

$$\bar{q}_j = -k_T\frac{\overline{\partial T}}{\partial x_j} \tag{3.47c}$$

$$\overline{Q}_j = c_p\left(\overline{\rho u_j T_t} - \bar{\rho}\tilde{u}_j\tilde{T}_t\right) = \bar{\rho}c_p\left(\widetilde{u_j T_t} - \tilde{u}_j\tilde{T}_t\right) \tag{3.47d}$$

where $\tilde{T}_t$ is the total or stagnation temperature.

The term associated with the work done by the viscous stress tensor $\overline{u_i\sigma_{ij}}$ can be expanded again using Reynolds variables and written as

$$\overline{u_i\sigma_{ij}} = \bar{u}_i\,\bar{\sigma}_{ij} + \overline{u'_i\sigma'_{ij}}. \tag{3.48}$$

The first term on the right can be approximated by the Favre mean velocity $\tilde{u}_i$ and the expression for the viscous stress tensor in Eq. (3.39). The second term enters into Eq. (3.44) as a turbulent viscous diffusion term that will be discussed further in Section 5.3 in connection with the turbulent stress closures. Alternatively, for filtered methods such as LES, the term $\partial(\tilde{u}_i\tilde{\sigma}_{ij})/\partial x_j$ can be added to both sides of Eq. (3.44) and subgrid-scale models developed just as would be the case for terms such as Eq. (3.47d). The viscous term appears in the total energy equation due to the kinetic energy contribution. Now the two contributions from the internal energy are considered.

Once again, a drawback associated with the density-weighted variables surfaces in the mean heat flux vector $\overline{q}_j$. If Reynolds variables are used, and fluctuations in the thermal conductivity are allowed, Eq. (3.47c) can be written as

$$\overline{q}_j = -\overline{k}_T \frac{\partial \overline{T}}{\partial x_j} - \overline{k'_T \frac{\partial T'}{\partial x_j}}. \tag{3.49}$$

If fluctuations in the thermal conductivity are neglected so that, $k_T = \overline{k}_T$, then a simple gradient diffusion model in terms of the Reynolds-averaged temperature is obtained. This Reynolds-averaged temperature can be linked to the required Favre-variable temperature through a relation analogous to Eq. (3.38), and given by

$$\frac{\widetilde{T}}{\overline{T}} = 1 + \frac{\overline{\rho'T'}}{\overline{\rho}\overline{T}}. \tag{3.50}$$

If the normalized density-temperature correlation $\overline{\rho'T'}/\overline{\rho}\overline{T}$ is assumed small, so that $\overline{T}$ can be approximated by $\widetilde{T}$, then the familiar simplified gradient transport model is obtained,

$$\overline{q}_j \approx -\overline{k}_T \frac{\partial \overline{T}}{\partial x_j} \approx -\overline{k}_T \frac{\partial \widetilde{T}}{\partial x_j}$$

$$\approx -\widetilde{k}_T \frac{\partial \widetilde{T}}{\partial x_j}, \tag{3.51}$$

where $\widetilde{k}_T$ is introduced consistent with the dependence on the density-weighted temperature and viscosity $\widetilde{\mu}$ (see also Favre, 1965b, p. 409). A quantitative measure of the validity of this assumption can be obtained from an evaluation of the $\overline{\rho'T'}/\overline{\rho}\widetilde{T}$ correlation. For the supersonic boundary layer flow with adiabatic wall temperature (Pirozzoli et al., 2004), the DNS results plotted in Figure 3.3 show a maximum variation of about 0.75% across most of the boundary layer. Even when the walls are cooled (Huang et al., 1995) similar deviation levels are predicted in channel flow simulations. In such simulations the maximum normalized mass flux can range from 0.3% to 1% with the maximum increasing with an increase in the ratio between centerline and wall temperature. At least for simulation data available for these simple boundary layer and channel flows, the substitution of the density-weighted variables for the Reynolds variables is certainly a valid approximation. More dynamically complex flow fields may yield different results so the reader should be cognizant of the various approximations that have been made in the formulation of the mean variable governing equations.

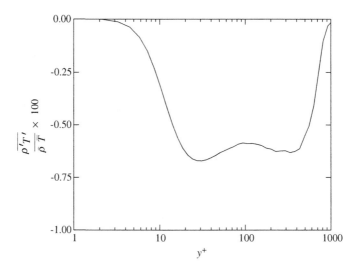

FIGURE 3.3 Scaled density–temperature correlation across supersonic boundary layer.

Substitution of the density-weighted decomposition for the velocity u_i and temperature T into Eq. (3.47d) yields

$$\overline{Q}_j = c_p \left(\overline{\rho \tilde{u}_j \tilde{T}_t} - \overline{\rho} \tilde{u}_j \tilde{T}_t + \overline{\rho u''_j \tilde{T}_t} + \overline{\rho \tilde{u}_j T''_t} + \overline{\rho u''_j T''_t} \right). \qquad (3.52)$$

Once again for an averaging approach, the procedure simply reduces Eq. (3.52) to the form

$$\overline{Q}_j = c_p \overline{\rho u''_j T''_t} = c_p \overline{\rho} \widetilde{u''_j T''_t}$$

$$= c_p \overline{\rho u''_j T''} + \frac{\overline{\rho u''_i u''_i u''_j}}{2} + \tilde{u}_i \overline{\rho u''_i u''_j}$$

$$= \overline{\rho} c_p \widetilde{u''_j T''} + \frac{\widetilde{\overline{\rho} u''_i u''_i u''_j}}{2} + \overline{\rho} \tilde{u}_i \tau_{ij}. \qquad (3.53)$$

With this background, the equation describing the evolution of the internal energy can be explored. The mean equation is obtained directly from Eq. (2.30) and written as

$$\frac{\partial (\overline{\rho} e)}{\partial t} + \frac{\partial (\overline{u_j \rho e})}{\partial x_j} = \overline{\left(-p \delta_{ij} + \sigma_{ij} \right) S_{ji}} - \frac{\partial \overline{q}_j}{\partial x_j} \qquad (3.54)$$

or, in Favre-variables, as

$$\frac{\partial (\overline{\rho} \tilde{e})}{\partial t} + \frac{\partial}{\partial x_j} \left(\tilde{u}_j \overline{\rho} \tilde{h} \right) = \overline{u_j \frac{\partial p}{\partial x_j}} + \overline{\sigma_{ij} S_{ji}} - \frac{\partial \overline{q}_j}{\partial x_j} - \frac{\partial \overline{Q}_{(h)j}}{\partial x_j}, \qquad (3.55)$$

where $h \, (= e + p/\rho)$ is the enthalpy, and

$$\tilde{h} = \tilde{e} + \frac{\overline{p}}{\overline{\rho}} \tag{3.56a}$$

$$\overline{Q}_{(h)j} = \overline{\rho u_j h} - \overline{\rho} \tilde{u}_j \tilde{h} = c_p \left(\overline{\rho u_j T} - \overline{\rho} \tilde{u}_j \tilde{T} \right)$$

$$= \overline{\rho} c_p \widetilde{u_j'' T''}. \tag{3.56b}$$

Even though the equation set that governs the mean motion of the turbulent flow has been cast using density-weighted dependent variables, additional higher-order correlations appear in the equations that require closure. In the incompressible, isothermal case, the only corresponding term requiring closure would be the turbulent stress $\overline{\rho} \tau_{ij}$ given in Eq. (3.41); however, the inclusion of the mean total energy equation necessitates models, or parameterizations, for the viscous diffusion term associated with $\overline{u_i' \sigma_{ij}'}$, and all the terms comprising the turbulent scalar flux $\overline{Q}_j$ given in Eq. (3.53). It should be noted that all these extra correlations requiring closure, with the exception of the heat flux vector $\overline{\rho u_j'' T_t''}$, appear due to the kinetic energy contribution to the total energy equation. Then, such terms also appear in the transport equations for the turbulent stress and turbulent kinetic energy. If these terms are retained in the solution of the mean total energy equation, then the effect is to provide a stronger coupling between the turbulence closure and the mean motion than in incompressible flows.

3.4 FLUCTUATION TRANSPORT EQUATIONS

With the derivation of the mean or resolved equations, it is now possible to derive equations for some fluctuating quantities that will be useful in the discussion to follow in the remainder of the book. Substitution of the decompositions Eq. (3.1), into the mass, momentum, and internal energy equations, Eqs. (2.3), (2.8), and (2.30), respectively, and subtracting the corresponding mean transport equations from the instantaneous equations, then yields the equations for the fluctuating density, velocity and energy fields.

Since the equation for the mean density takes the simple form given in Eqs. (3.30a,b), and the mass conservation equation, Eq. (2.3), can be written as

$$\frac{\partial \rho}{\partial t} + \tilde{u}_j \frac{\partial \rho}{\partial x_j} + \rho \frac{\partial \tilde{u}_j}{\partial x_j} + \frac{\partial}{\partial x_j} \left(\rho u_j'' \right) = 0, \tag{3.57}$$

the fluctuating density equation retains a form-invariance with Eq. (3.57) given by

$$\frac{\partial \rho'}{\partial t} + \tilde{u}_j \frac{\partial \rho'}{\partial x_j} + \rho' \frac{\partial \tilde{u}_j}{\partial x_j} + \frac{\partial}{\partial x_j} \left(\rho u_j'' \right) = 0. \tag{3.58}$$

Applying an averaging operation to Eq. (3.58) shows that the last term on the left must vanish, and which is consistent with the result shown in Eq. (3.19). The density variance equation will be discussed further in Section 5.3.4.

For the fluctuating momentum equation, Eqs. (3.32) and (2.8) are used

$$
\rho \frac{\partial u_i''}{\partial t} + \rho \tilde{u}_j \frac{\partial u_i''}{\partial x_j} + \rho u_j'' \frac{\partial \tilde{u}_i}{\partial x_j} + \rho u_j'' \frac{\partial u_i''}{\partial x_j}
$$
$$
+ \frac{\partial}{\partial x_j} \left[p' \delta_{ij} - \sigma_{ij}' - \overline{\rho} \tau_{ij} \right] - \frac{\rho'}{\overline{\rho}} \frac{\partial}{\partial x_j} \left[\overline{p} \delta_{ij} - \overline{\sigma}_{ij} + \overline{\rho} \tau_{ij} \right] = 0,
$$

(3.59)

with the fluctuating part, σ_{ij}', obtained by subtracting the mean from the total,

$$
\sigma_{ij}' = 2\overline{\mu} \left(s_{ij}' - \frac{1}{3} s_{kk}' \delta_{ij} \right) + 2\mu' \left(\overline{S}_{ij} - \frac{1}{3} \overline{S}_{kk} \delta_{ij} \right)
$$
$$
+ 2\mu' \left(s_{ij}' - \frac{1}{3} s_{kk}' \delta_{ij} \right) - \overline{2\mu' \left(s_{ij}' - \frac{1}{3} s_{kk}' \delta_{ij} \right)}.
$$

(3.60)

When coupled with Eq. (3.57), and again applying an averaging operation to the resultant equation for $\rho u_j''$ further validates the above equality. Such simple validations can be useful in attempting to assess the accuracy of numerical simulations. Equation (3.59) is the equation from which the derivation of the transport equations describing the motion of the velocity second-moment correlations, that is the compressible turbulent stresses, is obtained (see Section 5.3). The mass flux that arises in connection with the scalar flux model development (see Section 5.3.4) is also directly obtained.

For the energy field, the previous derivations for the mean and instantaneous fields were given in Eqs. (3.55) and (2.30), respectively. The fluctuating internal energy is then given by the difference as

$$
\rho \frac{\partial e''}{\partial t} + \rho \tilde{u}_j \frac{\partial e''}{\partial x_j} + \rho u_j'' \frac{\partial \tilde{e}}{\partial x_j} + \rho u_j'' \frac{\partial e''}{\partial x_j} - \left[-p S_{jj} + \sigma_{ij} S_{ji} - \frac{\partial q_j}{\partial x_j} \right]
$$
$$
+ \frac{\rho}{\overline{\rho}} \left[-\overline{p S_{jj}} + \overline{\sigma_{ij} S_{ji}} - \frac{\partial \overline{q}_j}{\partial x_j} - \frac{\partial \overline{Q}_{(e)j}}{\partial x_j} \right] = 0,
$$

(3.61)

where $\overline{Q}_{(e)j} = \overline{\rho} c_v \widetilde{u_j'' T''}$ or, in terms of the fluctuating enthalpy h'' as

$$
\rho \frac{\partial h''}{\partial t} + \rho \tilde{u}_j \frac{\partial h''}{\partial x_j} + \rho u_j'' \frac{\partial \tilde{h}}{\partial x_j} + \rho u_j'' \frac{\partial h''}{\partial x_j} - \left[\frac{Dp}{Dt} + \sigma_{ij} S_{ji} - \frac{\partial q_j}{\partial x_j} \right]
$$
$$
+ \frac{\rho}{\overline{\rho}} \left[\overline{\frac{Dp}{Dt}} + \overline{\sigma_{ij} S_{ji}} - \frac{\partial \overline{q}_j}{\partial x_j} - \frac{\partial \overline{Q}_{(h)j}}{\partial x_j} \right] = 0.
$$

(3.62)

In the case of a perfect gas, $h'' = c_p T''$, and Eq. (3.62) is the governing transport equation for the temperature fluctuations. From this equation, both the heat flux correlation equations and the temperature variance equations can be formed (see Section 5.3.4).

The fluctuating pressure field has long been investigated since the influence of compressibility on the fluctuating pressure has an important dynamic influence in controlling the turbulence energetics in compressible free shear flows such as mixing-layers (see Section 6.2). The equation governing the behavior of the fluctuating pressure is obtained from the divergence of the fluctuating momentum equation and can be written as

$$
\frac{\partial^2 p'}{\partial x_i \partial x_i} = \frac{D^2 \rho'}{Dt^2} - 2 \frac{\partial}{\partial x_i}\left[\left(\rho u_j''\right) \frac{\partial \tilde{u}_i}{\partial x_j} \right] - 2 \frac{\partial \tilde{u}_i}{\partial x_i} \frac{\partial (\rho u_j'')}{\partial x_j}
$$
$$
- \rho'\left[\frac{\partial \tilde{u}_i}{\partial x_i}\frac{\partial \tilde{u}_j}{\partial x_j} + \frac{\partial \tilde{u}_i}{\partial x_j}\frac{\partial \tilde{u}_j}{\partial x_i} \right] - \frac{\partial^2}{\partial x_i \partial x_j}\left[\rho u_i'' u_j'' - \overline{\rho u_i'' u_j''} - \sigma_{ij}' \right],
$$

$$(3.63)$$

where the Galilean invariant operator D^2/Dt^2 is given by (see Friedrich, 1998, 2007)

$$
\frac{D^2}{Dt^2} = \frac{\partial^2}{\partial t^2} + 2\tilde{u}_j \frac{\partial^2}{\partial x_j \partial t} + \tilde{u}_i \tilde{u}_j \frac{\partial^2}{\partial x_i \partial x_j}.
$$

$$(3.64)$$

As written, Eq. (3.63) is a Poisson equation for the fluctuating pressure field that includes acoustic contributions through the $D^2 \rho'/Dt^2$ term. If an isentropic relationship between the pressure and density fields is assumed so that Eq. (2.33b) holds, it can also be recast as a convective wave equation. This is discussed further in Section 5.3.3 and used to assess the behavior of the pressure–strain rate and pressure–dilatation correlations that appear in the turbulent stress tensor transport equations.

The fluctuating pressure equation just derived is valid for both incompressible and compressible flows. However, a fluctuating pressure field can also be extracted from the internal energy equation and the equation of state, that are assumed to describe the behavior of the compressible part of the fluctuating pressure field. Since the pressure is partitioned using Reynolds variables, the fluctuating pressure equation can be obtained by combining Eqs. (2.31) and (2.32b), which yields for a perfect gas,

$$
\frac{\partial p'}{\partial t} + \frac{\partial}{\partial x_j}\left(\tilde{u}_j p' \right) + \frac{\partial}{\partial x_j}\left[\overline{p}\left(u_j'' - \overline{u_j''} \right) \right] + \frac{\partial}{\partial x_j}\left(p' u_j'' - \overline{p' u_j''} \right)
$$
$$
+ (\gamma - 1)\left[p'\frac{\partial \tilde{u}_j}{\partial x_j} + \overline{p}\frac{\partial}{\partial x_j}\left(u_j'' - \overline{u_j''} \right) + \left(p'\frac{\partial u_j''}{\partial x_j} - \overline{p'\frac{\partial u_j''}{\partial x_j}} \right) \right.
$$
$$
\left. - \sigma_{ij}' \tilde{S}_{ji} - \overline{\sigma}_{ij}\frac{\partial}{\partial x_i}\left(u_j'' - \overline{u_j''} \right) - \left(\sigma_{ij}'\frac{\partial u_j''}{\partial x_i} - \overline{\sigma_{ij}'\frac{\partial u_j''}{\partial x_i}} \right) + \frac{\partial q_j'}{\partial x_j} \right] = 0.
$$

$$(3.65)$$

Associated with this is an equation that has proven important in the development of closure models, the pressure-variance equation,

$$
\frac{\partial \overline{p'^2}}{\partial t} + \tilde{u}_j \frac{\partial \overline{p'^2}}{\partial x_j} + 2\overline{p'u'_j}\frac{\partial \overline{p}}{\partial x_j} - \overline{p'^2 \frac{\partial u''_j}{\partial x_j}} + \frac{\partial}{\partial x_j}\overline{u''_j p'^2}
$$

$$
+ 2\gamma \left(\overline{p'^2 \tilde{S}_j} + \overline{p}\,\overline{p's'_{jj}} + \overline{p'^2 \frac{\partial u''_j}{\partial x_j}} \right) - 2(\gamma - 1)
$$

$$
\times \left[\overline{p'\sigma'_{ij}\tilde{S}_j} + \overline{p's'_{ij}\overline{\sigma}_{ij}} + \overline{p'\sigma'_{ij}\frac{\partial u''_j}{\partial x_i}} - \overline{p'\frac{\partial q'_j}{\partial x_j}} \right] = 0, \qquad (3.66)
$$

where the result from Eq. (3.25) showing the equality between the fluctuating variables $\overline{f'g''}$ and $\overline{f'g'}$ has been used in Eq. (3.66).

Although the equations for the fluctuation variables can be the starting point for developing the equations for some of the higher-order correlations that require closure in a RANS-type modeling formulation, they also form the basis for the development of some early ideas about the behavior of compressible turbulent flows. One of the earliest concepts to be developed was that of a mode decomposition proposed by Kovasznay (1953). This decomposition consisted of three modes: a vorticity mode, an acoustic (pressure) mode, and an entropy mode. It is linear perturbation theory, and it is useful in analyzing the structure and behavior of homogeneous turbulent flows. It was the basis for experimental studies for almost a decade after its introduction for unsteady supersonic flows. Morkovin (1962) in a classic paper used the concept to provide a visual guide to the energy exchange between large and small scales in a compressible turbulent boundary layer, and from which he introduced the strong reynolds analogy (see Section 3.5) and and Morkovin's hypothesis (see Section 3.5.2).

The most fundamental turbulent motion to analyze is homogeneous, isotropic turbulence. Recall that in homogeneous turbulence there is no difference between the density-weighted and Reynolds mean variables, and the fluctuations are centered about the mean. For this reason and simplicity of notation, Reynolds variables will then be used. In this "flow," the mean velocity vanishes, $\tilde{u}_i = \overline{u}_i = 0$, and it will be assumed here that the fluctuating quantities are small so that linear theory can be used. From Eq. (3.30a), $\overline{\rho}$ is constant in time (and space), and if both molecular diffusion and viscous dissipation effects are neglected the flow is isentropic, and mean pressure $\overline{p}$ is constant. Equations (3.58), (3.59) and (3.65) reduce to

$$
\frac{\partial \rho'}{\partial t} = -\overline{\rho}\frac{\partial u'_j}{\partial x_j} \qquad (3.67a)
$$

$$
\overline{\rho}\frac{\partial u'_i}{\partial t} = -\frac{\partial p'}{\partial x_i} \qquad (3.67b)
$$

$$\frac{\partial p'}{\partial t} = -\gamma \overline{p} \frac{\partial u'_j}{\partial x_j}, \tag{3.67c}$$

where only the linear terms have been retained. The motion of the vorticity mode is easily obtained by taking the curl of Eq. (3.67b) and described by,

$$\frac{\partial \omega'_i}{\partial t} = 0. \tag{3.68}$$

As shown, the fluctuating vorticity field is independent of any fluctuations in the thermal and pressure fields.

The equation for the acoustic or pressure mode, is obtained from a combination of the momentum and energy equations, Eqs. (3.67b) and (3.67c), respectively. This yields a wave equation for the pressure of the form

$$\frac{\partial^2 p'}{\partial t^2} - \overline{c}^2 \frac{\partial^2 p'}{\partial x_j \partial x_j} = 0, \tag{3.69}$$

with the density, pressure and thermal fields related through the isentropic relations

$$\frac{\rho'}{\overline{\rho}} = \frac{p'}{\gamma \overline{p}} = \frac{1}{(\gamma - 1)} \frac{T'}{\overline{T}}, \tag{3.70}$$

and the velocity field u'_i is then a function of $p'/\overline{p}$.

The entropy mode can be obtained using a combination of Eqs. (3.67a) and (3.67c) as shown in Eq. (2.32b) for a perfect gas. Consistent with the initial assumptions where the viscous dissipation and molecular diffusion were neglected, the entropy mode is given by

$$\frac{\partial s'}{\partial t} = 0. \tag{3.71}$$

Equations (3.68), (3.69), and (3.71) are clearly decoupled and show that the vorticity, acoustic and entropy modes are independent of one another, and that the vorticity and entropy modes are time independent. In Section 7.1 this modal decomposition will be used to analyze some aspects of shock–turbulence interactions.

Viscous effects can be introduced into the analysis but the one-to-one relationship between the vorticity, acoustic, and entropy modes and vorticity, pressure, and entropy fields disappears. The pressure and entropy fields each need to be split in order to identify three distinct modes; although, it still holds that the vorticity mode is independent of both the acoustic and entropy modes (Blaisdell, Mansour, & Reynolds, 1991; Blaisdell et al., 1993; Kovasznay, 1953). Nonlinear effects were introduced into the analysis by Chu and Kovasznay (1958) who attempted to analyze mode interactions. Of course, in order to make any analytical analysis tractable several assumptions were required that limited the general applicability of such formulations.

Nevertheless, these early studies provided both the stimulus and insights necessary to understand some of the fundamental interactions of high speed compressible flows.

3.5 MOMENTUM AND THERMAL FLUX RELATIONSHIPS

As was the case in Section 2.5 in the analysis of the equations governing compressible flows, additional insights into the underlying dynamics of turbulent compressible flows can be gained by an analysis of the mean conservation equations for mass, momentum and energy. These ideas were first exploited by Morkovin (1962) from an analysis of the two-dimensional supersonic ($M_\infty < 5$) boundary layer data at moderate values of wall heat transfer $\bar{q}_w$ (cf. Young, 1953). Since this original analysis pre-dates the utilization of filtered equations for simulations, the focus was necessarily on the Reynolds-averaged equations with time-independent mean variables and stationary turbulence statistics. In addition, usage of Favre or density-weighted averages had, as yet, not been introduced as a common means of compressible flow description. This partitioning in terms of Reynolds-averaged variables will be used here; although, it will now be possible to contrast some of these derivations with the form of these equations in Favre variables derived earlier in this chapter. Even today, these relations provide a valuable link between experimental measurements and both direct numerical simulations and Reynolds-averaged type computations (see Chapter 4).

3.5.1 Strong Reynolds Analogy

Since it is possible to initially frame the discussion and derivations in a somewhat general fashion before imposing the conditions of a two-dimensional mean flow, that was originally considered, the averaged mass conservation equation for a stationary flow can be obtained from Eq. (3.29), and written as

$$\frac{\partial}{\partial x_j}\left(\bar{\rho}\,\bar{u}_j + \overline{\rho'u'_j}\right) = 0. \tag{3.72}$$

Since Morkovin's intent was to examine the effects of Reynolds stress transport and the total enthalpy transport, peripheral effects such as mass transport were necessarily neglected in the original formulation. This allowed Eq. (3.72) to be written as (cf. Morkovin, 1962)

$$\overline{\rho\frac{D}{Dt}\left(\frac{1}{\bar{\rho}}\right)} = \left(\bar{\rho}\,\bar{u}_j + \overline{\rho'u'_j}\right)\frac{\partial}{\partial x_j}\left(\frac{1}{\bar{\rho}}\right) = \frac{\partial\bar{u}_j}{\partial x_j}, \tag{3.73}$$

where the mass transport term $\partial(\overline{\rho'u'_j}/\bar{\rho})/\partial x_j$ was neglected. This then represents a mean continuity equation with zero mass transport effects.

The averaged momentum conservation equation in Reynolds variables can be obtained from Eq. (3.31),

$$\left(\overline{\rho}\,\overline{u}_j + \overline{\rho'u'_j}\right)\frac{\partial \overline{u}_i}{\partial x_j} = -\frac{\partial \overline{p}}{\partial x_i} + \frac{\partial}{\partial x_j}\left(\overline{\sigma}_{ij} - \overline{\rho}\,\overline{u'_i u'_j}\right)$$

$$-\frac{\partial}{\partial x_j}\left(\overline{\rho'u'_i u'_j}\right) - \overline{\rho}\,\overline{u}_j \frac{\partial}{\partial x_j}\left(\frac{\overline{\rho'u'_i}}{\overline{\rho}}\right)$$

$$+\left(\frac{\overline{\rho'u'_i}}{\overline{\rho}}\right)\frac{\partial\left(\overline{\rho'u'_j}\right)}{\partial x_j}. \tag{3.74}$$

When both mass transport effects and higher-order mass flux correlations are neglected, Eq. (3.74) can be written as

$$\left(\overline{\rho}\,\overline{u}_j + \overline{\rho'u'_j}\right)\frac{\partial \overline{u}_i}{\partial x_j} \simeq -\frac{\partial \overline{p}}{\partial x_i} + \frac{\partial}{\partial x_j}\left(\overline{\sigma}_{ij} - \overline{\rho u'_i u'_j}\right). \tag{3.75}$$

With the Reynolds-averaged mean flow stationary, the total energy equation in Eq. (3.42) becomes an equation for the total enthalpy, or total temperature, and can be written in terms of Reynolds variables as

$$\left(\overline{\rho}\,\overline{u}_j + \overline{\rho'u'_j}\right)\frac{\partial \overline{H}}{\partial x_j} = \frac{\partial}{\partial x_j}\left[-\overline{q}_j - \overline{\rho u_j H'} + \overline{u_i \sigma_{ij}}\right]$$

$$= \frac{\partial}{\partial x_j}\left[\frac{\overline{k}_T}{c_p}\frac{\partial \overline{H}}{\partial x_j} - \overline{\rho u_j H'}\right.$$

$$\left. + \overline{u_i \sigma_{ij}} - \frac{\overline{k}_T}{c_p}\frac{\partial}{\partial x_j}\left(\frac{\overline{u}_i \overline{u}_i}{2} + \frac{\overline{u'_i u'_i}}{2}\right)\right], \tag{3.76}$$

where the mean Fourier conduction law $-\overline{k}_T(\partial \overline{T}/\partial x_j)$ is used with (assumed) constant mean thermal conductivity $\overline{k}_T$. The mean temperature $\overline{T}$ is also replaced with the mean total enthalpy using the following relation (cf. Eqs. (2.25) and (2.28)) valid for a perfect gas,

$$\overline{H} = c_p\overline{T} + \frac{\overline{u_i u_i}}{2}$$

$$= c_p\overline{T} + \frac{\overline{u}_i\overline{u}_i}{2} + \frac{\overline{u'_i u'_i}}{2} = c_p\overline{T}_t, \tag{3.77}$$

with c_p constant. Both the total enthalpy flux $\overline{\rho u_j H'}$ and the term representing the work done on the fluid by the viscous stress $\overline{u_i \sigma_{ij}}$ can be expanded to yield

$$\overline{\rho u_j H'} = \overline{\rho u'_j H'} + \overline{u}_j \overline{\rho' H'} + \overline{\rho' u'_j H'}$$

$$\simeq \overline{\rho u'_j H'} \tag{3.78a}$$

$$\overline{u_i \sigma_{ij}} = \overline{u}_i \overline{\sigma}_{ij} + \overline{u'_i \sigma'_{ij}}$$

$$\simeq \frac{\partial}{\partial x_j} \left(\frac{\overline{u}_i \overline{u}_i}{2} + \frac{\overline{u'_i u'_i}}{2} \right), \tag{3.78b}$$

where in Eq. (3.78a) the mass flux term $\overline{\rho' H'}$ and higher-order correlations have been discarded, and in Eq. (3.78b) a boundary layer approximation to the viscous stress tensor $\sigma_{ij} \approx \mu(\partial u_i / \partial x_j)$ has now been introduced. Equation (3.76) can then be written as

$$\left(\overline{\rho}\,\overline{u}_j + \overline{\rho' u'_j} \right) \frac{\partial \overline{H}}{\partial x_j} = \frac{\partial}{\partial x_j} \left[\frac{\overline{k}_T}{c_p} \frac{\partial \overline{H}}{\partial x_j} - \overline{\rho u'_j H'} \right.$$

$$\left. + \mu \left(1 - \frac{1}{Pr} \right) \frac{\partial}{\partial x_j} \left(\frac{\overline{u}_i \overline{u}_i}{2} + \frac{\overline{u'_i u'_i}}{2} \right) \right], \tag{3.79}$$

where Pr is the Prandtl number $\overline{\mu} c_p / \overline{k}_T$.

These same equations can now be expressed in terms of Favre variables. Both the Reynolds-averaged momentum conservation equation in Favre variables Eq. (3.32), with the corresponding Reynolds stresses defined in Eq. (3.41),

$$\overline{\rho} \widetilde{u}_j \frac{\partial \widetilde{u}_i}{\partial x_j} = - \frac{\partial \overline{p}}{\partial x_i} + \frac{\partial}{\partial x_j} \left(\overline{\sigma}_{ij} - \widetilde{\rho u''_i u''_j} \right), \tag{3.80}$$

and the total enthalpy (total temperature) equation in Favre variables Eq. (3.44), with the corresponding thermal fluxes defined in Eqs. (3.51) and (3.53),

$$\overline{\rho} \widetilde{u}_j \frac{\partial \widetilde{H}}{\partial x_j} = \frac{\partial}{\partial x_j} \left[\frac{\overline{k}_T}{c_p} \frac{\partial \widetilde{H}}{\partial x_j} - \overline{\rho u''_j H''} \right.$$

$$\left. + \mu \left(1 - \frac{1}{Pr} \right) \frac{\partial}{\partial x_j} \left(\frac{\widetilde{u}_i \widetilde{u}_i}{2} + \frac{\widetilde{u''_i u''_i}}{2} \right) \right], \tag{3.81}$$

retain the same form as the mean momentum and total enthalpy equations given in Eqs. (3.75) and (3.79). A comparison of Eq. (3.80) with Eq. (3.75), and Eq. (3.81) with Eq. (3.79) shows that differences are related to the defining relations given in Eqs. (3.20b) and (3.24).

With this background, now the case of the two-dimensional flow of a perfect gas is considered. Analogous to the derivations in Section 2.5, the flow is in the (x, y) plane with x the streamwise direction, y the direction of the mean shear, and the corresponding Reynolds-averaged velocities are $(\overline{u}, \overline{v})$. The mean mass, momentum and total enthalpy equations, Eqs. (3.72), (3.75) and (3.79), can then be approximated as

$$\frac{\partial}{\partial x}\left(\overline{\rho}\,\overline{u} + \overline{\rho' u'}\right) + \frac{\partial}{\partial y}\left(\overline{\rho}\,\overline{v} + \overline{\rho' v'}\right) = 0 \tag{3.82}$$

$$\left(\overline{\rho}\,\overline{u} + \overline{\rho' u'}\right)\frac{\partial \overline{u}}{\partial x} + \left(\overline{\rho}\,\overline{v} + \overline{\rho' v'}\right)\frac{\partial \overline{u}}{\partial y} = -\frac{\partial \overline{p}}{\partial x}$$
$$+ \frac{\partial}{\partial y}\left(\overline{\sigma}_{xy} - \overline{\rho u' v'}\right) \tag{3.83}$$

$$\left(\overline{\rho}\,\overline{u} + \overline{\rho' u'}\right)\frac{\partial \overline{H}}{\partial x} + \left(\overline{\rho}\,\overline{v} + \overline{\rho' v'}\right)\frac{\partial \overline{H}}{\partial y} = \frac{\partial}{\partial y}\left[\frac{\overline{k}_T}{c_p}\frac{\partial \overline{H}}{\partial y} - \overline{\rho v' H'}\right.$$
$$+ \mu\left(1 - \frac{1}{Pr}\right)\frac{\partial}{\partial y}$$
$$\left. \times \left(\frac{\overline{u}\,\overline{u}}{2} + \frac{\overline{u' u'}}{2}\right)\right], \tag{3.84}$$

in terms of Reynolds-averaged variables with $\overline{\sigma}_{xy} \approx \mu(\partial \overline{u}/\partial y)$, and as

$$\frac{\partial}{\partial x}(\overline{\rho}\widetilde{u}) + \frac{\partial}{\partial y}(\overline{\rho}\widetilde{v}) = 0 \tag{3.85}$$

$$\overline{\rho}\widetilde{u}\frac{\partial \widetilde{u}}{\partial x} + \overline{\rho}\widetilde{v}\frac{\partial \widetilde{u}}{\partial y} = -\frac{\partial \overline{p}}{\partial x} + \frac{\partial}{\partial y}\left(\widetilde{\sigma}_{xy} - \overline{\rho u'' v''}\right) \tag{3.86}$$

$$\overline{\rho}\widetilde{u}\frac{\partial \widetilde{H}}{\partial x} + \overline{\rho}\widetilde{v}\frac{\partial \widetilde{H}}{\partial y} = \frac{\partial}{\partial y}\left[\frac{\overline{k}_T}{c_p}\frac{\partial \widetilde{H}}{\partial y} - \overline{\rho v'' H''}\right.$$
$$+ \mu\left(1 - \frac{1}{Pr}\right)\frac{\partial}{\partial y}\left(\frac{\widetilde{u^2}}{2} + \frac{\widetilde{u''^2}}{2}\right)\right], \tag{3.87}$$

in terms of Favre variables with $\widetilde{\sigma}_{xy} \approx \mu(\partial \widetilde{u}/\partial y)$.

If, additionally, a zero mean pressure gradient is considered at a sufficiently high Reynolds number so that turbulent processes dominate, Eqs. (3.83) and (3.84) and Eqs. (3.86) and (3.87) will yield a solution of the form

$$\overline{H} = a_0\overline{u} + a_1, \tag{3.88a}$$
$$\widetilde{H} = b_0\widetilde{u} + b_1, \tag{3.88b}$$

respectively, and where it is assumed that the Prandtl number is unity. For a boundary layer flow, the coefficients a_i and b_i can be identified with parameter values at the wall that are assumed constant. Equation (3.88) can be written for the Reynolds and Favre mean total enthalpy as

$$\overline{H} - \overline{H}_w \equiv c_p \overline{T}_t - c_p \overline{T}_{tw} = Pr_w \left. \left(\frac{\overline{q}_y}{\overline{\sigma}_{xy}} \right) \right|_w \overline{u} \tag{3.89a}$$

$$\tilde{H} - \tilde{H}_w \equiv c_p \tilde{T}_t - c_p \tilde{T}_{tw} = Pr_w \left. \left(\frac{\tilde{q}_y}{\tilde{\sigma}_{xy}} \right) \right|_w \tilde{u}. \tag{3.89b}$$

Since the conditions at the wall are assumed constant, the linear relation between mean total enthalpy and mean streamwise velocity can be readily extended to the corresponding fluctuating quantities so that

$$H' \equiv c_p T_t' = Pr_w \left. \left(\frac{\overline{q}_y}{\overline{\sigma}_{xy}} \right) \right|_w u' \tag{3.90a}$$

$$H'' \equiv c_p T_t'' = Pr_w \left. \left(\frac{\tilde{q}_y}{\tilde{\sigma}_{xy}} \right) \right|_w u''. \tag{3.90b}$$

Morkovin (1962) (see also Gaviglio, 1987) termed the solutions in Eqs. (3.89a) and (3.90a) as the strong Reynolds analogy (SRA). In addition, following Morkovin (1962), the mean and fluctuating total, or stagnation, temperature has been introduced (cf. Eq. (2.28)).

While there exists a formal equivalence between the solutions given in Eqs. (3.89a) and (3.90a) and Eqs. (3.89b) and (3.90b), it needs to be recalled that the SRA in Reynolds-averaged variables resulted from equations where terms associated with the mass flux and higher-order correlations were neglected (cf. Eq. (3.78a) and (3.78b)). In this context, it follows that the SRA relations in Reynolds-averaged variables are a weaker set of conditions than the SRA relations in Favre variables. However, these SRA relations in Reynolds variables provide an important link between experimentally measured values and results determined from averaged (RANS) methods. For this latter reason the additional SRA relations will be derived in the original way in terms of the Reynolds–averaged variables.

3.5.1.1 Morkovin's Relations
If the case of an adiabatic wall is now considered, Eqs. (3.89a) and (3.90a) can be written as

$$\overline{T}_t = \overline{T}_{tw} = \overline{T}_w \tag{3.91a}$$

$$T_t' = 0, \tag{3.91b}$$

where the absence of total temperature fluctuations implies the absence of both temperature and velocity fluctuations which corresponds to neglecting fluctuations due to the entropy mode. Even with the limited amount of experimental data available at the time Morkovin (1962) questioned the validity of assuming negligible total temperature fluctuations as dictated by Eq. (3.91b). Nevertheless, the original SRA conditions were derived and expressed in *rms* forms using the zero total temperature fluctuations condition. Here the condition $T_t' \neq 0$ will be used throughout, but the added constraint of zero total temperature fluctuations will be also imposed at times for comparative purposes with the original SRA relations.

For the two-dimensional boundary layer flow considered here, the mean total temperature can be written as

$$\overline{T}_t \simeq \overline{T} + \frac{\overline{u}^2}{2c_p}, \tag{3.92}$$

where terms quadratic in the fluctuations are neglected in the definition. The fluctuating total temperature can then be approximated as

$$T_t' \simeq T' + \left(\frac{\overline{u}}{c_p}\right) u', \tag{3.93}$$

where it has additionally been assumed that $\overline{u}u' \gg \overline{v}v'$. Equations (3.92) and (3.93), when coupled with Eqs. (3.89a) and (3.90a), evaluated at the boundary-layer edge, can be combined to yield an expression for the fluctuating temperature,

$$\frac{T'}{\overline{T}} = -(\gamma - 1)\,\overline{M}^2 \left[1 + \frac{c_p(\overline{T}_w - \overline{T}_{te})}{\overline{u}\,\overline{u}_e}\right] \left(\frac{u'}{\overline{u}}\right), \tag{3.94}$$

where the local Mach number $\overline{M} = \overline{u}/\sqrt{\gamma \mathcal{R} \overline{T}}$ has been introduced. In the absence of any total temperature fluctuations, Eq. (3.94) must reduce to the simpler expression

$$\frac{T'}{\overline{T}} = -(\gamma - 1)\,\overline{M}^2 \left(\frac{u'}{\overline{u}}\right) \tag{3.95}$$

or, using Eq. (3.92) evaluated at the boundary-layer edge (with $\overline{T}_t = $ const),

$$\frac{T'}{\overline{T}_w - \overline{T}_e} = -2\frac{\overline{u}}{\overline{u}_e}\left(\frac{u'}{\overline{u}_e}\right). \tag{3.96}$$

Equation (3.94) was first introduced by Cebeci and Smith (1974) in the instantaneous form shown as an attempt to account for wall heat transfer effects. Gaviglio (1987) then referred to the *rms* form of this relation as an extended strong Reynolds analogy (ESRA).

It is possible to obtain the corresponding correlation relation to Eq. (3.95) under a condition less restrictive than $T_t' = 0$. Consider the temperature variance relationship formed from Eq. (3.93),

$$\overline{T'^2} = \left(\frac{\overline{u}}{c_p}\right)^2 \overline{u'^2} + 2\overline{T'T_t'} - \overline{T_t'^2}, \tag{3.97}$$

which, when coupled with an assumption that

$$\overline{T'^2} \gg \overline{T_t'^2} - 2\overline{T'T_t'}, \tag{3.98}$$

can yield a correlation expression analogous to the relation given in Eq. (3.95),

$$\frac{\sqrt{\overline{T'^2}}}{\overline{T}} \simeq (\gamma - 1)\,\overline{M}^2\,\frac{\sqrt{\overline{u'^2}}}{\overline{u}}, \tag{3.99}$$

and the relation given in Eq. (3.96),

$$\frac{\sqrt{\overline{T'^2}}}{\overline{T}_w - \overline{T}_e} \simeq 2\left(\frac{\overline{u}}{\overline{u}_e}\right)\frac{\sqrt{\overline{u'^2}}}{\overline{u}_e}. \tag{3.100}$$

In addition to Eq. (3.93), Eqs. (3.99) and (3.100) comprise three of the original five forms of the SRA obtained by Morkovin (1962) under the more stringent condition of zero total temperature fluctuations. The instantaneous form of Eq. (3.99) given in Eq. (3.95) has been referred to as the "fluctuation form" of the SRA (Smith & Smits, 1993). It should be noted that the instantaneous form given in Eq. (3.95) has not been verified either experimentally or numerically, thus this relation should only be taken as an approximation.

It is worth emphasizing here that results obtained from Eq. (3.97) will contrast with results obtained from alternate expressions of similar form, such as

$$\overline{T''^2} = \left(\frac{\tilde{u}}{c_p}\right)^2\overline{u''^2} + 2\overline{T''T_t''} - \overline{T_t''^2}, \tag{3.101}$$

as well as forms extracted from the Favre-averaged variables. Such complexities have only arisen from the availability of simulation data where various numerical averaging/filtering procedures are easily implemented. What is important, is the consistency of the approach and assumptions used in deriving the relevant relations. Then, it is possible to use the derived relations to analyze and validate both numerical and experimental results.

The remaining two relations that comprised the original SRA relations were based on the correlation coefficient $R_{u'T'}$, defined as

$$R_{u'T'} \equiv \frac{\overline{u'T'}}{\sqrt{\overline{u'^2}}\sqrt{\overline{T'^2}}}, \tag{3.102}$$

and the turbulent Prandtl number, defined as

$$Pr_t \equiv \frac{-\overline{\rho u'v'}(\partial\overline{T}/\partial y)}{-\overline{\rho v'T'}(\partial\overline{u}/\partial y)}. \tag{3.103}$$

Relations for these quantities have also been extended from their original form to more accurately account for the non-negligible total temperature fluctuations.

The correlation coefficient $R_{u'T'}$ can be formed from the streamwise scalar heat flux obtained from Eq. (3.93),

$$\overline{u'T'} = -\left(\frac{\overline{u}}{c_p}\right)\overline{u'^2} + \overline{u'T_t'}.$$ (3.104)

If Eqs. (3.93) and (3.99) are used, the correlation coefficient $R_{u'T'}$ is given by

$$R_{u'T'} = -1 + \frac{\overline{T_t'^2}}{2\overline{T'^2}}.$$ (3.105)

When total temperature fluctuations can be neglected, Eq. (3.105) simplifies to the original SRA form

$$R_{u'T'} = -1.$$ (3.106)

In order to obtain a relation for the turbulent Prandtl number Pr_t, both the vertical heat flux correlation and the turbulent shear stress are needed. The vertical scalar heat flux vector can be obtained from Eq. (3.93) and written as

$$\overline{v'T'} = -\left(\frac{\overline{u}}{c_p}\right)\overline{u'v'} + \overline{v'T_t'}.$$ (3.107)

With this, it is easy to show that the corresponding correlation coefficient $R_{v'T'}$ is related to the shear stress correlation coefficient $R_{u'v'}$ by the relation

$$R_{u'v'} = -R_{v'T'}\left[1 - \frac{\overline{v'T_t'}}{\overline{v'T'}}\right],$$ (3.108)

where

$$R_{u'v'} \equiv \frac{\overline{u'v'}}{\sqrt{\overline{u'^2}}\sqrt{\overline{v'^2}}}$$ (3.109a)

$$R_{v'T'} \equiv \frac{\overline{v'T'}}{\sqrt{\overline{v'^2}}\sqrt{\overline{T'^2}}}.$$ (3.109b)

If the linearized equation for the total temperature, Eq. (3.92), and the equation for the vertical heat flux, Eq. (3.107), are used, Eq. (3.103) can be rewritten in the form

$$Pr_t = \left[1 - \frac{\overline{v'T_t'}}{\overline{v'T'}}\right]\left[1 - \frac{\partial\overline{T_t}}{\partial\overline{T}}\right]^{-1}.$$ (3.110)

When the turbulent shear stress and vertical scalar flux correlations are anti-correlated, $R_{u'v'} = -R_{v'T'}$, Eq. (3.110) reduces to

$$Pr_t = \left[1 - \frac{\partial\overline{T_t}}{\partial\overline{T}}\right]^{-1},$$ (3.111)

and if the total temperature fluctuations are neglected, with $\overline{T}_t = \overline{T}_w$ from Eq. (3.91a), Eq. (3.111) further simplifies to the original SRA form

$$Pr_t = 1. \tag{3.112}$$

Both Eqs. (3.106) and (3.112) are the same form proposed in the original SRA relations.

3.5.1.2 Extended Forms

Recall that the SRA relations were intended for adiabatic conditions and that Cebeci and Smith (1974) attempted to account for wall heat flux effects through the relation given in Eq. (3.94). Gaviglio (1987) and Rubesin (1990) further examined flows under such non-adiabatic conditions and used mixing-length arguments to arrive at

$$\frac{T'}{\overline{T}} = -c_0 \left(\gamma - 1 \right) \overline{M}^2 \left[1 - \frac{\partial \overline{T}_t}{\partial \overline{T}} \right]^{-1} \left(\frac{u'}{\overline{u}} \right), \tag{3.113}$$

where c_0 is a constant. Gaviglio (1987) assumed the characteristic spatial scales of the velocity and thermal fluctuations to be the same and arrived at a value of $c_0 = 1$. Rubesin (1990) attempted to calibrate his relation from experimental and simulation data and arrived at a value of $c_0 \approx 0.75$. Later, these analyses were revisited (Huang et al., 1995) and it was shown that the value of c_0 could be related to the turbulent Prandtl number Pr_t.

It is possible to adapt these relations for temperature to corresponding relations for the density fluctuations. From the perfect gas law and for small fluctuations, the pressure fluctuations can be related to both the density and temperature fluctuations through

$$\frac{p'}{\overline{p}} = \frac{\rho'}{\overline{\rho}} + \frac{T'}{\overline{T}}. \tag{3.114}$$

If the pressure (acoustic) fluctuations are negligible, the density and temperature fluctuations are related through

$$\frac{\rho'}{\overline{\rho}} = -\frac{T'}{\overline{T}}, \tag{3.115}$$

where the temperature fluctuations are essentially non-isentropic from Eq. (1.32). Since the density and temperature fluctuations are now related by Eq. (3.115), it is possible to obtain a set of relations between the density and velocity correlations. A practical example of the usefulness of such relationships can be found by first considering the velocity–temperature correlation coefficient given in Eq. (3.102) and rewritten, using Eq. (3.99), as

$$\frac{\overline{u'T'}}{\overline{u}\,\overline{T}} = R_{u'T'} \left(\gamma - 1 \right) \overline{M}^2 \frac{\overline{u'^2}}{\overline{u}^2}. \tag{3.116}$$

If the pressure fluctuations can be neglected as just discussed, and Eq. (3.115) is used, Eq. (3.116) can be rewritten in terms of the velocity–density correlation coefficient as

$$\frac{\tilde{u}}{\bar{u}} - 1 = \frac{\overline{\rho'u'}}{\bar{\rho}\,\bar{u}} = R_{u'\rho'}\,(\gamma - 1)\,\overline{M}^2\frac{\overline{u'^2}}{\bar{u}^2}. \tag{3.117}$$

Recall that in Section 3.3, it was shown in deriving the governing equations in terms of density-weighted, or Favre, variables approximations were made requiring that an equality between Reynolds-averaged and Favre-averaged variables be assumed. Figure 3.2 showed that the maximum error incurred across the boundary layer was less than 2%. If direct simulation data had not been available, a similar estimate could have been obtained from Eq. (3.117). If the SRA is assumed to hold in its original form (negligible total temperature fluctuations), that is $R_{u'\rho'} \simeq 1$ (recall $R_{u'T'} \simeq -1$), then for the Mach 2.25 boundary layer flow, shown in Figure 3.2, with a 10% *rms* fluctuation level, the difference between the two mean values is of order of 2%. This is consistent with the direct estimate from DNS data obtained in Section 3.3.

As was just discussed in extending some of the SRA relations for temperature and velocity to density and velocity, for example, an underlying assumption was the neglect of the normalized pressure fluctuations relative to the normalized density fluctuations. The applicability of these relations to both wall-bounded and free shear flows then depends on how well this assumption can hold. As will be seen in Chapter 6, the turbulence in free shear flows is much more susceptible to compressibility effects than the turbulence in wall-bounded flows. In free shear flows, the pressure fluctuations increase in importance as the Mach number increases and often are comparable to the corresponding density fluctuations. It is possible to show, however, that the relative magnitude of the pressure and density fluctuation levels in itself may not negate the application of these SRA relations to such flows, and that a more relevant comparison is based on a (linear) mode analysis.

Consider the linearized relation between the pressure, density and entropy for a perfect gas extracted from Eqs. (1.32) and (1.35) where the assumption of small amplitude turbulent fluctuations is made. (Recall the earlier discussion of the mode decomposition of Kovasznay (1953).) For example, the mean square temperature and density fluctuations can be written in terms of the acoustic (pressure) and entropic modes as

$$\frac{\overline{T'^2}}{\overline{T}^2} = (\gamma - 1)^2\,\frac{\overline{p'^2}}{\gamma^2\overline{p}^2} + \frac{\overline{s'^2}}{c_p^2} \tag{3.118a}$$

$$\frac{\overline{\rho'^2}}{\overline{\rho}^2} = \frac{\overline{p'^2}}{\gamma^2\overline{p}^2} + \frac{\overline{s'^2}}{c_p^2}, \tag{3.118b}$$

where for (inviscid) small amplitude pressure and entropy fluctuations are uncorrelated, $\overline{p's'} = 0$ (Blaisdell et al., 1993). This shows that the difference between the normalized temperature and density mean square fluctuations can be expressed in terms of the acoustic mode and given by (Barre, private communication)

$$\frac{\overline{\rho'^2}}{\overline{\rho}^2} - \frac{\overline{T'^2}}{\overline{T}^2} = \left(\frac{2-\gamma}{\gamma}\right)\frac{\overline{p'^2}}{\overline{p}^2}. \tag{3.119}$$

If it is assumed that the *rms* turbulent thermal fluctuations can be related in a generalized polytropic manner (Rubesin, 1976),

$$\frac{\sqrt{\overline{p'^2}}}{\overline{p}} = n_{p\rho}\frac{\sqrt{\overline{\rho'^2}}}{\overline{\rho}} \tag{3.120a}$$

$$\frac{\sqrt{\overline{\rho'^2}}}{\overline{\rho}} = \left(\frac{1}{n_{\rho T}-1}\right)\frac{\sqrt{\overline{T'^2}}}{\overline{T}}, \tag{3.120b}$$

where $n_{p\rho}$ and $n_{\rho T}$ are polytropic (modeling) coefficients, then $n_{p\rho} = 0$ means isobaric, $n_{\rho T} = 1$ means isothermal, and $n_{p\rho} = n_{\rho T} = \gamma$ means isentropic. Using these relations, a normalized difference between density and temperature intensities, Γ_B, can be defined and written as (Barre, private communication)

$$\Gamma_B \equiv \frac{\sqrt{\overline{\rho'^2}}/\overline{\rho} - \sqrt{\overline{T'^2}}/\overline{T}}{\sqrt{\overline{\rho'^2}}/\overline{\rho}} = \left(\frac{n_{p\rho}^2}{\gamma\, n_{\rho T}}\right)\Gamma_B^{(isen)}, \tag{3.121}$$

where $\Gamma_B^{(isen)} = (2-\gamma)$. From Eq. (3.121), it is easily shown that for the isobaric, isentropic and isothermal cases, the values obtained for Γ_B are 0, $2-\gamma$, and 1, respectively. This suggests that a quantitative criterion of the relative importance of the pressure fluctuations to the other modes can be obtained by comparing the value of Γ_B with its isentropic value $\Gamma_B^{(isen)}$, with the condition that

$$\Gamma_B \ll \Gamma_B^{(isen)} \quad \text{or} \quad \frac{n_{p\rho}^2}{\gamma\, n_{\rho T}} \ll 1, \tag{3.122}$$

should be satisfied in order to justify the neglect of the pressure or acoustic mode. As will now be shown, this criterion is then distinct from a simple comparison of pressure fluctuation intensity and density fluctuation intensity. These points will be further utilized in Section 7.1.1 when the linear interaction analysis (LIA) theory is discussed.

Consider first the case of a purely solenoidal turbulent field passing through a normal shock at a Mach number of 3 (Jamme, 1998). Simulation data yields values for $n_{p\rho}$ and $n_{\rho T}$ of 0.7 and 1.9, respectively. This results in a value for Γ_B of $0.18\Gamma_B^{(isen)}$, which corresponds to an acoustic mode contribution of only 18% when the pressure fluctuations are 70% of the density fluctuations (cf. Eq. (3.120a)). Thus, as this example shows, the importance of the normalized pressure fluctuation intensity is not necessarily based on its relative magnitude to the normalized density fluctuation intensity obtained from Eq. (3.120a,b), but rather on a parameter that is a measure of the importance of the fluctuating pressure intensity to its value relative to the other fluctuation modes.

The fundamental relations derived from the strong or fluctuating Reynolds analogies have highlighted some turbulent correlations that are either unique to the compressible regime or that can be used to better characterize such flows. They can also be used to validate both experimental and numerical simulation data as well as the predictive capability of statistical average (RANS) predictions. In turn, experimental and numerical simulation data can be used to validate the assumptions used in the derivation of the relations comprising the Reynolds analogies.

3.5.2 Morkovin's Hypothesis

From the discussion in Section 3.5.1, Morkovin (1962) further deduced from the solution of the boundary layer averaged momentum and total enthalpy equations, that the "...*essential dynamics of these supersonic shear flows will follow the incompressible form...*" This observation was quantified by the proposal that

$$\left(\frac{\overline{\rho}}{\overline{\rho}_w}\right)\frac{\overline{u'v'}}{u_\tau^2}, \qquad \left(\frac{\overline{\rho}}{\overline{\rho}_w}\right)\frac{\overline{u'^2}}{u_\tau^2} \tag{3.123}$$

"*would depend little on Mach number ...*" This hypothesis asserts that compressibility or Mach number only affects the turbulence (at least the velocity correlations) through variations in mean density, and that fluctuations in density ρ' have little effect on the turbulence. These observations, as well as the previously discussed strong Reynolds analogies (see Section 3.5), were originally focused on turbulence structure, specifically, the large-scale turbulent structure. Both experimental measurements and numerical simulations have now largely confirmed (at least qualitatively) the hypothesis. In the experimental investigations (e.g. Dussauge & Gaviglio, 1987; Smits & Dussauge, 2006), difficulties in measuring the various velocity correlations and wall scaling quantities provide a less clear situation than the numerical studies. Nevertheless, the trends over a rather wide Mach number range tend to confirm the hypothesis.

The numerical simulations provide a more detailed picture although over a narrower range of Mach numbers. The most relevant comparison is with the incompressible simulation data of Spalart (1988) at values of Re_θ of 670 and 1410. In a temporal simulation at $M_\infty = 2.5$ that used a method that transformed the spatially evolving boundary layer into a parallel, streamwise homogeneous flow (Guarini, Moser, Shariff, & Wray, 2000), both the scaled turbulent intensities, $\sqrt{\overline{\rho}/\rho_w}(\sqrt{\overline{u_i'^2}}/u_\tau)$, and the scaled shear stress component, $\sqrt{\overline{\rho}/\rho_w}(\overline{u'v'}/u_\tau^2)$, of the velocity second-moments were examined. The simulations were performed at a free-stream Re_θ of 1577

Compressibility, Turbulence and High Speed Flow

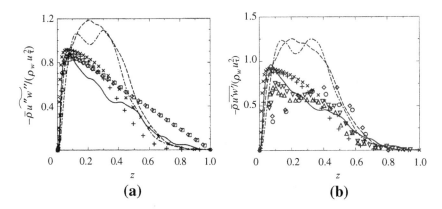

FIGURE 3.4 Comparison of density-weighted and Reynolds averaged shear stress distributions across boundary layer: (a) —— $M_\infty = 3$, --- $M_\infty = 4.5$, -·- $M_\infty = 6$, symbols $+$, $\times$, $\diamond$ represent incompressible DNS data (see Figure 13b of Maeder et al., 2001); (b) —— $M_\infty = 3$, --- $M_\infty = 4.5$, -·- $M_\infty = 6$, symbols $+$, $\times$, represent incompressible DNS data and open symbols represent experimental data (see Figure 15 of Maeder et al., (2001)). From Maeder et al. (2001) with permission.

which corresponded to an equivalent incompressible Re_θ of 902. Overall, for this simulation, both the scaled normal and shear components compared favorably with the incompressible data when scaled with the local mean density. An analogous assessment of the hypothesis was carried out later (Maeder, Adams, & Kleiser, 2001), where a different temporal DNS was used that also accounted for streamwise mean flow variation at Mach numbers of $M_\infty = 3, 4.5$ and 6 with free-stream Re_θ values of 3028, 3305, and 2945, respectively, and with corresponding equivalent incompressible Re_θ values of 1456, 1032, and 652. The scaled longitudinal normal stress component $\overline{\rho u'' u''}/(\rho_w u_\tau^2)$ and their shear stress components $\overline{\rho u'' w''}/(\rho_w u_\tau^2)$ and $\overline{\rho u' w'}/(\rho_w u_\tau^2)$ were examined (see Figure 3.4). While the case of $M_\infty = 6$ was clearly outside the range of Morkovin's hypothesis, the other two cases fell within the Mach number range where the hypothesis is considered to apply. In addition, the corresponding equivalent incompressible Re_θ of all the cases was either within or near the limits of the incompressible simulation data. Although the peak values in both the $M_\infty = 3$ and 4.5 cases tended to match the incompressible simulation data, the overall variation across the boundary layer was less satisfactory than the study of Guarini et al. (2000).

In contrast to these temporal-based simulations, a fully spatial simulation at $M_\infty = 2.25$ was performed (Pirozzoli et al., 2004). The flow field was allowed to evolve along the flat plate from a compressible laminar flow to a fully turbulent flow. The free-stream momentum thickness Reynolds number at the start of the fully turbulent regime was $Re_\theta = 4263$. This corresponded to an equivalent incompressible Re_θ of 1967 at the start of the fully turbulent region, and at the end of the simulated turbulent region the value was 2283. Even though these values were outside the range of values used in the incompressible simulations of Spalart (1988), the assessment of Morkovin's hypothesis was in very good agreement with the incompressible simulations for both the normal $(\overline{\rho}/\rho_w)(\overline{v'^2}/u_\tau^2)$ and spanwise components $(\overline{\rho}/\rho_w)(\overline{w'^2}/u_\tau^2)$, and slightly less favorable agreement with the longitudinal component $(\overline{\rho}/\rho_w)(\overline{u'^2}/u_\tau^2)$. The shear stress component $(\overline{\rho}/\rho_w)(\overline{u'v'}/u_\tau^2)$ was slightly over-predicted relative to the incompressible simulations.

As suggested in the original hypothesis put forth by Morkovin, this density scaling should apply at moderate values of wall heat transfer. Huang et al. (1995) analyzed the DNS data of Coleman, Kim, & Moser (1995) for the case of fully developed (isothermal) cooled-wall channel flow at Mach numbers of 1.5 and 3 (based on bulk velocity), and a friction Reynolds number Re_τ of 150 for both. They compared the shear stress $\overline{\rho \widetilde{u'' w''}}/(\rho_w u_\tau^2)$ and kinetic energy $\overline{\rho \widetilde{u_i''^2}}/(2\rho_w u_\tau^2)$ distributions of two cooled-wall cases against the incompressible channel flow simulation data of Mansour, Kim, and Moin (1988) at Re_τ of 180. The case with the mean centerline to wall temperature ratio of 1.38 compared very well with the incompressible data; whereas, the stronger mean centerline to wall temperature case of 2.49 comparison was less favorable. This latter case appeared to be outside the range where simple density scaling could account for the change in dynamics.

These experimental and numerical results certainly provide confirmation for the original hypothesis of Morkovin. It is the basis on which compressible, Reynolds-averaged Navier–Stokes strategies have used variable density extensions of their incompressible counterparts to compute compressible subsonic and supersonic flow fields with a high degree of success. Nevertheless, limitations to such strategies exist, such as in the presence of shocks or strong scalar fluxes and, therefore, require closures that account differently for the compressibility of the turbulence.

Chapter 4

Experimental Measurement and Analysis Strategies

This chapter focuses on the unique constraints imposed by compressibility on measurement and analysis strategies for turbulent flows at high speeds when compared to incompressible situations. Indeed, the major characteristics arising from high speed compressibility are the coexistence of several fluctuating fields (having both dynamic and thermodynamic properties) and the associated modes, the presence of shocks, and the wide bandwidths. None of the conventional available measurement systems can handle all these characterizing features in total. More importantly, these characteristics can impose severe limitations of the measurement capabilities of the methods more widely developed, optimized, and used in incompressible regimes. It is not the intent here to provide an exhaustive description of the different measurement methods and analysis procedures (there is a large literature on these topics), but to highlight the unique influence compressibility has on these methods and procedures.

Although the topics of measurements and analyses may appear focused to physical experimentalists, the material is applicable to both laboratory (physical) and numerical experiments. For the laboratory experiments, it is important in their design that the orders of magnitude of the flow quantities, that characterize the physics, be established, and their impact on the measurement methods identified. For the numerical experiments, which are often used in either the prediction or validation of the physical experiment, it is crucial that the physical experiment is properly replicated. In this context, there are three factors that should be considered in formulating the numerical problem: (i) constraints imposed on the experimental study; (ii) limitations on the data collected; and (iii) accurate replication of boundary conditions and parameterizations.

Compressibility, Turbulence and High Speed Flow. http://dx.doi.org/10.1016/B978-0-12-397027-5.00004-6

4.1 EXPERIMENTAL CONSTRAINTS FOR SUPERSONIC FLOWS

Just as in subsonic flows, wind tunnels are an essential tool in the study of supersonic flows. However, the supersonic regime imposes some unique constraints on the tunnels themselves and on the measurement techniques used in obtaining the data. While some of the constraints are simply related to tunnel size and run times or to limitations on measurement apparatuses, other constraints relate to underlying flow physics. In this section, these various constraints are identified and discussed.

4.1.1 Constraints on Wind Tunnel Testing

As noted, boundary conditions and parameterizations, play an important role in both laboratory and numerical experiments. This makes it necessary to consider characteristics of wind tunnels where the experiments are performed and that define the typical space and time scales of the fluctuating fields. For laboratory experiments in general, the higher the flow velocity, the smaller are the dimensions of the wind tunnel, since the required power to drive a wind tunnel is related to the size of the wind tunnel (L) and speed (U), which is proportional to $\rho L^2 U^3$. The result is that the size of the test section L of the tunnel is generally less than in the subsonic cases. With the exception of the large high velocity wind tunnels of research agencies, university wind tunnel sizes are of the order of a few tens of centimeters at most. This limited size of tunnels is particularly relevant when models are tested since blockage effects can create parasitic shocks or modify the entire flow regime. When shock waves are present (either intrinsic to the flow or imposed by the boundary conditions), the influence of the walls becomes predominant, inducing shock reflections and possible expansion fans, parasitic separations, etc. These conditions strongly depend on wall characteristics (shock absorbing, adaptive walls) of the wind tunnels. Thus, the boundary conditions of any transonic or supersonic experiment have to be carefully specified so that the flow field is free of the unwanted parasitic effects, and that the real physics is well represented. In addition, for computational fluid dynamic comparisons it is necessary to provide a complete and detailed boundary condition specification in order to have the same physical (laboratory) and numerical experiment.

For most practical applications of compressible, high speed flows, the Reynolds number is very large, and this is also the case for most experimental studies provided the velocities are sufficiently high. However, it should be noted that a unique feature of supersonic wind tunnels, when compared to subsonic ones, is that the pressures can be different. The influence of the pressure is noticeable in jets discharging into ambient surroundings where the pressure difference can prevent the jet from being fully expanded. In tunnels with a closed test section, the static pressure can be modified and this has an effect on the Reynolds number. Two different categories of tunnels exist that allow for such pressure modifications but do so in a different manner.

The first category (WT1) corresponds to suction or return-circuit wind tunnels, in which the settling chamber pressure is of the same (or lower) order as (than) the ambient. Values on the order of 0.5×10^5 Pa are often encountered and are designated as WT1 conditions (values as low as 10×10^5 Pa can be also encountered (e.g. Tedeschi, Gouin, & Elena, 1999)). The second category (WT2) corresponds to pressurized wind tunnels, in which the settling chamber pressure for moderate Mach numbers can be higher, with typical values up to 10×10^5 Pa. The result is that the Reynolds numbers can vary about an order of magnitude between different facilities for the same Mach number. These pressure conditions can have a large influence on the molecular length scale of motion, the mean free path ξ, given by

$$\xi = \frac{1}{\sqrt{2}\pi d_m^2}\left(\frac{m_m \mathcal{R} T}{p}\right), \tag{4.1}$$

with m_m the molecular mass and d_m the diameter of a molecule. Thus, the mean free path is inversely proportional to the pressure so that at atmospheric conditions, for example, ξ is $\mathcal{O}(10^{-7}$ m). From the kinetic theory of gases, under the approximation of hard sphere gas, the viscosity is proportional to the mean free path and the average molecular velocity, $\bar{v}_m = \sqrt{8\mathcal{R}T/\pi}$, and can be written as

$$\mu = 0.499\rho\bar{v}_m\xi, \tag{4.2}$$

(see p. 98, Chapman & Cowling, 1970). If the speed of sound is introduced ($c = \sqrt{\gamma \mathcal{R} T}$), Eq. (4.2) can be rewritten as

$$\mu \simeq \rho\sqrt{\frac{2}{\gamma\pi}}c\xi, \tag{4.3}$$

or for the mean free path as

$$\xi = \left(\frac{\mu}{\rho}\right)\sqrt{\frac{\pi}{2\mathcal{R}T}}. \tag{4.4}$$

From Eq. (4.3), ξ can be expressed in terms of a unit Reynolds number, $Re^* = \rho U_{\text{ref}}/\mu$, and a characteristic Mach number $M_r = U_{\text{ref}}/c$, so that

$$\xi = \sqrt{\frac{\pi\gamma}{2}}\left(\frac{M_r}{Re^*}\right). \tag{4.5}$$

As will be seen later, the choice of the velocity U_{ref} that defines the Mach and Reynolds numbers needs to be correctly associated with the particular physical aspect of the problem under study. Typically the local mean velocity is chosen if the macroscopic analysis has to be performed, and when the turbulent scales are of interest, then a reference velocity characterizing the turbulent motions is needed.

For air flow at $M = 2$ and a plenum temperature of 300 K, the low pressure condition for WT1 yields a mean free path of $\approx 4.5 \times 10^{-7}$ m, which is almost an order of magnitude larger than the value of 2.4×10^{-8} m that is obtained for the high pressure condition of the WT2 wind tunnel at the same Mach number and temperature conditions. This has a major impact on the thicknesses of the shocks—which are typically of the order of a few mean free paths. It also affects some measurement systems that are associated with the physical scales of the flow, such as the hot-wire anemometer (HWA) or the particle-based (non-intrusive) optical methods. In addition, mesh requirements for numerical simulations can also be affected where the exact or filtered conservation equations are solved in flow fields with (weak) shocks.

The effect just described can be parameterized through the Knudsen number Kn ($= \xi/L$, where L is a characteristic spatial length). The simplest situation occurs when Kn $\lesssim 0.01$ since the continuum hypothesis is assumed valid, but if Kn $\gtrsim 10$, the condition corresponds to free molecular flow. Between these two limits, there is a transition regime where Kn $= \mathcal{O}(1)$ and a regime where a slip flow condition applies Kn < 0.1. The choice of the relevant values used to compute Kn is not straightforward, however. As mentioned before, the velocity scale used to compute ξ can be either the local velocity or the reference velocity, and the length scale L can be, for example, either the physical dimension of an obstacle or the thickness of the boundary layer that develops on the obstacle.

The Knudsen number can also be useful for the analysis of the turbulent scales. Turbulent eddies can be associated with the small scale motion of turbulent flows with the smallest of these being the Kolmogorov scale, η. In the subsonic case, these scales are removed from the mean free path limit; however, the supersonic character of the flow can have a strong influence on this limit, and it is important to consider whether these smallest scales are within this continuum limit. For the turbulent scales of size η, the relevant turbulent velocity scale to be considered is $\sqrt{u_i'^2}$. From Eq. (4.3), the relevant Knudsen number is simply given by

$$\mathrm{Kn}_\eta = \frac{\xi}{\eta} = \sqrt{\frac{\pi \gamma}{2}} \left(\frac{\mu}{\rho c \eta} \right), \qquad (4.6)$$

which can be related to the more useful turbulent Reynolds and Mach number parameters. Since the ratio of the Kolmogorov scale η to the large, energetic spatial (integral) scales L is proportional to $Re_t^{-3/4}$ (Tennekes & Lumley, 1972), the corresponding Reynolds and Mach numbers are, $Re_t = \rho \sqrt{u_i'^2} L / \mu$ and

$M_t = \sqrt{\overline{u_i'^2}}/c$, respectively.[1] Equation (4.6) can then be expressed in terms of these parameters as

$$\mathrm{Kn}_\eta = \sqrt{\frac{\pi \gamma}{2}} \left(\frac{M_t}{Re_t^{1/4}} \right). \tag{4.7}$$

Kn_η is sometimes called the "micro-structure Knudsen number" (Tennekes & Lumley, 1972). The weak dependency on Reynolds number due to the $-1/4$ power dependence suggests that the continuum hypothesis will generally hold over a wide range of Reynolds numbers. As an example, consider the two categories of tunnels, WT1 and WT2, referred to above. For these two wind tunnels, the unit Reynolds numbers based on a Mach 2 regime are respectively of order of 6×10^6 m^{-1} and 1.2×10^6 m^{-1} for WT1 and WT2. If a turbulence level of 20% is considered, with a typical integral scale of $L \simeq 8 \times 10^{-3}$ m (an example from the typical value observed in the mixing-layer of Barre, Quine, and Dussauge (1994)), the turbulent Reynolds numbers are respectively 10^4 and 2×10^5. The magnitude of the micro-structure Knudsen numbers Kn_η then lies between 6×10^{-2} and 3×10^{-2}, respectively. In the WT1, low pressure tunnel case, the Kn number lies within the slip-flow regime, while in the WT2 high pressure wind tunnel case, the continuum assumption is most likely applicable.

Supersonic wind tunnels can be operated either continuously (long time running) or intermittently. As with subsonic wind tunnels, continuous blowing supersonic wind tunnels pose no particular experimental problems provided the temperature is properly adjusted. This is not the case for the intermittent, or so-called blowdown, wind tunnels in which the running times can be as long as a few minutes, or even as short as hundreds of milliseconds. As an example, the Shock Wind Tunnel of the Institute of Aerodynamics and Gasdynamics (IAG) at Stuttgart University in Germany has a typical quasi-steady state duration of 120 ms (Weiss, Knauss, & Wagner, 2003). Some hypersonic wind tunnels can have even shorter running durations. In these wind tunnels, the time required to reach a stabilized state for both the dynamic and thermal fields has to be considered, since the time required to obtain converged statistics can also limit the amount of data collected during each run. Fortunately, this is in general easy to account for, provided the typical time scales of the flows are also sufficiently short (see Section 4.1.2). The time constraints are more demanding when seeding is required. The seeding location is often located in the upstream, subsonic part of the flow, and so the response times of the flow itself, and of the seeding transport, can then be different.

---

[1] It should be noted that the definition of the turbulent Mach number is not unique. For example, the turbulent velocity scale used for the above definition of M_t is twice the turbulent kinetic energy. However, the velocity is sometimes chosen as $\sqrt{1/3\overline{u_i'^2}}$, that is, referring to one component of the velocity field coupled with an isotropy hypothesis (Tennekes & Lumley, 1972).

4.1.2 Constraints on Data Collection and Measurement Apparatus

For supersonic turbulent shear flows, space, and time scales need to be specified, or at least estimated, to assess the quality of measurements (and computations), since the inherent three-dimensionality and unsteadiness of the turbulence requires well-resolved (in both space and time) measurements. The range of scales in any turbulent flow depends on the (physical and numerical) experimental conditions, and this specification becomes more difficult in high speed compressible flows where different thermodynamic properties evolve at high frequencies and small spatial scales when compared with the equivalent subsonic, incompressible situation. As pointed out in the last section, the ratio of the Kolmogorov scale η to the large, energetic spatial (integral) scales L is proportional to $Re_t^{-3/4}$. Due to the very large values of the Reynolds numbers generally encountered in compressible high speed flows, the dynamic range of the fluctuations can be considerably larger than the equivalent incompressible situation.

If measurements are to be applicable to numerical experiments, it is necessary that as complete a specification as possible be provided consistent with the independent variables of the numerical problem. For numerical simulations such as direct numerical simulation (DNS) and large eddy simulation (LES), the instantaneous details of the flow structure are important for a sufficiently accurate physical interpretation of the turbulent phenomena with a level of consistency with the numerical procedure used. The experimental methods used in obtaining the data required for such use are mostly based on imaging methods that have to be resolved in time (see Tropea, Yarin, & Foss, 2007a, for a review). Optimally, the entire flow field should be resolved, ranging from the smallest Kolmogorov scales (or larger ones if LES type filtering is used) up to the (often coherent) largest scales. For numerical computations involving the averaged equations such as Reynolds-averaged Navier-Stokes (RANS), correlations (one-point, one-time) of the velocity second moments and scalar fluxes (see Chapter 3) are required. It then becomes essential to know the extent of the energetic part of the frequency or wavenumber spectra that has to be accurately measured. Since many flows of relevance involve statistically stationary variables, the discussion to follow will be on the time domain, that is the frequency characteristics. In order to give some order of magnitude estimates of these characteristics, two (generic) supersonic flows, the supersonic boundary layer and supersonic mixing-layer will be used as examples. (The physical characteristics of these flows will be presented in more detail in Chapter 6.) Typical values for the boundary layer and mixing-layer can be taken from Weiss, Chokani, and Comte-Bellot (2005) and Barre et al. (1994), respectively.

Typical spectral distributions are commonly represented by the power spectral density, PSD(f), in a log–log framework for easy assessment and evaluation of the relevant scaling laws. In Figure 4.1a, such a spectrum is shown for the longitudinal mass flux fluctuations for a $M = 2.5$ supersonic

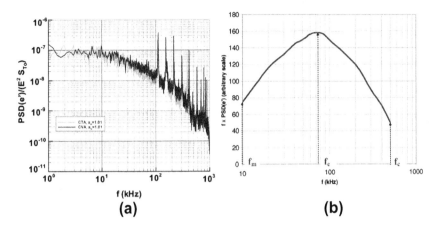

FIGURE 4.1 Typical frequency spectra of longitudinal mass flux fluctuations in a $M = 2.5$ boundary layer at $y/\delta = 0.36$: (a) conventional log–log plot (from Weiss et al., 2005, with permission); (b) "energetic plot" $f \times \text{PSD}(e')$ (arbitrary vertical scale).

boundary layer flow ($U_\infty = 570$ m/s and a boundary layer thickness $\delta = 7.5$ mm). (The PSD of the output signal fluctuations e' is scaled in the figure by the relative total temperature sensitivity S_{T_t}, and the total energy E^2 (Weiss et al., 2005).) Alternative representations using linear scales are also sometimes used to highlight the energy-containing frequencies, but then the precise estimation of the energy repartitioning is not straightforward as illustrated in Figure 4.2. An alternative representation has been proposed (Dumas, 1976) where $f \times \text{PSD}(f)$ is plotted in a linear scale against $\log f$ (see Figure 4.1b). The area under the curve is then proportional to the energy $E^2_{f_a - f_b}$ of the signal in a bandwidth lying between f_a and f_b, that is,

$$E^2_{f_a - f_b} = \text{PSD}(f_a < f < f_b) = \int_{f_a}^{f_b} \text{PSD}(f)\,\mathrm{d}f$$

$$= \int_{f_a}^{f_b} \left[f \times \text{PSD}(f) \right] \mathrm{d}(\log f). \quad (4.8)$$

With this "energetic" scaling, the frequency corresponding to the energy maximum, f_e, can be easily identified when compared to Figure 4.1a, and from Figure 4.1b lies around 76 kHz which roughly corresponds to U_∞/δ.

For the turbulent supersonic mixing-layer, it is expected that the corresponding energetic frequencies will be centered around $f_e = St \times U_c/\delta_\omega$, where St is the Strouhal number, δ_ω the (vorticity) mixing-layer thickness, and U_c the convection velocity (see Section 6.2). The Strouhal number is only weakly dependent on compressibility effects (e.g. Barre, 1993; Debisschop, 1993) with values $St \approx 0.2 - 0.25$. For a mixing-layer developing between free streams of Mach

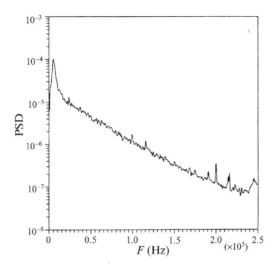

FIGURE 4.2 Power spectral density in a $M_c = 0.62$ supersonic mixing-layer (supersonic side) at $y/\delta_\omega = 1.56$. From Barre et al. (1994) with permission.

number of 1.8 and 0.3, respectively (Barre et al., 1994), the convection velocity U_c is 290 m/s (convective Mach number $M_c = 0.62$), with a typical vorticity thickness of 8 mm. From the PSD plot in Figure 4.2, the frequency of the maximum of energy, f_e, is $\simeq 8$ kHz, and corresponds to a Strouhal number St of 0.22.

An indirect consequence of such a PSD analysis of the mixing-layer as well as the boundary layer, is that the shear layer thickness (either δ or δ_ω) is the appropriate spatial scaling parameter, at least as a first approximation in estimating the maximum frequency. It is possible, from an estimation of f_e, to determine the lower, f_m, and upper, f_c, characteristic frequency limits. These frequencies represent either the limits beyond which no significant fluctuation energy is present or the frequency cutoffs of the measurement apparatus itself.

In order to properly measure the kinetic energy (or other second-order moments), it is necessary that the upper frequency limit, f_c, be sufficiently high to ensure the resolution of the most energetic scales in the flow. In the pioneering work of Kistler (1959), it was first proposed that f_c should be 4.5–5 times the most energetic frequency f_e which, for turbulent boundary layers, can be estimated as $f_e \simeq U_\infty/\delta$. This estimate of $f_c \geqslant 5f_e$, established initially for supersonic boundary layers, has been successfully extrapolated to the mixing-layer (Barre et al., 1994) where a usable bandwidth of up to $f_{c1} = 5f_e \approx 40$ kHz has been obtained. An alternative estimate (Barre et al., 1994) has been obtained by a direct analogy with the boundary layer results, that is $f_{c2} = 5U_c/\delta_\omega \approx 200$ kHz. From Figure 4.2, either value is applicable to the mixing-layer case, and from Figure 4.1 an acceptable value of $f_c \approx 40$ kHz is obtained for the boundary layer case.

In estimating the lower bound, f_m, of the resolved frequency range, the criterion used is dependent on whether the data are time resolved or are discretely sampled. For time-resolved series processing, such as obtained from HWA (sometimes laser-Doppler velocimetry, LDV), and numerical simulations (DNS or LES), a minimum time duration of observation, T_{obs}, is imposed in order to ensure the statistical convergence of the moments. The relative error of the second-order statistics (Bendat & Piersol, 2000) can be estimated (for a Gaussian process) as $(\sqrt{f_c \times T_{obs}})^{-1}$. Thus, for a 1% error on the mean square energy in the supersonic boundary layer example, this value can be estimated to be $\sim 2.5 \times 10^{-2}$ s, and for the supersonic mixing-layer example, the time can be estimated to be $\sim 5 \times 10^{-2}$ s. A consequence of this for laboratory experiments is that for several blowdown wind tunnels, the number of observations during a run is limited, and for numerical experiments using DNS or LES, the simulations need to be run over a sufficiently long length of time. In the numerical experiments, it is possible to achieve more rapid statistical convergence by exploiting any conditions of homogeneity in the flow. This often happens in simulations of flows that are two-dimensional in the mean so that the third spatial direction can be assumed statistically homogeneous.

If the data are discretely sampled, such as by particle image velocimetry (PIV), statistical independence is required. In laboratory experiments, the time between two successively collected velocity fields is generally $\mathcal{O}(10^{-1})$ s. For the boundary layer and mixing-layer estimates of T_{obs} just obtained, the $\mathcal{O}(10^{-1})$ s time interval should be sufficient to obtain independent flow realizations. (Note that high speed PIV data collection is still not sufficiently resolved in time to capture the entire spectral domain $f_m - f_c$ in such shear layers.) For discretely sampled signals (George, Beuther, & Lumley, 1978, see also Albrecht, Borys, Damaschke, & Tropea, 2003) a condition for assuming statistically independent samples can be introduced, that is $N = T_{obs}/2T_x$, where $T_x (\approx f_e^{-1})$ is the integral time scale and the time between two successive samples should be greater than $2T_x$.

It should be cautioned that these orders of magnitude estimates are applicable to measurements of turbulent second-order moments, and that these values should be carefully checked for each individual case by testing, for example, the influence of the data acquisition parameters. It should also be mentioned that these characteristics apply to velocity fields. For other turbulent fields, corresponding estimates need to be made. For example, the spectral extent of the temperature fluctuations can be larger than the velocity fluctuations (Gaviglio, 1976), and for the pressure fluctuations, the required bandwidths are quite different than the velocity or thermal fields and should be assessed on a different basis. If higher moments need to be obtained, wider bandwidths will be required (Lumley & Panofsky, 1964).

The wide range of turbulent fluctuation frequencies in high speed flows induce spatial scales smaller than those in subsonic flows. This means that turbulence measurements may need to be an order of magnitude better resolved in space and time than the corresponding subsonic, incompressible case. In such situations, dissipative scales are then, in general, out of the range of the common measurement techniques in use in high speed flows.

Added complexity also comes from the possible presence of shock waves that may be induced by either boundary conditions, high turbulence levels, flow distortion, pressure effects, etc. or geometry, such as parasitic shock waves introduced by intrusive sensors. These shock waves induce very rapid distortions in space, at mean free path scales, and time so that the measurement apparatus should be able to take into account very high pressure gradients, and supersonic/subsonic regime transitions during the flow survey. Clearly, significant care needs to be exercised if probes are to be embedded in the flow or if particles are to be injected (seeding). Of course, analogous constraints exist in numerical experiments as well. For numerical simulations, the constraints are inherent in the meshing and time resolution, in particular for (weak) shock resolution or for (strong) shock capturing, when the shock waves are neither steady nor localized in space.

4.2 MEASUREMENT METHODS

Measurement methods for high speed turbulent flows can be organized into three categories: (i) intrusive methods; (ii) non-intrusive methods—with particles; and (iii) non-intrusive methods—without particles. Intrusive methods involve the placement of solid probes in the flow, so that in transonic and supersonic flows the major problem of these methods is the presence of probe induced shocks. This means that account needs to be taken of the fact that the quantities are actually measured behind the induced shocks, and that additional fluctuations may have been introduced into the flow by both the probe and probe support. Another problem with such methods is the limitation on space and time scale resolution. The application of non-intrusive methods—with particles to supersonic flows imposes increased demands on this method not usually found in subsonic flows. In high speed flows, the particles are required to both follow the large accelerations related to the velocity fluctuations and behave correctly as they traverse (decelerate through) a shock. Such limits require that the method have a broad dynamic range of accuracy when applied to the supersonic regime. The third category of measurement method is the non-intrusive—without particles method. Such methods are, in principle, not susceptible to any of the above-mentioned high speed flow limitations, and are based on variations of the thermodynamics properties of the (compressible) flows.

The first two categories of methods are popular in subsonic flows. In the following sections, the discussion of these two categories will focus on the specific constraints linked to velocity fields near the speed of sound, and show how such measured fields are coupled with the Reynolds or

Favre decompositions used in obtaining second-moment correlations and more generally closure. For some purposes, the intrusive and non-intrusive (with particles) methods can be combined. For example, hot-wire anemometry (intrusive method) can be combined with LDV (non-intrusive with particles) to take advantage of both systems in supersonic flows. The X-wire probes used with HWA can be used with a two-dimensional LDV system to obtain three-dimensional measurements (Chambres, Barre, & Bonnet, 1997), or simultaneous measurements of temperature/velocity (Bowersox, 1996). In Section 4.2.2 the third method, non-intrusive without particles, will be discussed along with some recent advances in their development.

4.2.1 Intrusive Method: Hot-Wire Anemometry

HWA is the pioneering and most popular measurement system for obtaining information on both the velocity and temperature fields in supersonic flows. The method was introduced in the 1950s and immediately applied to supersonic flows (Kistler, 1959; Kovasznay, 1950; Morkovin, 1956). The major advantage of the method is its sensitivity to the thermodynamics fluctuations. For more details, the reader is referred to the review by Smits and Dussauge (1989) and the more recent review by Tropea, Yarin, and Foss (2007b, chap. 5).

4.2.1.1 Anemometers and Probes
The principle behind the hot-wire anemometer can be simply described. As is well known, a sensor composed of a thin wire (typically 2.5–5 μm in diameter and 0.5–1 mm long) is soldered or welded between prongs and placed in the flow. For supersonic flows, the prongs have to be designed for minimizing parasitic effects of shock waves, and as described in Section 4.1.2, the bandwidth of the system has to be particularly high when compared to low speed flows. The time constant of the heat transfer of the HWA depends on several parameters: the wire (sensor) temperature, the aerothermodynamic conditions, and the operating electronics. Three operating modes are possible that can affect the characteristics of the sensor:

- The current passing through the wire is maintained constant for the constant current anemometry (CCA). The current level sets the temperature of the wire in the flow, and this temperature difference is related to the overheat ratio which is a key parameter for the resolution of the operating mode. Since the natural frequency of the wire is quite low, the output signal needs to be amplified and carefully post-processed to enhance the low bandwidth. This is performed by appropriate electronic and/or numerical amplification of a high-pass filtered signal.
- The temperature of the wire is maintained constant by a closed loop control and the anemometer operates as a constant temperature anemometer (CTA). Appropriate closed loop electronics with wide bandwidths and

signal-to-noise ratio have to be used. The bandwidth of this apparatus is better adapted to measurements when the overheat ratio is high.

- The voltage of the wire is kept constant and the anemometer operates as constant voltage anemometer (CVA). This method has only been recently introduced (Comte-Bellot & Sarma, 2001; Sarma, 1998). The electronics is optimally suited for obtaining the necessary measurement bandwidths. This is a very promising method with many advantages related to bandwidth and ease of use.

Although the CCA and CTA methods are complementary, it appears that the CTA is probably the simpler to use (Bestion, Gaviglio, & Bonnet, 1983). It should be emphasized, however, that the signal-to-noise ratio is an important issue that has to be addressed whatever the electronic circuit (operating mode). In supersonic boundary layers, CTA (and CVA) can have bandwidths well suited for moderate supersonic regimes, with typical values going up to 300–500 kHz (Kegerise & Spina, 2000; Weiss et al., 2005).

For each operating mode, it is necessary to measure and adjust the time constant of the anemometer. Since this will strongly depend on the thermodynamic conditions at the wire location, the adjustment is more difficult to perform in supersonic flows. As in the subsonic regime, test electrical signals are often used and consist of a sine or square wave signal being input to the circuit creating an unsteady heating of the wire. However, this heating is not equivalent to the unsteady heat transfer variation in the flow, and in particular for high frequency characteristics, the electrical test can be ambiguous. Direct heating of the wire can be performed using an appropriate modulated laser beam focused on the wire placed in the flow. Although this procedure is less ambiguous, it is more delicate to setup, but it has been used in supersonic flows (Bonnet & Alziary de Roquefort, 1980; Kegerise & Spina, 2000).

In the supersonic regime, there is a detached shock of finite thickness in translational equilibrium ahead of the wire. The determination of this shock thickness δ_s is not straightforward. For the case of a weak shock, an estimate of the characteristic thickness in the laminar case can be obtained (Thompson, 1988). A low-order estimate of the thickness for a normal shock is obtained from the ratio of the velocity jump across the shock to an estimate of the rate of change in velocity. The slope of the velocity can be estimated by considering the locations where the velocity is in defect or excess of the upstream or downstream velocities, respectively. This leads to the relationship for the shock Knudsen number (Thompson, 1988)

$$\frac{\delta_s}{\xi} = \frac{1}{\mathrm{Kn}_{\delta_s}} = \frac{3}{4}\left(\frac{A_s}{M_\mathrm{u} - 1}\right), \tag{4.9}$$

where A_s depends on the magnitude of the velocity defect selected, and M_u is the shock Mach number (Mach number computed with the upstream velocity normal to the shock and the upstream speed of sound, c_u). For example, a 1%

deficit/excess yields a value for A_s of 9.19 with a limiting value of $A_s = 4$ for a maximum slope. Then for a normal shock at $M = 2$ in air, the shock thickness can be estimated to be between 3 and ≈ 7 times the mean free path, with equivalent Knudsen numbers of 0.33 and ≈ 0.15. For the low pressure tunnel (WT1) example, this thickness is of order of 2 μm, and thinner for the high pressure wind tunnels (WT2) example where the thickness is ≈ 0.1 μm depending on the percent deficit/excess selected.

For a turbulent flow, the relevant scale is the Kolmogorov scale (Moin & Mahesh, 1998). By combining Eqs. (4.7) and (4.9), the ratio of Kolmogorov length scale to shock thickness can be written as

$$\frac{\eta}{\delta_s} \approx \frac{1}{Kn_\eta} \frac{M_u - 1}{A_s} \simeq 0.1 Re_t^{1/4} \left(\frac{M_u - 1}{M_t} \right), \qquad (4.10)$$

where $\gamma = 1.4$ has been taken and a 1% deficit/excess value is assumed. Alternately, this expression can be written in terms of a Reynolds number associated with the Taylor microscale λ (Tennekes & Lumley, 1972)

$$Re_\lambda = \frac{\rho \sqrt{u_i'^2} \lambda}{\mu} = Re_t \frac{\lambda}{L} \approx \sqrt{15} Re_t^{1/2}, \qquad (4.11)$$

so that

$$\frac{\eta}{\delta_s} \approx 5.1 \times 10^{-2} Re_\lambda^{1/2} \left(\frac{M_u - 1}{M_t} \right). \qquad (4.12)$$

(The difference in numerical coefficient in Eq. (4.12) and that in Moin and Mahesh (1998) can be rectified by including the factor $\sqrt{2/\gamma\pi}$ in their derivation of the η/δ_s ratio.) Dramatic changes in the shock thickness can occur at higher Mach numbers and with other gases. For helium, at $M_\infty = 10$, the larger thickness was found to be $\delta_s \sim \mathcal{O}(5 \ \mu m)$ (Spina & McGinley, 1994).

The HWA measurements occur behind this shock wave, and an estimate of the distance between the shock and the wire can be obtained from data, for example, shown in Figure 4.3. For the moderate Mach number of 2 used in the boundary layer and mixing-layer examples above, the shock occurs typically at 0.6 wire diameters upstream of the wire. This value can vary from 0.3 for a Mach number of 4 up to 0.5 when approaching the transonic regime, with sensitivity to Mach number increasing with decreasing Mach number.

It is useful to determine the relevant Knudsen number. The conditions upstream of the detached shock have to be chosen to determine the mean free path, and the applicable length scale is the wire diameter (assuming the system formed by the wire and the shock arrangement is of order of the wire diameter). The mean free path estimates used previously are once again valid, so that if a 2.5 μm diameter hot-wire is considered operating at Mach 2, the Kn numbers are equal to 0.2 and 0.01, for the WT1 and WT2 tunnels, respectively. This means that in the high pressure tunnel (WT2) a continuum can probably be

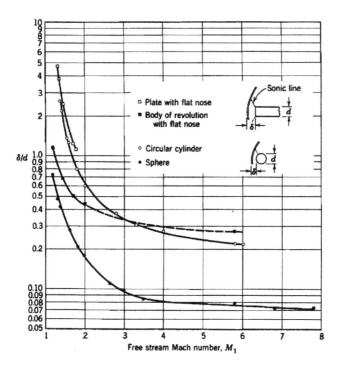

FIGURE 4.3 Typical shock detachment ahead of different bodies in supersonic flows. From Liepmann and Roshko (1957) with permission.

assumed, but in the low pressure tunnel (WT1), the conditions are within the transition to the slip-flow regime.

Another constraining factor that arises in high speed flow is the presence of a transonic regime delimited by the Mach number range $0.8 < M < 1.2$. Figure 4.3 clearly shows the sensitivity of the shock system when approaching the transonic regime. This regime is present in most shear flows such as mixing-layers between supersonic and subsonic streams, supersonic jets in still air, and boundary layers in supersonic flows. In this regime, the shock system becomes unsteady, depending on the probe geometry, with a high sensitivity to prong design. *In situ*, careful calibrations have to be made to take into account the flow around the probe in this case (Barre, Dupont, & Dussauge, 1992; Dupont & Debiève, 1992).

Finally, it should be noted that in addition to the constraints imposed on the sensors due to flow dynamics, there are related issues associated with probe and wire (sensor) construction that also need to be mentioned. One is the parasitic wire vibrations due to the high dynamic loads on the wire that can induce spurious effects known as "strain gage effects." Fortunately, such effects can be avoided by appropriate slacking of the wires. Another is the ever increasing need to measure several velocity or velocity–temperature correlation components. In doing this, inclined or multiple wire designs are used, with the ultimate arrangement for the transonic regime in a shock tube being designed by Briassulis, Agui, and Andreopoulos (2001) with 12 wires. This allows for access to three velocity components, spatial gradients, and vorticity. In the supersonic regime, however, less wires can be used but new problems arise from shock interaction between prongs and wire requiring careful design and calibration (Bonnet & Knani, 1988; Bowersox, 1996).

4.2.1.2 Data Reduction

For supersonic flows, the sensitivity of the anemometer can be adapted to different fluctuating quantities depending on the choice of primary flow variables. Morkovin (1956) initially discussed the sensitivity of the HWA in terms of primary variables with the output expressed in terms of the variables ρ, u_i, and T_t, or alternately in terms of (ρu), (v and w in case of cross-wire), T_t and M (local Mach number). It is possible from these measured quantities to extract the fluctuating vorticity, entropy, and acoustic modes proposed by Kovasznay (1953) and discussed in Section 3.4. In the absence of shear and molecular dissipation effects, the three modes are independent which would be the case in free stream flows. Even in flows with shocks, such as downstream of a shock wave/homogeneous turbulence interaction, the modes are correlated but useful in the description and understanding of the flow (Chambres et al., 1997).

If second-order moments are of interest, the method with adjusted sensitivity to the primary variables $(\rho u), T_t$ and M is preferred since it is easier to use due to a simpler calibration procedure. Just as in Section 3.5.1 where a variety of linearized velocity–thermal relationships were obtained, several additional linearized relationships can be obtained that aid in the analysis of experimental data. Within an appropriate range of Mach numbers, Reynolds numbers, and Knudsen numbers, the instantaneous HWA output normalized by its mean value $(e'/\bar{e})$ of the HWA can be written for a normal-wire as (Morkovin, 1956)

$$\frac{e'}{\bar{e}} = F_{\rho u} \frac{(\rho u)'}{\overline{\rho u}} + F_{T_t} \frac{T_t'}{\overline{T_t}} + F_M \frac{M'}{\overline{M}}, \qquad (4.13)$$

where $F_{(\rho u)}, F_{T_t}$, and F_M are sensitivity coefficients to the mass flux, total temperature, and Mach number fluctuations, respectively. These coefficients are functions of the wire temperature or of the overheat ratio (defined as the difference between hot-wire temperature at operating conditions and without heating normalized by the latter value). In order to reduce the number of sensitivity coefficients, it is possible to develop a linearized relationship between the fluctuating Mach number and the fluctuations arising from the mass flux and total temperature.

The fluctuating Mach number M' can be related to both the fluctuating velocity and temperature fields (see Section 3.5.1) by the linearized relation

$$\frac{M'}{\overline{M}} = \frac{u'}{\overline{u}} - \frac{1}{2}\frac{T'}{\overline{T}}. \tag{4.14}$$

However, what is needed are relationships with $(\rho u)'$ and T_t'. For the mass flux fluctuations, the usual Reynolds decomposition, truncated at second-order, yields

$$\frac{(\rho u)'}{\overline{\rho}\,\overline{u}} = \frac{u'}{\overline{u}} + \frac{\rho'}{\overline{\rho}}$$

$$= \frac{u'}{\overline{u}} + \frac{p'}{\overline{p}} - \frac{T'}{\overline{T}}, \tag{4.15}$$

where Eq. (3.114) has been used. The remaining task of relating the ratio $T'/\overline{T}$ to the ratio $T_t'/\overline{T}_t$ is achieved using Eqs. (3.92) and (3.93)

$$\frac{M'}{\overline{M}} = \frac{1}{(\alpha + \beta)}\left[\frac{(\rho u)'}{\overline{\rho}\,\overline{u}} - \frac{p'}{\overline{p}} + \frac{1}{2}\frac{T_t'}{\overline{T}_t}\right], \tag{4.16}$$

where, to simplify notation, $\alpha = \left[1 + (\gamma - 1)\overline{M}^2/2\right]^{-1}$ and $\beta = (\gamma - 1)\overline{M}^2\alpha$. Within this approximation, two flow regimes can be considered that would further reduce the number of variables. The first regime occurs inside most turbulent shear layers, where the acoustic mode can be neglected (Morkovin, 1956). In this regime, the Mach number fluctuations can be expressed in term of $(\rho u)'$ and T_t' directly by neglecting the p' term in Eq. (4.16).

The second regime that can be considered is exterior to the shear layer, where the pressure fluctuations come essentially from the sound radiated from the turbulent, rotational parts of the flow. This acoustic mode then dominates (Laufer, 1961), and the pressure, density, and temperature can be related through the isentropic relationships $p'/\overline{p} = \gamma\rho'/\overline{\rho} = \gamma/(\gamma - 1)T'/\overline{T}$. With this assumption, the fluctuations of the Mach number can be expressed as (see Barre, 1993)

$$\frac{M'}{\overline{M}} = \frac{1}{\alpha(1 - M^2)}\left[\frac{(\rho u)'}{\overline{\rho}\,\overline{u}} - \frac{1}{2}\left(\frac{\gamma + 1}{\gamma - 1}\right)\frac{T_t'}{\overline{T}_t}\right]. \tag{4.17}$$

By introducing Eq. (4.17) into (4.13), the fluctuating Mach number term can be eliminated, then for both flow regimes just described, the instantaneous hot-wire output can be simply expressed in terms of $(\rho u)'$ and T_t' as

$$\frac{e'}{\overline{e}} = F\frac{(\rho u)'}{\overline{\rho}\,\overline{u}} + G\frac{T_t'}{\overline{T}_t}, \tag{4.18}$$

where F and G are the (new) sensitivity coefficients to the mass flux and total temperature.

Since the coefficients F and G can be adjusted through the overheat ratio, the HWA should be operated at different temperatures in order to obtain the primary variables. Some attempts to extract instantaneous values have been made (Smith & Smits, 1993), but these measurements are quite complex and generally the HWA output is written in an average form as

$$\overline{\left(\frac{e'}{\overline{e}}\right)^2} = F^2 \frac{\overline{(\rho u)'^2}}{(\overline{\rho}\,\overline{u})^2} + 2FG \frac{\overline{(\rho u)'T_t'}}{(\overline{\rho}\,\overline{u}\,\overline{T_t})} + G^2 \frac{\overline{T_t'^2}}{\overline{T_t}^2}. \qquad (4.19)$$

In order to solve the system, at least three wire temperatures have to be used, and in general, more than three for redundancy and better precision in the measurement of $\overline{(\rho u)'^2}/(\overline{\rho}\,\overline{u})^2, \overline{T_t'^2}/\overline{T_t}^2$, and $\overline{(\rho u)'T_t'}/(\overline{\rho}\,\overline{u}\,\overline{T_t})$. As might be expected, variation of the overheat can be done easily in continuous wind tunnels, but is more difficult in short running time tunnels. However, by using the appropriate electronics, it is possible to vary the heating condition of the wires during each run. As an example, for multiple wire operating ratios in a very short running time wind tunnel such as the Shock Wind Tunnel, only six observations of 20 ms each (value obtained with the expression of T_{obs} given above) can be achieved during a run (Weiss et al., 2003).

With these three normalized variables known, it is possible to construct other turbulent correlations in terms of Reynolds variables that can be useful in validating results from numerical experiments. In a shear layer where the pressure fluctuations can be neglected, correlations involving $\overline{u'^2}, \overline{u'T'}$, and $\overline{T'^2}$ can be obtained, and are given by

$$\frac{\overline{u'^2}}{\overline{u}^2} = (\alpha + \beta)^{-2} \left[\frac{\overline{T_t'^2}}{\overline{T_t}^2} + 2\alpha \frac{\overline{(\rho u)'T_t'}}{(\overline{\rho}\,\overline{u}\overline{T_t})} + \alpha^2 \frac{\overline{(\rho u)'^2}}{(\overline{\rho}\,\overline{u})^2} \right]$$

$$\frac{\overline{u'T'}}{\overline{u}\overline{T}} = (\alpha + \beta)^{-2} \left[\frac{\overline{T_t'^2}}{\overline{T_t}^2} + (\alpha - \beta) \frac{\overline{(\rho u)'T_t'}}{(\overline{\rho}\,\overline{u}\overline{T_t})} - \alpha\beta \frac{\overline{(\rho u)'^2}}{(\overline{\rho}\,\overline{u})^2} \right]$$

$$\frac{\overline{T'^2}}{\overline{T}^2} = (\alpha + \beta)^{-2} \left[\frac{\overline{T_t'^2}}{\overline{T_t}^2} - 2\beta \frac{\overline{(\rho u)'T_t'}}{(\overline{\rho}\,\overline{u}\overline{T_t})} + \beta^2 \frac{\overline{(\rho u)'^2}}{(\overline{\rho}\,\overline{u})^2} \right]. \qquad (4.20)$$

An additional simplification to Eq. (4.20) can be easily made by assuming negligible total temperature fluctuations. Corresponding relations can also be derived when pressure fluctuations are non-negligible, although the relationships become more complex.

These same correlations can be obtained from direct numerical and large eddy simulation data. Although in the case of LES, it is important to note that such fluctuating quantities can only be obtained from the computed resolved field and not from the entire turbulent field as with the DNS. For the Reynolds-averaged computations, where the entire turbulent field is modeled, the results cannot yet be directly applied since the averaged equations were formulated in terms of Favre variables. In Section 4.4, additional relations are presented providing the necessary link to the averaged equation or RANS results.

4.2.2 Non-Intrusive Methods

As the name suggests, non-intrusive methods involve data collection systems where the sensor is not embedded into the flow field. Within this category of measurement method are two subcategories that include the methods with and without seeding. The method with seeding does involve the injection of particles into the flow that provide a sensing link between the measuring device and the instantaneous flow field. The method with no seeding senses light pattern changes due to thermodynamic variations in the instantaneous flow. Each of these approaches are now briefly discussed with respect to their application to compressible turbulent flows.

4.2.2.1 With Particles: LDV, PIV, and DGV

The now common non-intrusive LDV, PIV, and Doppler global velocimetry (DGV) or planar Doppler velocimetry (PDV, when Mie scattering is used) methods all require the presence of seeding particles in the flow. The data processing for supersonic flows is identical to the incompressible regime, provided the signal characteristics for the three methods are the same, that is, succession of Doppler signals, succession of images, or Doppler shift in the scattered light, respectively. Several references deal with these methods, and among them some are devoted to high speed flows (Riethmuller & Scarano, 2005; Samimy & Wernet, 2000; Scarano, 2008; Tropea et al., 2007). In these methods, the quality of the signal depends on the intensity of the light collected on the receivers. For high velocity flows, this becomes more crucial due to the shorter time constants (short residence time inside the observation volume of the LDV, short illumination time of the illumination needed for frozen images in PIV) compared to the equivalent subsonic flows. In contrast, the DGV is better adapted to supersonic flows provided the Doppler shift of the frequency of the scattered light is larger. The power of the laser source required is then higher and an increase in the sensitivity of the detectors or cameras is necessary. For LDV, forward diffusion, although more difficult to adjust, is often preferred for better scattering efficiency. In addition, since the sampling rate in high speed flows is large, the electronics should have smaller time constants than for an equivalent subsonic flow. Fortunately, all these issues have been adequately addressed by most of the commercial apparatuses available.

The common, major difficulty of these methods lies in the behavior of the seeding particles. Due to inertial effects, the particles act as low velocity filters since they are only able to follow limited accelerations. In high speed flows, the absolute value of the accelerations seen by the particle are indeed higher than in subsonic flows, with the relative level of velocity fluctuations being of the same order of magnitude in both regimes. These inertial effects are equivalent to low pass filtering with a given (particle) cutoff frequency f_{cp}. The turbulent energy captured by these methods then should be analyzed with respect to the energy-containing spectral distributions discussed in Section 4.1.2 in order to ascertain the quality of the measured data. In other terms, the particle-cutoff frequency f_{cp} has to be compared with the expected cutoff frequency f_c, and the effect quantified by some statistical or spectral measure.

A different, although analogous, effect of inertia is observed when particles traverse shock waves. This, of course, is unique to supersonic flows or shock tube configurations. In this case, the velocity gradient is a step function, and the particles are subjected to a deterministic velocity jump. Since the particles have non-negligible inertia due to their mass being far higher than in ambient air, a relaxation time, or distance is required before recovering to the local velocity downstream of the shock. A typical example of the evolution of the particle velocity passing through a normal shock wave is given in Figure 4.4. (This figure will be discussed in more detail shortly.) The thickness of the shock (of order of 3–7 mean free paths) should be compared directly to the particle diameter. For the conditions used as examples for the two wind tunnels (WT1 and WT2) discussed in Section 4.2.1, particles of 1 μm diameter (the usual order of magnitude for PIV and LDV) will be of the same order or larger than the shock thicknesses, that are respectively of order of 2 μm for WT1 and

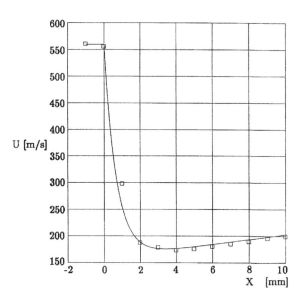

FIGURE 4.4 Typical evolution of particle velocity through a normal shock wave. Symbols: LDV measurements; solid line: from Eq. (4.23) (Lamarri, 1996).

0.1 μm for WT2 (see Section 4.1.1). In this case, the particles are only weakly (indirectly) sensing the shock itself.

Whether these inertial effects are observed in spectral or physical space, the particle's response to acceleration is a function of several parameters including its relative density and drag coefficient (related to the (assumed) spherical particle diameter) (Johnson, 1989, chap. 6). The determination of the particle's response time (or cutoff frequency f_{cp}) is not simple, even in subsonic flows, and requires either a direct or indirect method be followed to estimate it.

The direct method is specific to the supersonic regime. This is a rather unique way of estimating the particle inertial effect and takes advantage of the (well known) speed jump across the shock that is a function of the flow conditions. Indeed, downstream of a shock the mean velocities are uniquely determined by the flow conditions (temperature, incident Mach number, and shock angle or angle of the shock generating wedge), and is one of the few "advantages" of supersonic flows when compared to subsonic ones. In Figure 4.5a, an oblique shock wave is shown being generated by a (known) wall deflection (wedge). For a $M = 2.3$ flow, for example, with a wall deviation of $8°$, the velocity variation seen by the particle through the shock is of order of 10% (e.g. Tedeschi et al., 1999). Great care needs to be taken in ensuring the stability of the shock system since it can be unsteady due to possible effects of turbulence in the incoming boundary layer on the wedge that generates the shock. In most cases, the wedges are detached from the wind tunnel floor to avoid such shock unsteadiness. For larger velocity gradients, normal shocks can be generated by converging two oblique shocks resulting in a Mach effect (shown schematically in Figure 4.5b). This configuration is somewhat more difficult to set up (and particularly difficult to stabilize) but it has been achieved (Lamarri, 1996) for an incident Mach number of 2.82. In this case the velocity jump is far greater—on the order of 350 m/s or 70%. Once either the oblique or normal shock configuration is established, the particle velocity is then easily obtained by LDV or PIV. It is important to recognize that in these configurations, mixing-layers develop at the edges of the shock surface. These layers induce, after an initial decrease, an acceleration of the subsonic flow downstream of the shock location with, for

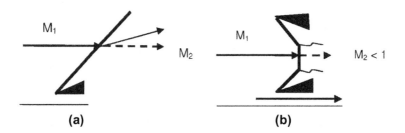

FIGURE 4.5 Schematic view of two configurations for particle drag measurements: (a) oblique shock and (b) normal shock.

the example given, a typical velocity gradient in the streamwise direction on the order of 50 m/s within 10 mm (Barre, Alem, & Bonnet, 1996). This is the reason for the increase of the particle velocity observed on Figure 4.4 downstream of the shock location. Downstream of the shock the particle velocity exceeds the local gas velocity due to inertia, but subsequently relaxes to the local fluid velocity. The particle relaxation, or response, time τ_p can be estimated by a simple relaxation formula (Melling, 1997),

$$\left| \frac{U_2 - u_p}{U_2 - u_{pi}} \right| = \exp\left[-t/\tau_p\right], \tag{4.21}$$

where u_{pi} and u_p are the particle velocities before and after the shock, respectively, U_2 is the gas velocity after the shock, and $t = x/U_2$ (with x the distance from the shock location where the velocity is measured). The corresponding frequency response can be directly deduced from this relaxation (response) time. The maximum value of the lag velocity $u_p - U_2$ being the velocity difference between the particle condition upstream of the shock (the maximum possible velocity of the particle) and the local fluid velocity downstream of the shock (asymptotic velocity of the particle).

The other method, the indirect method, can be used if the shock system is not available. For this method, an expression for the response time from a simple balance equation between acceleration and viscous (Stokes law) forces can be deduced (Melling, 1997). The resulting relaxation equation for the particle velocity yielded a response time constant τ_p given by

$$\tau_p = \frac{4}{3} \frac{\rho_p d_p^2}{C_d Re_p \mu}, \tag{4.22}$$

where ρ_p is the fluid density, Re_p the Reynolds number based on the fluid density ρ_f, the particle velocity lag ($V = u_p - U_2$) and diameter, d, and C_d the drag coefficient. Here, the drag coefficient can be estimated from the Stoke's relation $C - d = 24/Re_p$.

A more precise relation for the drag coefficient has been proposed (Tedeschi et al., 1999) for spherical particles in regimes ranging from continuum to free molecular flow for Reynolds numbers $\lesssim 200$ and Mach numbers $\lesssim 1$ (based on particle velocity and diameter). As discussed previously in Section 4.1.1, the Knudsen number is an important parameter that may need to be taken into account in high speed flows, and requires the specification of the mean free path as well as a relevant length scale L. Correspondingly, these Mach and Reynolds numbers are then introduced requiring the specification of a relevant velocity scale into the parameterization. The velocity needed to determine the mean free path is the velocity of the fluid on the surface of the particle. For Knudsen numbers corresponding to a continuum, the particle velocity is the same as the velocity of the fluid on the particle surface; however, for a slip flow the two velocities are different. The values of length scale L needed in the specification of Kn_p are based on values on the flow regime around the particle.

For $Re_d \gg 1$, there is high speed flow around the particle, and a boundary layer develops on the particle itself. The relevant length scale is now the thickness of the boundary layer which, as a first approximation, is proportional to $d/\sqrt{Re_d}$ (with $L = d$ being the particle diameter). From Eq. (4.5), the Knudsen number ($= \xi/d$) is proportional to M/Re_d so that the parameter Kn_p is then proportional to $M/\sqrt{Re_p}$. When Re_d is small, the particle diameter d is the only relevant spatial scale. This is the case in most supersonic wind tunnels where Re_d is $\mathcal{O}(1)$ for micron sized particles (the value decreases downstream of the shock as the lag velocity decreases), and yielding $Kn_p = Kn_d = (M/Re_d)\sqrt{\gamma\pi/2}$ (e.g. Humble, Scarano, & van Oudheusden, 2007a). Recall that the Knudsen number is strongly affected by the different wind tunnel pressures. As an example, Kn_d can vary by a factor of 20 between the low pressure WT1 tunnel (settling chamber pressure of 1.5×10^4 Pa), and the high pressure WT2 tunnel (settling chamber pressure of 1.0×10^6 Pa). The corresponding flow regimes in each case can then be quite different.

Depending on the flow regime (Kn number), a slip velocity U_s may exist between the particle surface and the surrounding fluid. To account for this effect and retain the proper relation for a continuum flow, a relation for the drag coefficient is given by (Tedeschi et al., 1999)

$$C_d = \frac{24 k_d}{Re_d} \left[1 + 0.15(k_d Re_d)^{0.687} \right] \xi(Kn_p), \qquad (4.23)$$

where

$$k_d \approx \left(1 + \frac{9}{2} Kn_p \right)^{-1} \qquad (4.24a)$$

$$\xi(Kn_p) = 1.177 \left[1 + \frac{0.851 Kn_p^{1.16} - 1}{0.851 Kn_p^{1.16} + 1} \right], \qquad (4.24b)$$

with $Re_d = \rho u_p d/\mu$ and Kn_p the particle Knudsen number. A compressibility correction factor can, in general, be also included in Eq. (4.23) (Tedeschi et al., 1999). The function $\xi(Kn_p)$ allows Eq. (4.23) to be valid from slip to free molecular flows which should include all the Kn_p number regimes encountered in most wind tunnels. The relation given in Eq. (4.23) has been validated for several particle diameters and wind tunnel conditions (e.g. Scarano & van Oudheusden, 2003; Tedeschi et al., 1999). The velocity distribution downstream of the shock is a function of the drag, and can be calculated by integration of the equation of motion once the drag coefficient is known. An example of such a calculation can be seen for the case of a strong velocity jump across a normal shock with an upstream Mach number of 2.83, and a downstream Mach number of 0.487 (SiO_2 particles are used in this example). Integration of Eq. (4.23) yields a velocity distribution that is plotted in Figure 4.4 and compared with LDV measurements, confirming the validity of the previous hypotheses and formulas. It is worth noting (Tedeschi et al., 1999) that inertial effects influencing the lag velocity (as discussed previously) are most often of primary importance when compared to these rarefaction effects.

Once the drag coefficient is estimated, the time constant is known from Eq. (4.22) and the frequency cutoff can be estimated. The estimates of time constants, through either the direct or indirect methods, combined with the values of the required frequencies established in Section 4.1.2, provide the necessary information in choosing the particle to be used for seeding. Characteristics (diameters and mass) of the available particles can be found in the literature (Albrecht et al., 2003). It should be emphasized, however, that for practical applications, the particle diameter is never simple to obtain nor unique, since most seeding media are polydisperse (as is the case with incense smoke), and others can agglomerate. Assuming that the drag law given in Eq. (4.23) is valid for most of the moderate supersonic laboratory experiments, it is possible to obtain an "effective" particle diameter from the measured particle velocity evolution downstream of the shock. This evolution can be described by a relation obtained by the integration of the equation of motion using the expression for C_d with the diameter now being considered as an unknown. The effective diameter that is obtained in this way can be quite different than the one expected from nominal particle specification. This is illustrated by an experiment (Schrijer, Scarano, & van Oudheusden, 2006) in which titanium dioxide (TiO_2) particles were supposed to be of (nominal) 50 nm diameter. Electron microscopy shows single agglomerates of typically 400 nm in diameter. The estimation of the diameter from the integration of Eq. (4.23), in a Mach 2 experiment, gives an effective diameter of ≈ 900 nm (Humble et al., 2007a).

When complex flows are investigated, a measurement bias can be introduced that is associated with a velocity bias that is related to the origin of the particle that is observed. A typical illustration is in the mixing-layer where the (particle) velocities can depend on the stream from which they emanated. This seeding bias is present in any turbulent flow, but when compared with subsonic flows, is much more pronounced in supersonic regimes due to the low mixing efficiency (e.g. see Figure 6.12).

4.2.2.2 Without Particles: Rayleigh-Scattering Methods

There are several non-intrusive, non-seeding or quasi non-seeding methods applicable to compressible, high speed flow in inert gases (see Bonnet, Grésillon, & Taran, 1998; Miles & Lempert, 1997; Samimy & Wernet, 2000, for reviews). Most of them are based on Rayleigh scattering—elastic scattering of an electromagnetic wave by neutral particles. Such a regime exists when the particles that are illuminated are far smaller than the wavelength λ of the light, that is the particle diameter d is such that $\pi d / \lambda \ll 1$. This is the case in direct scattering by gas molecules or from condensation clusters of CO_2, H_2O, or O_2 molecules that occurs in the Rayleigh regime. In this case, the problem of inertia no longer exists, and the method can be considered as a non-particle method. The efficiency of Rayleigh scattering is quite low, and high sensitivity detectors have to be used.

The basic optical arrangement consists of illuminating the flow field by a sheet of laser light in a manner comparable to PIV methods. However, in such a basic configuration the background scattering (from walls, windows or models) is collected as well as the light scattered by the flow which then hides the true phenomenon. Miles and Lempert (1990) introduced the filtered Rayleigh scattering (FRS) method that takes advantage of the fact that the frequency of the light scattered by the flow field particles, or molecules, is Doppler shifted by the bulk velocity. Since the background scattered light is not Doppler shifted, two main goals can be achieved by properly adjusting the filter. Of course, the first is that the background scattering is avoided so that camera or detector amplifiers can be used that are not saturated by parasitic signals. The second is that by adjusting the filter characteristics and the laser wavelength in such a way that the absorption depends on the frequency, the amplitude of the light collected can be made proportional to the frequency of the scattered light. This is then proportional to the velocity of the small particles/molecules through the conventional Doppler frequency relationships,

$$\Delta f_D = \mathbf{v} \cdot \frac{\mathbf{k}_d - \mathbf{k}_l}{\lambda}, \qquad (4.25)$$

where $\mathbf{v}$ is the bulk velocity of the (scattering) observed medium, and $\mathbf{k}_d$ and $\mathbf{k}_l$ the unit vectors of the observation direction and of the laser propagation, respectively (see Figure 4.6). The principle is then the same as the DGV, or PDV (Samimy & Wernet, 2000) in which the appropriate filtering is obtained by using a spectral molecular (or atomic) filter (iodine filter). Indeed, high velocity flows exhibit larger Doppler frequency shifts and ranges, and they are well suited for FRS (or DGV/PDV) methods. The optical arrangement, schematically illustrated in Figure 4.6, is the same for the two methods, the only difference being the size of the particles (if any) for either Rayleigh or Mie scattering. The FRS and the DGV/PDV are excellent flow imaging methods in high velocity regimes for the large scale characterization of turbulent flows (Arnette, Samimy, & Elliott, 1998; Smits & Dussauge, 2006; Wu, Lempert, & Miles, 2000). Quantitative measurements of density, temperature, pressure, and velocity can be obtained with the frequency scanning technique, with the total intensity of the collected light related to the density. The width of the spectra is related to the temperature, and the shape can be analyzed in terms of pressure and temperature. Finally, the velocity can be measured through the frequency of the Doppler shift. Turbulence statistics can be obtained as from PIV data, with the same data processing and then with the same criteria for convergence, etc. It should be noted that the method requires careful calibration, is not trivial to adjust, and the precision is not easy to determine. The interested reader can find reviews on the method (Boguszko & Elliott, 2005; Miles, Lempert, & Forkey, 2001) which is still under development.

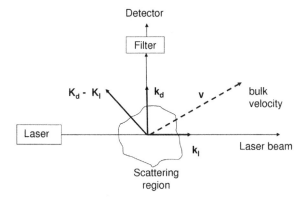

FIGURE 4.6 Schematic of the Doppler shift notations and FRS arrangement.

An alternative method to simultaneously measure velocity and density fluctuations has been proposed (Panda & Seasholtz, 2006). In this method, instead of scattered light filtering, by using a Fabry–Perot interferometer, the Doppler shift is translated in terms of a fringe pattern. The pattern is analyzed through two photomultiplier outputs and compensated to take into account temperature influence. The density fluctuations are measured through the intensity variation of the Rayleigh scattered light obtained from a third photomultiplier. The schematic of the optical arrangement is given in Figure 4.7. With this unique arrangement, simultaneous fluctuations can be measured in small volumes (of order of 0.2 mm in diameter for 1 mm in length). With this apparatus, it is possible to determine (Panda & Seasholtz, 2006) velocity and density spectra, *rms*, and cross-correlations in supersonic ($M = 0.95$, 1.4, and 1.8) jets with a 90 kHz bandwidth. Special attention must be given to the quality of the air in the wind tunnel, to the high sensitivity electronics, and to the delicate use of the interferometer.

A third method to enhance the Rayleigh scattering uses an optical arrangement comparable to the conventional LDV method. In the collective light scattering (CLS) technique, light scattered by density fluctuations in the medium are collected. It was initially developed within the context of the Tokamak phenomenon, and then extended to supersonic turbulent flow studies in wind tunnels (Bonnet, Grésillon, Cabrit, & Frolov, 1995; Grésillon, Gémaux, Cabrit, & Bonnet, 1990). This method is still under development at a few research centers (Aguilar, Azpeitia, Alvarado, & Stern, 2005), and so in the following only some salient features are given. In a gas, the total scattered light is the sum of the light re-emitted by all the molecules. The characteristics of the scattered light thus reflect the characteristics of the structure and motion of the gas. If the density is uniform, the scattering from the medium is of very low energy; however, in Section 4.1.2 it was shown that in supersonic wind tunnels the

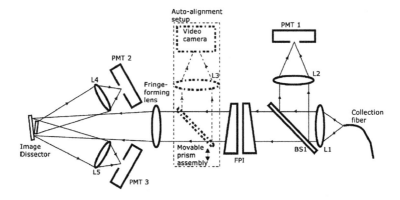

FIGURE 4.7 Schematic of optical arrangement for the Rayleigh scattering method: PMTi, photomultipliers; FPI, Fabry–Perot interferometer; Li, lenses; BS1, beam splitter. From Panda and Seasholtz (2006) with permission.

density fluctuations can be large so that the scattering from the medium can be of high energy. In addition, the characteristic spatial scales of the fluctuating fields (of order of millimeters, Section 4.1.2) are large when compared to laser light wavelengths (lower than 10 μm). Then, if the observed scales of the system λ are of the order of the scales of the turbulent flow, a much larger scattered light power can be detected. This effect is termed a "collective" or enhanced scattering effect (Grésillon et al., 1990). Figure 4.8 gives a schematic of the method. The optical arrangement is reminiscent of regular LDV systems; however, the angle θ is far smaller in the CLS method—with angles typically of the order of a few to tens of milliradians. The CLS apparatus uses an heterodyne detection process, beating the scattered light with a reference beam, called a local oscillator. This method not only allows for the observed signal to be amplified, but also for the frequencies to be shifted around the frequency difference $\Delta\omega$ imposed by the acousto-optic deflector. This then allows the data to be collected with a conventional apparatus.

In short, the resulting *instantaneous* output signal of the CLS $i(\mathbf{k},t)$ is proportional to the Fourier transform of the instantaneous density inhomogeneities $\rho'(\mathbf{r}',t)$ at the location $\mathbf{x}$,

$$i(\mathbf{k},t|\Delta\omega,\mathbf{x}) \propto \int_{\mathcal{V}} \rho'(\mathbf{x}+\mathbf{r}',t)e^{-\iota(\mathbf{k}\cdot\mathbf{r}'-\Delta\omega t)}\mathrm{d}^3\mathbf{r}', \qquad (4.26)$$

where the wavevector $\mathbf{k}$ is $|\mathbf{k}| = 4\pi \sin(\theta/2)/\lambda_0$ (λ_0 is the wavelength of the laser source), and $\mathcal{V}$ the intersection volume (see Grésillon, 1994, for details). The orientation of the vector $\mathbf{k}$ is defined by the apparatus. CLS performs a real three-dimensional analysis at wavelengths $\lambda = 2\pi/|\mathbf{k}|$. Note that the wavelength has to be chosen according to the flow physics and particularly the

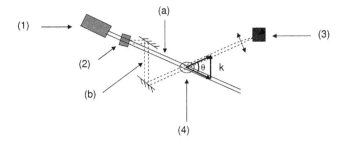

FIGURE 4.8 Schematic of the collective light scattering apparatus: (1) laser source; (2) acousto-optic-deflector (AOD); (3) detector; (4) interrogation volume, (a) incident beam, and (b) local oscillator beam.

spatial scales as described in Section 4.1.2, in such a way that the corresponding spectral component has enough energy in the density fluctuations. As an example, values of λ can be of order of 0.1–2 mm (Bonnet, Grésillon, & Taran, 1998; Stern, Aguilar, & Alvarado, 2006). As may be expected, there are several factors that need to be taken into account in setting the CLS measurement apparatus.

The usefulness of the CLS method is illustrated by the fact that instantaneous density and velocity data, and associated statistics, can be obtained from the spectrum of the signal output. The method is sensitive to the density fluctuations and the corresponding velocity fluctuations. It is then able to discriminate between density fluctuations convected with the local instantaneous fluid flow, and those propagating with the local velocity *plus or minus the speed of sound*. This is a unique feature for compressible high speed flows. Lastly, the speed of sound being determined, the mean temperature can be determined from the CLS measurements if acoustic waves are present.

There are some limitations to the system, however, with the major limitation being linked to the volume needed to perform the space Fourier transform (see Eq. (4.26)). The interrogation volume of the optical arrangement is the intersection of the two laser beams, and can be considered as a cylinder with a diameter the size of the beam width, and a longitudinal length one or two orders of magnitude greater than the diameter. The method performs an integration in this volume and is thus better adapted to flows with a marked two-dimensional character (see also Cabrit, 1992).

Recall that in Section 3.2 it was shown that the density-weighted average could be written in terms of a joint probability density function (see Eqs. (3.26) and (3.27)). It is only possible to obtain this joint *pdf* through a few experimental methods that can sense density fluctuations, and which are in turn linked with the corresponding velocity field. The collective light scattering method is potentially one such method and allows for the determination of both the density and associated velocity fields.

4.3 ANALYSIS USING MODAL REPRESENTATIONS

The intent here is not to provide an exhaustive list of all the various analysis tools available in assessing turbulent, incompressible or compressible, flows. There are however methods, such as the proper orthogonal decomposition (POD) and wavelet decomposition, that have a unique role to play. Such methods provide more than statistical single-point correlation distributions that are used, for example, in the RANS or conventional experiments to characterize the flow in a global manner. They also provide information on both modal structure and amplitude. The POD is based on the two-point correlation that contains non-local information not only on the energy of the fluctuations, as is the case for one-point correlations, but also on the flow organization in space and time. From the original ideas of Townsend (1976), an illustration of the information that is contained in the space-time correlation can be obtained (Bonnet & Delville, 1996). While such methods have been used extensively in the incompressible regime, extension to the compressible regime has been less widespread. In this section, the focus is on the POD approach and its extension to the compressible regime. Although the material is presented somewhat generally so that the non-specialist can have an intuitive understanding of the approach, it can also provide those who are familiar with the method in the incompressible regime, with an overview of specific aspects linked to compressible high speed flows.

The POD is a polynomial expansion which is a generalization of the Karhunen–Loève expansion that was adapted to analyze the structure of inhomogeneous turbulent flows (Lumley, 1967, 1970). An extensive amount of mathematical formalism can be invoked to both solve the resultant problem and to prove a variety of properties related to the representation. The reader is referred to the extensive literature that has evolved in this area (Berkooz, Holmes, & Lumley, 1993; Cordier, Delville, & Bonnet, 2007, chap. 22.4; Holmes, Lumley, & Berkooz, 1998; Lumley, 1970, and associated references). It is an optimal (here energy based), polynomial representation of a (vector) variable $q_i(x)$ in terms of eigenfunctions of the autocorrelation function of $q_i(x)$. (For notational convenience, (x) is simply used to designate the $(\mathbf{x}, t)$ pair.) Of course, the method is simply a mathematical tool that can be applied to any scalar or vector quantity depending on the interest and data availability.

Formally, the representation can be written as

$$q_i(x) = \sum_{n=1}^{\infty} a_n \phi_i^n(x), \qquad (4.27)$$

with the a_n random coefficients and basis functions $\phi_i^n, n = 1, \dots, \infty$ obtained by solving the optimization problem

$$\lambda = \max_{\phi_i \in H(D)} \frac{\overline{(q_i, \phi_i)^2}}{(\phi_i, \phi_i)}, \qquad (4.28)$$

where $H(D)$ is a Hilbert space ($D = \Omega + [0,T]$, Ω is the spatial domain and T is the time of observation), and inner product $(\cdot, \cdot)$. Solving this optimization problem can be performed through a classical calculus of variations approach leading to a countable number of solutions given by the eigenvalues λ_n and eigenfunctions $\phi_i^n(\mathbf{x})$ ($n = 1,2,3,\ldots,N_{POD}$), where N_{POD} is the total number of POD modes. There are a variety of properties that can be associated with these results (see Cordier et al., 2007, chap. 22.4, for a summary). The corresponding expansion coefficients can be related to the eigenvalue set λ_n and are given by

$$\overline{a_n a_m^*} = \delta_{mn}\lambda_n. \tag{4.29}$$

This decomposition or representation can be applied to the analysis of turbulent flows in two successive ways. First an analysis associated with a *filtering* of the flow structure, can be applied so that relevant information on the large (energetic) scales or organized structures can be obtained. A second, follow-on, analysis associated with *low-dimensional modeling* of the flow can be applied. This latter step introduces dynamics into the analysis and allows for subsequent flow stability analysis and control (Bergmann, Cordier, & Brancher, 2005). A variant of the standard POD approach has recently been proposed (Jordan, Schlegel, Stalnov, Noack, & Tinney, 2007) for aeroacoustic applications by relating the dynamics of a low Reynolds number jet (computed by LES) to the radiated sound by means of a Linear Stochastic Estimation (LSE) (see Adrian, 2007, chap. 22.5, for a description of the method). Rather than using an inner-product based on the turbulent kinetic energy for example, an inner-product constructed from the radiated sound field is used to compute the modes.

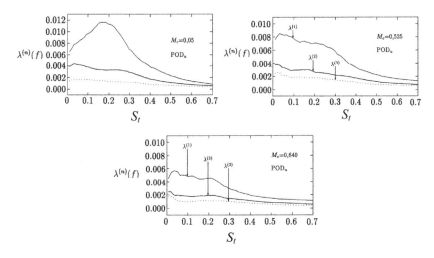

FIGURE 4.9 First three POD eigenvalue spectra for three convective Mach numbers (vertical axis arbitrary scale). From Debisschop et al. (1993) with permission.

Within the context of flow structure filtering, the inner product associated with the eigenfunctions is important since these functions contain structure information—the so-called "building blocks" of the turbulent field (Lumley, 1967). One of the major advantages of the POD is its rapid and optimal convergence in inhomogeneous flows. If large scale structures dominate the flow (mean field being subtracted), it is not unusual to have 40% of the energy in the first mode, and 80% in the first 10% of the total number of modes (Delville & Bonnet, 2008). An example of such a rapid convergence is observable in Figure 4.9 in which the first three POD modes in a supersonic mixing-layer are represented. The third mode contains only a small part of the energy when compared with the first two modes. By using a limited number of modes, the flow field can be reconstructed from this truncated set. As an example, the velocity can be estimated from a truncated series of N_t modes

$$\hat{u}_j^{N_t}(\mathbf{x},t) = \sum_1^{N_t} a_n(t)\phi_j^n, \qquad (4.30)$$

with the limit of $\hat{u}_j^{N_t}$ when $N_t \to N_{\text{POD}}$ being u_j. If the flow is time resolved, it is possible to reconstruct the instantaneous velocity field using the "filtered" field obtained from the POD. This is readily achievable if the flow is measured through HWA rakes (Delville & Bonnet, 2008), time resolved LDV or DNS/LES simulations. A typical application to mixing-layers measured with hot-wire rakes is shown in Figure 6.7.

Debisschop, Sapin, Delville, and Bonnet (1993) were the first to apply such a method to the turbulent supersonic mixing-layer by means of a rake apparatus of 12 hot-wires, using HWA (see Section 4.2.1) and defining a transverse slice of the flow. They collected instantaneous mass flux data, $(\rho u)'$ and applied the POD to the longitudinal component (scalar POD) in order to educe the (large scale) coherent structures. Figure 4.9 shows the spectral (frequency) distribution of the first three eigenvalue modes at three different convective Mach numbers, $M_c = 0.05$ (subsonic), 0.525 and 0.64, as a function of Strouhal number $St = f\delta_\omega/U_c$. The POD clearly shows the M_c influence on the spectral distribution. A broad distribution of energy is shown, and the sharp peak corresponding to Strouhal number around 0.18 observed in the incompressible mixing-layer is no longer observed in the supersonic one. This behavior corresponds to the effect of compressibility effects discussed in Section 6.2. In contrast, the percentage of energy in the first three modes is essentially independent of M_c being 79%, 76%, and 81% for $M_c = 0.05, 0.525$, and 0.64 respectively.

Unfortunately, the use of rakes of hot-wires is particularly difficult in supersonic flows due to probe holder design, HWA calibration, etc. With the increased use of non-intrusive techniques such as PIV, it is possible to obtain information on the other velocity components. As an example, a POD analysis of a supersonic bluff body wake has been performed (Humble, Scarano, & van

Oudheusden, 2007b), and two and three component velocities obtained in a plane have been obtained.

Once the POD eigenfunction and eigenvalue analysis is completed, and has shown that a limited number of modes is sufficient to represent a large part of the turbulent energetics, it is then possible to further exploit the POD by performing low-dimensional modeling of the flow. This can be done through the POD-Galerkin method that projects the Navier–Stokes equations onto the POD modes ϕ_j^n (e.g. Holmes et al., 1998). The Galerkin projection on the POD subspace then transforms the partial differential equations for the fluctuating fields into ordinary differential equations for the expansion coefficients (the a_n). The evolution of the truncated expansion coefficients Eq. (4.30) coupled with the corresponding basis eigenfunctions can then potentially represent the instantaneous flow field. The relevance of any low-dimensional modeling is heavily dependent on the choice of variables comprising the vector $q_i(x)$ and should be performed so that the number of modes required is as small as possible. The POD is also used in numerical methods computational fluid dynamics (CFD) modeling: Vigo, Dervieux, Mallet, Ravachol, & Stoufflet, 1998, for example) or inlet conditions (for incompressible cases, see Perret, Delville, Manceau, & Bonnet, 2006).

In order to effectively perform the filtering and/or low-dimensional modeling just discussed, there are three important factors that need to be considered. These include the: (i) definition of the function space and associated inner product; (ii) selection of the average operator in Eq. (4.28); and (iii) selection of the elements of **q** that best represent the flow characteristics to be studied. Each of these factors will be discussed briefly.

In most cases, the function space (i) consists of square-integrable functions L^2 and the associated norm. Other choices can be made for the space in which the eigenfunctions are defined. The Sobolev H^1 space is sometimes used (Iollo, Lanteri, & Désidéri, 2000) which then allows for the introduction of both function and gradient values into the norms and then subsequently into the POD average. At least for some compressible flow analyses, this H^1 inner product can be more stable than the L^2 one. Since the generalization of this approach is still under study (Cordier, 2008), the remainder of the discussion will focus on the L^2 space.

The choice, (ii), to be made next is the type of averaging procedure $\overline{(\cdot)}$ to be used. Whichever the choice, ensemble, time, or space averages, within the L^2 space, Eq. (4.28) can be written as

$$\lambda = \max_{\phi_i \in L^2(D)} \frac{(\Re\phi_i, \phi_i)}{(\phi_i, \phi_i)}, \tag{4.31}$$

with the operator $\Re$ given by

$$\Re\phi_i(x) = \int_\Omega R_{ij}(x, x^\dagger)\phi_j(x^\dagger)dx^\dagger, \tag{4.32}$$

and the kernel $R_{ij}(x,x^\dagger) = \overline{[q(x) \otimes q^*(x^\dagger)]}_{ij}$ is the two-point correlation tensor ($*$ denotes complex conjugate). (The corresponding one-point correlation is given by $R_{ij}(x,x)$ and is the basis for the RANS model development and closure.)

The conventional (or direct) POD method consists of time averages (see Eq. (3.5)) where the assumption of ergodicity is used. The kernel $R_{ij}(x,x^\dagger)$ is then a two-point spatial correlation defined over the spatial domain Ω. This kind of average is performed, for example, with hot-wires (rakes for two-dimensional analysis) or LDV (with sufficient data rate for high speed flows). These measurement techniques allow for long time histories and, therefore, well converged (temporal) statistics to be obtained. Since the POD degenerates to a Fourier decomposition if the process is stationary or the turbulence is homogeneous, it is possible to mix a Fourier (harmonic) decomposition and POD in such flow fields (e.g. Delville, Ukeiley, Cordier, Bonnet, & Glauser, 1999). This approach is illustrated in Figure 4.9 where the (statistically steady) data is collected from hot-wire rakes (12 hot-wires) in a supersonic ($M_c = 0.525$) turbulent mixing-layer, and the energy of the modes is plotted in terms of the (time) frequency (or Strouhal number). In contrast to the conventional (or direct) POD method, the snapshot method (Sirovich, 1987) utilizes averages in space, $\overline{(\,\cdot\,)} = \int_\Omega (\,\cdot\,) dx$. Thus, the kernel, $R_{ij}(x,x^\dagger)$, is now the two-point temporal correlation computed from all the flow realizations and averaged in the space of observation. This method is well suited for PIV or DGV methods that are well resolved in space, and for high speed flows where it is difficult to produce long time histories. For simulation data from either DNS or LES, the space resolution is usually high and converged spatial statistics relatively easy to obtain; whereas, the length of the time histories may not be sufficient for adequate converged statistics. As discussed in Section 4.1, care has to be taken if statistically independent snapshots are required. Such independence is obviously the case for PIV, since the time between successive images is far greater than the typical time scales of high velocity flows, but is less obvious, and often not possible, to ensure for DNS/LES results due to short time histories. In any event, it should be recognized that the simulation data processed from DNS and LES are distinctly different—with the LES data being associated with the resolved scale motions. Thus, modal decompositions can yield different results depending on the method.

Now that the choices associated with (i) function space and (ii) averaging operator have been considered, there remains only the choice of (iii) flow variable(s) to be analyzed. In the incompressible case, the variables most often analyzed are the velocity components and the points x and $x^\dagger$ are sufficiently separated so as to encompass the underlying dynamics of the large scales. In this case, the POD decomposition is an optimal kinetic energy-based representation of the flow structure given by

$$(q,q^\dagger) = \int_\Omega u_i' u_i'^\dagger \, dx. \tag{4.33}$$

Another quantity which is sometimes analyzed, and of dynamical significance, is the vorticity with the corresponding correlation tensor associated with the enstrophy.

For the application of the POD to compressible flows, a fundamental question to be addressed at the outset is the choice of suitably normalized quantities for the vector $\mathbf{q}$. However, this choice of the primary variables and the associated equations that have to be used is not as straightforward. The choice is important and should be performed in such a way that the system obtained is polynomial and the number of modes required is as small as possible (in some situations a departure from polynomial representation can be obtained (Vigo et al., 1998)). Note that the procedure can be applied to either density-weighted or Reynolds variables extracted using averaged or filtered methods. Whichever variable or averaging method is chosen, it is known that for compressible turbulence the velocity and thermal variables are strongly coupled. It is then not rational to attempt the decomposition on vector fields q_i that contain either the velocity or thermal field variables only. However, vector fields of the form $\mathbf{q} = \mathbf{q}(\rho', u_i')$ or $\mathbf{q} = \mathbf{q}(u_i', p')$ are problematic, since forming the various inner products, as defined in Eq. (4.33), immediately leads to inconsistent dimensional variables. It is clearly necessary in formulating the problem that the vector field $\mathbf{q}$ be composed of variables properly normalized so that relations such as Eq. (4.33) have a rational meaning

$$(\mathbf{q}, \mathbf{q}^\dagger) = \int_\Omega \left(\frac{q_i q_i^\dagger}{\overline{q_i^2}} \right) d\mathbf{x}. \tag{4.34}$$

Lumley & Poje (1997) also addressed this point by associating velocity and density for an homogeneous turbulent flow (in a low velocity regime), this method sometimes being called the normalized inner product. In this, the compressibility arises essentially from density variations. Each fluctuating variable is in this case normalized by its *rms* value so the form of the inner product involving density and velocity would be defined as

$$(\mathbf{q}, \mathbf{q}^\dagger) = \int_\Omega \left(\frac{\rho' \rho'^\dagger}{\overline{\rho'^2}} + \frac{u_i' u_i'^\dagger}{\overline{u_i'^2}} \right) d\mathbf{x}. \tag{4.35}$$

It appears that the first application of POD to compressible flows was to a two-dimensional mixing-layer (Kirby, Boris, & Sirovich, 1990). The decomposition was applied to the vector variable $\mathbf{q}$ consisting of *rms* values of the fluctuating horizontal and vertical momentum and pressure normalized by their respective mean values over a plane. Other proposals have been made where the specific volume ρ^{-1} is added to the primitive variables (u, v, p) in two-dimensional computations (Iollo et al., 2000; Vigo et al., 1998); so that the relevant variables are (ρ^{-1}, u, v, p). An alternative normalization of the primitive variables has been proposed (Bourguet, Braza, & Dervieux, 2007):

$$(\mathbf{q}, \mathbf{q}^{\dagger}) = \int_{\Omega} \left(\frac{q_i q_i^{\dagger}}{\overline{q_i^2} + \epsilon_q} \right) d\mathbf{x}, \qquad (4.36)$$

where ϵ_q is a parameter, introduced for numerical stability considerations and assumed to be small. In general, one of the obvious choices for the compressible form of the vector field $\mathbf{q}$ involves the primitive variables (ρ, u_i) or (u_i, p); however, as has been shown (Rowley, Colonius, & Murray, 2004), the resulting equations for the expansion coefficients are too complex to be easily analyzed.

An alternative choice was considered (Rowley et al., 2004) using only a single thermodynamic variable. If the flow is isentropic, the speed of sound c can be used since it is related to the enthalpy h by $c^2 = (\gamma - 1)h$. The primary variables are then the velocity variables (u_i, c), and the relevant governing equations are the (isentropic) Navier–Stokes equations. This formulation leads to an energy-based norm of the form

$$(\mathbf{q}, \mathbf{q}^{\dagger}) = \int_{\Omega} \left(u_i' u_i'^{\dagger} + \frac{2}{(\gamma - 1)} c' c'^{\dagger} \right) d\mathbf{x}, \qquad (4.37)$$

the kernel being twice the total enthalpy. This approach was used to compute laminar supersonic cavity flows. The same approach has been followed in experiments (Moreno, Krothapalli, Alkislar, & Lourenco, 2004) in a rectangular, turbulent supersonic jet. In that study, use was made of planar snapshots of two components of the velocity to obtain the POD modes. In order to establish a low-dimensional model, the speed of sound was calculated from the stagnation temperature. The same experimental approach has been followed in a study of an embedded cavity in a high subsonic turbulent boundary layer flow (Samimy et al., 2007). These solutions seem to be well suited to moderate compressible flows without the influence of shocks, but these methods may not be as well suited to higher turbulent Mach number flows.

Recently, the vector variable $\mathbf{q} = \mathbf{q}(\rho, u_i, T)$ has been proposed (Yang & Fu, 2008). The resulting normalized inner product used in the analysis was

$$(\mathbf{q}, \mathbf{q}^{\dagger}) = \int_{\Omega} \left[\frac{w_q}{2} \left(\frac{\rho' \rho'^{\dagger}}{\overline{\rho'^2}} + \frac{u_i' u_i'^{\dagger}}{\overline{u_i'^2}} \right) + (1 - w_q) \frac{T' T'^{\dagger}}{\overline{T'^2}} \right] d\mathbf{x}, \qquad (4.38)$$

where w_q was a weighting factor used to adjust the relative importance to the temperature, and a value of $w_q = 0.5$ was used (Fu, private communication); however, the results do not seem to be very sensitive to the exact value.

In light of the amount of data the decomposition method requires, it is only feasible to conduct extensive validation studies using results from numerical simulations. From plane mixing-layer simulation data, with convective Mach numbers ranging from 0.4 to 1.2 (Yang & Fu, 2008), a comparison of results from the inner products defined in Eqs. (4.33), (4.35), (4.37), and (4.38) can be performed. As seen in Figure 4.10, such a study shows that the inner products defined in Eqs. (4.35) and (4.38) capture more energy than the incompressible or enthalpy forms given in Eqs. (4.33) and (4.37). The convergence is rapid, and only subtle differences can be observed in the eigenfunctions.

From an experimental perspective, and as just suggested, the major limitation of the POD method is the amount of data needed to determine the eigenvalues and eigenfunctions. Nevertheless, the POD has become an accepted and common analysis tool in both subsonic and moderate Mach number supersonic flows. For the velocity and vorticity fields it provides energy and enstrophy distributions useful in understanding (large scale) flow structure. The exact choice of variables and inner product does not appear to be crucial; whereas, for low-order systems, the approach proposed by Rowley et al. (2004) has generally been adopted (Moreno et al., 2004; Samimy et al., 2007). The best choice of inner product space and analysis variables still remains an open question; particularly in flow situations where no obvious choice is apparent to judge the respective performance of the POD.

Overall, in the application of the POD there are two distinct yet important considerations. The first deals with the proper representation of the physical variables that is linked to the underlying physics of the problem. The choices are many and range from sole velocity or thermal variable to various combinations of both, and subject to different constraints (such as isentropic flow for example). The second consideration deals with numerical robustness in which the numerical stability of the POD-Galerkin method is driving the choice. Nevertheless, the POD application to compressible, supersonic flows appears

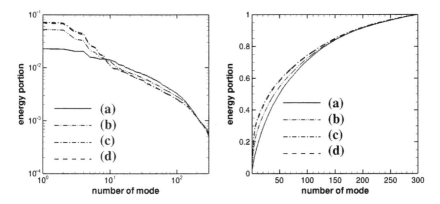

FIGURE 4.10 Energy distribution for several inner products in a supersonic mixing-layer from a DNS at convective Mach number 0.4. (a) Incompressible form Eq. (4.33), (b) total enthalpy Eq. (4.37), (c) normalized density/velocity Eq. (4.35), (d) mixed normalized density/velocity/temperature Eq. (4.38). Left is energy in each mode, and right is convergence. From Yang and Fu (2008) with permission.

to be an analysis tool that binds together experimental, numerical and applied mathematical disciplines.

4.4 REYNOLDS- AND FAVRE-AVERAGED CORRELATIONS

As was shown in Chapter 3, the governing averaged equations used in flow calculations involve density-weighted or Favre averages and corresponding moments or correlations. Some of these correlations have incompressible counterparts while others associated with the thermal field, appeared for the first time. However, as has been extensively discussed in this chapter, experimental measurements necessarily yield fluctuating fields corresponding to the Reynolds fluctuations. In addition, some of the currently popular experimental methods, such as LDV and PIV, only provide information on the velocity field. Relationships between the density-weighted, or Favre, variables and correlations, and the Reynolds variables and correlations are needed in order to facilitate the utilization of the experimental data in Reynolds-averaged type computations. In Eq. (4.20), relationships were established for the streamwise fluctuating velocity variance, heat flux, and temperature variance in terms of Reynolds variables. These relations were necessary for the HWA since the streamwise velocity is not directly measured and has to be extracted from a relation involving the mass flux. In the other measurement methods, Reynolds variables can be obtained. In order for the measurement methods to generate data useful in numerical computations such as RANS, it is necessary to now show how the density-weighted correlations can be related to the correlations involving Reynolds variables.

Consider the relation between the Favre fluctuating velocity, u'' and temperature, T'' and the Reynolds fluctuating velocity and temperature, u' and T', discussed previously in Sections 3.2 and 3.5. The difference between the two fluctuating quantities is equivalent to the difference between the two means and is given by

$$u'' - u' = \overline{u} - \tilde{u} = -\frac{\overline{\rho'u''}}{\overline{\rho}} = -\frac{\overline{\rho'u'}}{\overline{\rho}} \tag{4.39a}$$

$$T'' - T' = \overline{T} - \tilde{T} = -\frac{\overline{\rho'T''}}{\overline{\rho}} = -\frac{\overline{\rho'T'}}{\overline{\rho}}. \tag{4.39b}$$

From these two relationships, it is possible, through the application of the strong Reynolds analogy in Section 3.5, to obtain equivalent relationships for correlations involving these variables. It should be cautioned that the assumptions underlying the derivations in Section 3.5 were based on two-dimensional, slowly evolving flow fields. Additionally in some cases, the assumption of negligible total temperature fluctuations was also invoked. In the absence of such assumptions, simple relationships between the density-weighted correlations and correlations involving the Reynolds variables would probably not be possible. Nevertheless, even with such constraints, the relationships to follow are often applicable, and do provide an important link between measurements and computations.

From Eqs. (3.95), (3.97), and (3.115), one can relate the different mass flux terms to the corresponding velocity and temperature second moments. Equation (4.39) for the fluctuating velocity and temperature difference can then be written as

$$u'' = u' - (\gamma - 1) \overline{M}^2 \frac{\overline{u'^2}}{\overline{u}} \tag{4.40a}$$

$$T'' = T' + (\gamma - 1) \overline{M}^2 \frac{\overline{u'^2}}{\overline{u}} \left(\frac{\overline{u}}{c_p} \right). \tag{4.40b}$$

Since by definition the density-weighted fluctuations have non-zero mean, Eq. 4.40 can be used to estimate the magnitude of this mean value. The corresponding second moments can also be easily obtained. For example, the longitudinal component velocity second moment, heat flux and temperature variances can be written as

$$\frac{\overline{u''^2}}{\overline{u'^2}} = \frac{\overline{u''T''}}{\overline{u'T'}} = \frac{\overline{T''^2}}{\overline{T'^2}} = 1 + \left[(\gamma - 1)\overline{M}^2 \right]^2 \frac{\overline{u'^2}}{\overline{u}^2}. \tag{4.41}$$

With these equivalent relations, it is straightforward to estimate the difference between the correlations involving the density-weighted variables and the Reynolds-averaged variables. For the case of a 10% fluctuation level in a Mach 2.25 boundary layer flow, for example, there is only a difference of $\mathcal{O}(2\%)$ between the *rms* values.

Of course, the difference between the correlations can also be expressed directly in terms of second- and third-order moments of the density and velocity fluctuations (cf. Eq. (3.24), see also Eq. (5.35)). Simultaneously measured velocity and density fluctuations in supersonic jets with a Rayleigh scattering method described in Section 4.2.2 have been obtained (Panda & Seasholtz, 2006). The measurement of $\overline{\rho' u'}$ and $\overline{\rho' u'^2}$ yielded an estimation of the difference between Favre (density-weighted) and Reynolds energy to be typically less than 4%. This is consistent with the estimate obtained from the application of the strong Reynolds analogy given above.

Recall that for complete closure, turbulent correlations, such as $\overline{\rho u''^2}, \overline{\rho u'' T''}$, and $\overline{\rho T''^2}$ (or $\overline{\rho} \widetilde{u''^2}, \overline{\rho} \widetilde{u'' T''}$, and $\overline{\rho} \widetilde{T''^2}$), may be required. For the quantities, $\widetilde{u''^2}, \widetilde{u'' T''}$, and $\widetilde{T''^2}$, the relationships with the Reynolds variable correlations are

$$\frac{\widetilde{u''^2}}{\overline{u'^2}} = \frac{\widetilde{u'' T''}}{\overline{u'T'}} = \frac{\widetilde{T''^2}}{\overline{T'^2}} = 1 + \left[(\gamma - 1)\overline{M}^2 \right]^2 \frac{\overline{u'^2}}{\overline{u}^2} f(u'), \tag{4.42}$$

where Eq. (3.95) has been used along with the perfect gas law in Eq. (3.115), with

$$f(u') = \frac{S(u')}{(\gamma - 1)\overline{M}^2 \left(\sqrt{\overline{u'^2}}/\overline{u} \right)} - 1. \tag{4.43}$$

The streamwise skewness factor $S(u')$,

$$S(u') = \frac{\overline{u'^3}}{(\overline{u'^2})^{3/2}}, \tag{4.44}$$

has been introduced into Eq. (4.42) and is a measure of the deviation of the probability distribution function of the velocity fluctuations away from a Gaussian behavior (Lumley, 1970). In addition, the sign of the skewness factor is indicative of the sign of the velocity fluctuations and can be either positive or negative. Such a parameter can be useful in analyzing the dynamics of the inner layer region of a boundary layer and will be discussed later in Chapter 6. Not surprisingly, there is a close similarity with the relations shown in Eq. (4.41), with the differences now being attributed to this skewness factor which vanishes when the *pdf* of the (streamwise) velocity fluctuations is Gaussian. Although the algebra may be tedious, all the remaining higher-order correlation relationships could be constructed. It is clear that numerous simplifying assumptions (such as the SRA) have been introduced in this section to obtain the various relations. This inherently restricts the generality of the relations and further restricts the results to be a guide, albeit a potentially useful guide, in assessing the behavior of and similarity between the different averaging methods.

As was noted at the outset of this section, the relationships presented here can be both a useful and important link between measurements and numerical calculations. However, their extensions to numerical simulations where filtered rather than centered statistical moments are required for closure can be problematic since many of the relations no longer hold due to the fact that the filter operator is not idempotent. The relationships can then only be applied to fluctuations of the resolved field variables and these resolved field variables are density-weighted or Favre variables. Whether these relations are applicable to the fluctuating resolved field has not been validated as yet. In contrast, the added availability of direct numerical simulation databases can easily provide the necessary validation of these relationships. The limitation, unfortunately, lies in the fact that current DNS flow fields are far less complex than the practical engineering flow fields now being calculated by averaged methods. For averaged type methods such as RANS, the usefulness of the data is limited to validations of either closure models or computed correlations. For application to boundary condition specification, their usefulness may be limited due to the incomplete data sets since in order to accurately replicate the physical experiment with a CFD calculation, it is necessary to accurately and consistently specify all the conditions at the boundaries.

Chapter 5

Prediction Strategies and Closure Models

The overwhelming majority of numerical solutions of turbulent flow fields utilize either a direct numerical simulation (DNS), a filtered approach such as a large eddy simulation (LES), or an averaged approach such as the Reynolds-averaged Navier–Stokes (RANS) formulation. Other methodologies, such as the lattice Boltzmann method (LBM), are less common and have not been used in the study of more relevant engineering flow fields. The RANS approach has been used for over 40 years and obviously pre-dates the emergence of DNS and LES as a common engineering turbulence prediction tool. The direct numerical simulation approach emerged next in the early 1980s followed a decade later by large eddy simulations. It is an almost endless task to try to fully document the enumerable compressible flow studies that have been conducted. Rather the focus here will be on closure modeling since this is still by far the most popular, and utilized, methodology in studying compressible high speed flows. However, it would be remiss not to discuss the various contributions of the numerical simulation approaches since they have served as a guide for compressible research as well as an effective tool in probing details of the flow dynamics. It is in this, as well as an historical context that the results from both direct and large eddy simulations are presented in this chapter (as well as in Chapters 6 and 7). The remainder of the chapter will be devoted to the averaged, RANS method, and the discussion will focus on the formulation of various compressible closure models for both the velocity and thermal fields. It will be assumed that the reader is familiar with RANS approaches in the incompressible, subsonic flow regime. For those interested in reviewing or learning more about the topic, they are directed toward the excellent recent text by Hanjalić & Launder (2011).

Compressibility, Turbulence and High Speed Flow. http://dx.doi.org/10.1016/B978-0-12-397027-5.00005-8

5.1 DIRECT NUMERICAL SIMULATIONS

Direct numerical simulations have played, and continue to play, an important role in both the understanding of compressible turbulence dynamics and calibration of turbulence closure models. However, even in the simplest of turbulent flows, such as decaying isotropic turbulence, numerical simulations of compressible flows is dependent on additional parameters relative to the incompressible case. In the incompressible case, the only dependent variable that needs to be specified is the velocity; however, the dynamics of compressible flows additionally requires a description of the thermodynamic variables for each initial turbulence level and level of compressibility. As the flows become more dynamically and/or geometrically complex, additional parameters need to be considered in order to achieve accurate and representative simulations.

5.1.1 Homogeneous Turbulence

In the absence of a mean flow (shear), the dynamics of a turbulence field is in many ways simplified due to the absence of a (kinetic) energy cascade. It is then possible to better isolate (fluctuating) velocity and thermal field interactions, albeit in an environment that may not fully represent engineering type flows. Nevertheless, many of the early studies on compressible flow dynamics utilized simulation data from such homogeneous turbulence fields. Such shear-free turbulence field simulations are readily amenable to theoretical analysis and validation. In such simulations, (pseudo)spectral formulations are standard since the simulations are performed within computational domains with periodic boundary conditions. Resolution issues are focused on having a sufficient number of spectral modes to adequately resolve the dynamic range of motions. An inherent constraint, however, relates to the turbulent length scales which grow with time. It is necessary to insure in such simulations that the computational domain (box) be sufficiently large to have uninhibited growth of these scales. The compressibility level is usually quantified through either the turbulent Mach number M_t given by

$$M_t = \frac{\sqrt{2K}}{c}, \tag{5.1}$$

where $K = \widetilde{u_i'' u_i''}/2$ is the turbulent kinetic energy and $c \; (= \sqrt{\gamma R \widetilde{T}})$ is a mean speed of sound, or the *rms* Mach number M_{rms} given by

$$M_{rms} = \sqrt{\overline{\left(\frac{u_i''}{c}\right)^2}}. \tag{5.2}$$

(In this chapter unless noted, the density-weighted variable notation is adopted, since numerical solutions of filtered or averaged equations are usually based on equations written in these variables.)

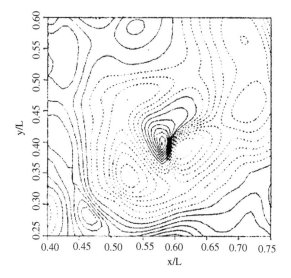

FIGURE 5.1 Snapshot of instantaneous pressure field from DNS of isotropic decaying compressible turbulence. From Lee et al. (1991) with permission.

These early simulations (Lee, Lele, & Moin, 1991; Passot & Pouquet, 1987) suggested the existence of eddy shocklets (see Figure 5.1) in whose vicinity transfer mechanisms existed that expedited an exchange between the turbulent kinetic energy and the internal energy as well as affected the kinetic energy decay. These mechanisms were identified as the pressure–dilatation and dilatation dissipation and were the focus of theoretical scrutiny and closure model development (Sarkar, Erlebacher, Hussaini, & Kreiss, 1991; Zeman, 1990, 1991a; Zeman, Blaisdell, Johansson, & Alfredsson, 1991). In the absence of a mean motion, the linear analysis presented in Section 3.4 readily applies. Recall the discussion in Section 3.4 of the linear analysis and physical fluctuation "modes" (the vorticity, acoustic, and entropy modes) (Kovasznay, 1953). In the absence of a mean motion, this linear analysis readily applies. These modes could be either coupled to or decoupled from one another, depending on the mean flow field. In a shear-free isotropic flow, the three modes are decoupled in the inviscid case, but coupled in the viscous case. (In a shear flow, the entropy mode is decoupled from the vorticity and acoustic modes, but the vorticity and dilatation fields are coupled.) Any resultant energy exchange between modes is important because it gives an indication of the modes' independence from initial conditions. In the isotropic decay case, for example, the decoupling between modes would suggest a strong dependence on the respective initial conditions. It was within this framework that truly compressible closure model development was initiated, and was in distinct contrast to the more traditional approach of variable (mean) density extensions to incompressible models or velocity–thermal relationships derivable from the various strong Reynolds analogies (SRAs). Of course the lack of any imposed deformation, such as compression or shear, limited the amount of information that could be obtained and in many ways precluded obtaining a complete picture of the flow dynamics.

An obvious limitation in examining a homogeneous decaying turbulence is the lack of any mean flow distortion. While a homogeneous shear is of obvious interest and relevance, another mean distortion of practical relevance is the case of mean flow compression and the subsequent investigation of rapidly compressed turbulence. Studies of such distortions were initially motivated by their applicability to flow situations in internal combustion engines, hypersonic flight, and supersonic combustion. Such extra-strain fields imposed on the turbulence had long been recognized (Bradshaw, 1974) as having an observable effect on the turbulence energy amplification. In addition, if the compressions produce a turbulence with an eddy turnover time of K/ϵ, and which is much greater than the inverse of the mean distortion rate, rapid distortion theory (RDT) can be used. For such rapid compressions, a parameter Δm was used (Cambon, Coleman, & Mansour, 1993; Durbin & Zeman, 1992; Jacquin, Cambon, & Blin, 1993) to quantify the effects of compressibility. In a subsequent RDT analysis of compressed turbulence, the limit $\Delta m \ll 1$ was investigated (Durbin & Zeman, 1992), which decoupled the fluctuation dynamics of the acoustic mode from the vortical mode (see Section 3.4). This decoupling allowed for the specification of separate acoustic and solenoidal contributions to the pressure–variance and subsequent model specification for the pressure–dilatation. Axial (one-dimensional) compressions were then investigated (Cambon et al., 1993; Jacquin, Cambon & Blin, 1993) as a guide in understanding and predicting the turbulent energy amplification in (isotropic) turbulence/shock interaction. For $\Delta m \gg 1$ (Cambon et al., 1993; Jacquin et al., 1993), both the acoustic and turbulent time scales are large compared to the mean inverse distortion and pressure–dilatation has little effect on the turbulence dynamics.

These early studies have clearly shown the usefulness of RDT as a tool in analyzing some homogeneous turbulence dynamics. It is applicable to flows in which the time scale of the turbulence is long in comparison with the time scale of the mean deformation, and the turbulence does not have time to interact with itself. When the turbulence is compressed by shocks and the distortion is rapid as just defined, other factors such as inhomogeneous effects and thermodynamic influences also enter, and are not accounted for with RDT. For such problems, the linear interaction analysis (LIA) can provide a more accurate description of the shock–turbulence interaction (see Section 7.1). Nevertheless, it is easy to see why direct simulations and analysis of homogeneous turbulence has been and can still be an important tool in understanding compressible turbulence physics.

5.1.2 Homogeneous Sheared Turbulence

Although turbulent simulations in the absence of mean deformations could be a natural starting point for analyzing compressible turbulence, it appears that the first DNS of a compressible turbulent flow was performed (Feiereisen, Reynolds, & Ferziger, 1981) for the case of homogeneous shear flow. By today's measure, the simulations were performed at low Reynolds number ($18 < Re_\lambda < 120$) and fluctuating Mach number ($0.06 < M_{rms} < 0.31$). Nevertheless, the results from this study clearly suggested that compressibility did alter the turbulence dynamics in some ways. In contrast to the isotropic case, the shear rate acts to couple the vorticity, acoustic, and entropy modes within the compressible turbulence. This coupling allows the compressible turbulence to evolve to (quasi)equilibrium states that are independent of the initial state of the turbulence. Feiereisen et al. (1981) also pointed out that in homogeneous turbulence the mass-flux term $\overline{\rho' u'_i}$ vanishes. As was shown in Section 3.2, this term represented the difference between the density-weighted and Reynolds variables so that in a homogeneous flow the two are equivalent (cf. Eq. (3.20a)); however, from Eq. (3.24) it is seen that the two turbulent stresses would still differ by a high-order factor $\overline{\rho' u'_i u'_j}$. The dynamical consequences of having the mass-flux correlation vanish apparently has not been addressed, but in these early homogeneous flow studies the role of fluctuating dilatation seemed to play a dominant role that did not exist in the inhomogeneous case.

An important Mach number parameter used to quantify the effects of compressibility under mean deformations is the product of the ratio between the mean deformation rate, D and turbulent time scale, K/ϵ (ϵ is the kinetic energy dissipation rate), and the turbulent Mach number M_t, that is DKM_t/ϵ. This is equivalently the ratio of the mean deformation rate D to the inverse sonic time scale c/l, where l is the turbulent eddy size. As shown in the previous section on homogeneous turbulence, it has been used in simulation and theoretical studies of rapidly compressed turbulence (where it was defined as $\Delta m = DKM_t/\epsilon$, and termed the deformation Mach number), and shear flow turbulence (where it is defined as $M_g = SKM_t/\epsilon$, with S a measure of the mean shear and termed the gradient Mach number). It is easily seen that even for low turbulent Mach numbers, under sufficiently rapid compression, or shearing, the flow can exhibit noticeable compressibility effects based on this parameter. Alternatively, for moderate values of both M_t and SKM_t/ϵ, it can yield deformation, or gradient Mach numbers >1.

In shear flows, for example, an increase in Mach number across an eddy limits the influence of the pressure waves to a zone in space within which the speed difference is subsonic. If this Mach number becomes too high, then the velocity difference will become too large relative to the sound speed, and this will result in a decrease in the communication across the eddy and thus restrict the structure size. This idea of sonic-delimited eddy size was used to formulate (Kim, 1990) a mixing-length model for supersonic shear layers (see Aupoix & Bézard, 2006 for a correction to the estimate of turbulent structure size), and to develop (Breidenthal, 1990, 1992) a "sonic-eddy" concept to qualitatively account for the free shear layer reduced spreading. A more quantitative assessment was provided (Papamoschou & Lele, 1993) by analyzing the direct simulation data of the disturbance field generated by the shearing of a vortex within a compressible shear layer. In all these studies, the underlying basis for the reduced spreading rate was attributed to an anisotropy in the propagation of information within the shear layer.

A consequence of the "sonic-eddy" concept highlights the discussion in Section 4.1.1 concerning the useful role the micro-structure Knudsen number, Kn_η can play. In Eq. (4.7) it was shown that $Kn_\eta \propto (M_t / Re_t^{1/4})$. This estimate was based on the (usual) relation that the ratio between the Kolmogorov microscale to the turbulent integral scale was $Re_t^{-3/4}$. However, instead of using the integral scale associated with the largest eddy size, it was assumed to be more appropriate to use (Breidenthal, 1992) a "sonic-eddy size" that is smaller than the size of the largest eddies and inversely proportional to the flow Mach number. It was further assumed that the ratio of the Kolmogorov scale to the scale of the "sonic-eddy" is proportional $Re_t^{-3/4}$. This then led to a micro-structure Knudsen number $Kn_\eta \propto (M_t^{1/2} / Re_t^{1/4})$, and an estimate for the reduction of shear layer spreading rate. As alluded to in Section 4.1.1 for the specification of the Mach number used in determining the Knudsen number, there is no universal nor unique definition for Kn, and it needs to be adapted to the physics of the phenomena under consideration as was done here.

Just as in the case of axial compressions, homogeneous sheared turbulence introduces anisotropies into the statistical correlations that have an impact on the turbulence dynamics. Unlike incompressible homogeneous shear, where the mean shear is arbitrary, the mean shear in compressible flow must satisfy a coupled system of nonlinear ordinary differential evolution equations (Blaisdell, Mansour, & Reynolds, 1993; Feiereisen et al., 1981). In addition to the turbulent or *rms* Mach numbers, the flow is also parameterized by a Mach number analogous to the deformation rate Mach number discussed in the case of rapid compressions. Although not extensively analyzed, it was recognized (Blaisdell, Mansour, & Reynolds, 1991) that such a parameter would be needed to quantify the shock structures that appeared in the homogeneous shear simulations since these structures were roughly aligned in the direction of shear. Sarkar (1995) utilized a similar Mach number parameter, M_g, based on the local mean shear and turbulent length scale. This gradient Mach number was useful in that it could be applied to both wall-bounded and free shear flows and provided an accurate relative measure of the effects of compressibility on each type of flow. As shown above, this Mach number can also be related to the turbulent Mach number through the scaled turbulent time scale. This variation between turbulent time (or length scale) in either wall-bounded or free shear flows provides the relative influence of compressibility on each type of flow. By isolating the effect of M_g, the turbulent energy growth rate was shown to decrease as M_g increased. This effect on the turbulent production mechanism, rather than on the earlier assumed dependence on direct dilatational effects, showed the stabilizing effect of compressibility on the turbulence. In addition, since the gradient Mach number associated with a boundary layer flow was much less than the corresponding value for a mixing-layer, the analysis led to an explanation for the reduced importance of compressibility corrections in boundary layer flows relative to free-shear flows such as mixing-layers.

5.1.3 Inhomogeneous Sheared Turbulence

Studies of homogeneous turbulence and homogeneous shear flows provide useful information about the turbulence dynamics at a fundamental level, and have played an influential role in RANS model development for incompressible and now compressible flows. In addition, these flows have also been amenable to theoretical analysis. Unfortunately, relevant compressible engineering flows are more complex often being influenced by impinging shocks and varying thermal conditions. Nevertheless, with both numerical algorithms and computational power now capable of direct simulations of free shear flows and wall-bounded flows (with/without shocks) at relatively low, but yet, dynamically relevant Reynolds numbers and Mach numbers, such flows are now being routinely investigated through direct simulations.

Recall that one of the earliest motivations for compressible research was the observed reduced spreading rates with increasing Mach number observed in compressible mixing-layers. Not surprisingly, such mixing-layers were the subject of simulation scrutiny with temporal planar and annular mixing-layers being the dominant focus (e.g. Freund, Lele, & Moin, 2000a, 2000b; Sandham & Reynolds, 1991; Vreman, Sandham, & Luo, 1996). For wall-bounded flows, recent simulation studies have primarily focused on channel flow simulations and have extended early DNS channel flow simulations (Coleman, Kim, & Moser, 1995; Huang, Coleman, & Bradshaw, 1995). Many previous DNS studies of supersonic channel and boundary layer flows (Guarini, Moser, Shariff, & Wray, 2000; Martin, 2007; Morinishi, Tamano, & Nakabayashi, 2004; Pirozzoli, Grasso, & Gatski, 2004, 2005; Shahab, Gatski, & Comte, 2009; Shahab, Lehnasch, Gatski, & Comte, 2011; Tamano & Morinishi, 2006) focused on adiabatic wall conditions; however, there have also been some with isothermal (hot/cold, $T_w > T_{aw}/T_w < T_{aw}$) wall conditions (Ghosh, Foysi, & Friedrich, 2010; Maeder, Adams, & Kleiser, 2001; Morinishi et al., 2004; Morinishi, Tamano, & Nakamura, 2007; Shahab et al., 2011; Tamano and Morinishi, 2006).

Since simulations of inhomogeneous flows are an ongoing and currently active topic of study, it is not possible to give a retrospective overview of the research being conducted. However, what is hopefully of more interest is to inject in the remaining chapters results from these simulations that both support existing theories and understanding of the flow dynamics and where new insights have been gleaned.

5.2 LARGE EDDY SIMULATIONS AND HYBRID METHODS

While the early DNS studies focused on investigating the underlying dynamics of homogeneous compressible flows, the early studies associated with large eddy simulations focused on suitable subgrid scale (SGS) models and were based on validations involving the homogeneous DNS results. Although the initial studies on compressible LES formulations are only about 25 years old, the last decade has seen a phenomenal growth in the application of such simulations to a wide variety of compressible engineering flows. As with the discussion on direct numerical simulations, a brief chronology of the early studies is presented. The intent is not to present a record of development for the LES methodology itself, but to highlight the various issues associated with the adaptation of the method to the compressible flow regime. In this context, within the last decade hybrid methods—mixed RANS and LES methodologies—have surfaced and have also been adapted to compressible flows. However, such models have not as yet developed closure, or subgrid scale, models unique to the hybrid method itself.

As was pointed out in Chapter 3, the starting point for both averaged and filtered methodologies is the same. Thus, the governing equations for the resolved field required in the LES methods are form-invariant with those in the RANS methods, and are given by the resolved mass and momentum equations, Eqs. (3.30a) and (3.32) respectively, and an energy equation given by either Eq. (3.55) for the internal energy, or Eq. (3.42) for the total energy. The earliest studies of LES subgrid scale models (Erlebacher, Hussaini, Speziale, & Zang, 1987; Speziale, Erlebacher, Zang, & Hussaini, 1988; Yoshizawa, 1986) utilized the internal energy (enthalpy) equation, Eq. (3.55). Whichever energy formulation is used, the equation system requires the specification of the subgrid scale stress (cf. Eq. (3.34)) and the subgrid scale heat flux (cf. Eq. (3.56b)),

$$\overline{R}_{ij} = \overline{\rho} \left(\widetilde{u_i u_j} - \tilde{u}_i \tilde{u}_j \right) \tag{5.3}$$

$$\overline{Q}_{(h)j} = \overline{\rho} \left(\widetilde{u_j h} - \tilde{u}_j \tilde{h} \right)$$

$$= \overline{\rho} c_p \left(\widetilde{u_j T} - \tilde{u}_j \tilde{T} \right). \tag{5.4}$$

The bar notation is intentionally kept on the left to emphasize that the filtered operation is applied and that the equation is simply written in density-weighted variables. If an internal energy or enthalpy equation is used (cf. Martin, Piomelli, & Candler, 2000), then subgrid scale models for pressure–dilatation and viscous energy dissipation rate terms,

$$\bar{\varepsilon}_v = \overline{\sigma_{ij} S_{ji}} - \tilde{\sigma}_S \tilde{S}_{ji}, \tag{5.5}$$

$$\overline{\Pi} = \overline{p S_{jj}} - \overline{p} \tilde{S}_{jj}, \tag{5.6}$$

are required; whereas, if the total energy equation is used, the terms in Equation (5.5) are not required, and are replaced with subgrid scale models for the triple-velocity correlation and viscous diffusion terms given by

$$\overline{\mathcal{J}}_j = \overline{\rho} \left(\widetilde{u_j u_k u_k} - \tilde{u}_j \widetilde{u_k u_k} \right) \tag{5.7}$$

$$\overline{\mathcal{D}}_j = \overline{u_i \, \sigma_{ij}} - \tilde{u}_i \, \tilde{\sigma}_{ij}. \tag{5.8}$$

Lesieur, Comte, and Normand (1991) and Normand and Lesieur (1992) appear to be the first to utilize the total energy equation for large eddy simulations of compressible flows. In this form (cf. Eq. (3.44)), all the terms can be written in conservative form; although, as Eq. (5.8) shows, models for SGS transport and viscous diffusion are required. In these early studies, however, such terms were not considered.

A compressible SGS model was proposed (Yoshizawa, 1986) that was based on an adaptation of the incompressible Smagorinsky model (Smagorinsky, 1963), and under the assumption of small density fluctuations. The tensorial SGS stress model was decomposed into deviatoric and isotropic parts and given by

$$\overline{R}_{ij}^d = \overline{R}_{ij} - \frac{\delta_{ij}}{3} \overline{R}_{kk} = -2\overline{\rho} \nu_t \left(\tilde{S}_{ij} - \frac{\delta_{ij}}{3} \tilde{S}_{kk} \right)$$

$$= -2\overline{\rho} (C_d \Delta)^2 |\tilde{S}| \tilde{S}_{ij}^d \tag{5.9a}$$

and

$$\overline{R}_{kk} = 2\overline{\rho} (C_i \Delta)^2 |\tilde{S}|^2, \tag{5.9b}$$

where $\nu_t = (C_d \Delta)^2 |\tilde{S}|$ is the eddy viscosity, Δ is the filter width, and the invariant $|\tilde{S}|$ is $(2\tilde{S}_{ij} \tilde{S}_{ji})^{1/2}$. For the subgrid scale heat flux in Eq. (5.4), a Prandtl number can be introduced so that an eddy-diffusivity model can be formed,

$$\overline{Q}_{(h)j} = -\overline{\rho} c_p \frac{\nu_t}{Pr_t} \frac{\partial \tilde{T}}{\partial x_j} = -\overline{\rho} c_p \frac{\nu_t}{Pr_t} \frac{\partial \tilde{T}}{\partial x_j}. \tag{5.10}$$

In such a specification of the SGS stress and heat flux, the models are simple variable density extensions of the incompressible forms, and the SGS kinetic energy, $\overline{R}_{kk}$, needs to be modeled as well with proposed values (Yoshizawa, 1986) of $C_d \approx 0.16$ and $C_i \approx 0.3$ for the closure coefficients. It has been shown (van der Bos & Geurts, 2006) that for the compressible mixing-layer the density fluctuations do make a significant contribution to the computational turbulent stress and may have to be explicitly taken into account in the subgrid scale modeling.

The isotropic contribution, $\overline{R}_{kk}$, can be further analyzed by introducing a subgrid scale Mach number, M_{sgs}, and defined as $M_{sgs}^2 = \overline{R}_{kk}/\overline{\rho}c^2$. For a perfect gas, it is then possible to relate the isotropic SGS stress to the pressure field through $\overline{R}_{kk} = \gamma M_{sgs}^2 \overline{p}$. Yoshizawa (1986), as noted above (see also Moin, Squires, Cabot, & Lee, 1991), and Erlebacher et al. (1987) initially modeled the term, although later, Erlebacher, Hussaini, Speziale, and Zang, (1992) in their study of decaying isotropic turbulence assumed that $\gamma M_{sgs}^2 \ll 1$ and neglected the term.

As in the incompressible case, the Smagorinsky model is overly dissipative, so Erlebacher et al. (1987) and Speziale et al. (1988) developed a mixed Smagorinsky and scale similarity model in terms of density-weighted variables (see also Erlebacher et al., 1992; Zang, Dahlburg, & Dahlburg, 1992). A Leonard decomposition (Leonard, 1974) was used, and the subgrid scale stress decomposed into the Leonard, cross, and Reynolds stresses based on density-weighted filtering. The Leonard stress was directly obtainable, with the cross stress and Reynolds stress closed by a scale similarity model and a Smagorinsky model, respectively. This mixed model then yielded the functional forms given in Eqs. (5.9a) and (5.9b) plus a scale-similarity term given by $-\overline{\rho}(\widetilde{\widetilde{u}_i \widetilde{u}_j} - \widetilde{\widetilde{u}}_i \widetilde{\widetilde{u}}_j)$. The coefficients C_d and C_i were calibrated against DNS data and assumed the values 0.012 and 0.0066, respectively. In addition, a fixed value of 0.5 for the turbulent Prandtl number was assumed.

Further improvements were made (Lilly, 1992; Moin et al., 1991) and compressible, dynamic models were developed for both the SGS stress and the heat flux. Additional validations on homogeneous and inhomogeneous flows were later conducted (e.g. Spyropoulos & Blaisdell, 1996; Vreman, Geurts, & Kuerton, 1994; Zang, Street, & Koseff, 1993) using the Smagorinsky and dynamic models. Fundamental issues of backscatter were also investigated (Piomelli, Cabot, Moin, & Lee, 1991) with behavior similar to the incompressible case and with little dependence on Mach number. The SGS stress models retain the same functional form as in Eqs. (5.9a) and (5.9b), but the coefficients were then given by

$$(C_d \Delta)^2 = \frac{\langle L_{ij} M_{ji}^d \rangle}{\langle M_{ij}^d M_{ji}^d \rangle} \tag{5.11a}$$

$$(C_i \Delta)^2 = \frac{\langle L_{kk} \rangle}{\langle M_{kk} \rangle} \tag{5.11b}$$

with

$$L_{ij} = \overline{\widehat{\rho} \, \widetilde{u}_i \widetilde{u}_j} - \frac{\widehat{\overline{\rho u_i}} \; \widehat{\overline{\rho u_j}}}{\widehat{\overline{\rho}}} \tag{5.12}$$

$$M_{ij}^d = 2\left[r \widehat{\overline{\rho}} |\widehat{\widetilde{S}}| \widehat{\widetilde{S}}_{ij}^d - \overline{\overline{\rho} |\widetilde{S}| \widetilde{S}_{ij}^d} \right] \tag{5.13a}$$

$$M_{kk} = 2\left[r \widehat{\overline{\rho}} |\widehat{\widetilde{S}}|^2 - \overline{\overline{\rho} |\widetilde{S}|^2} \right], \tag{5.13b}$$

where $r = (\widehat{\Delta}/\Delta)^2$ is the ratio (squared) of the test filter-scale to the filter-scale ($r = 4$ is often assumed), and the notation $\widetilde{\widehat{f}}$ is defined as the filter operation $\widehat{\overline{\rho f}}/\widehat{\overline{\rho}}$.

In an analogous way for the subgrid scale heat flux Eq. (5.10), the turbulent Prandtl number can be specified or also extracted from a dynamic procedure and given by

$$Pr_t = \frac{(C_d\Delta)^2 \langle \Theta_k \Theta_k \rangle}{\langle \varkappa_l \Theta_l \rangle},$$
(5.14)

where

$$\Theta_k = -r \,\widehat{\overline{\rho}}|\widehat{\widetilde{S}}|\frac{\partial \widehat{\widetilde{T}}}{\partial x_k} + \widehat{\overline{\rho}|\widetilde{S}|\frac{\partial \widetilde{T}}{\partial x_k}}$$
(5.15a)

$$\varkappa_l = \widehat{\overline{\rho}\,\widetilde{u}_l\widetilde{T}} - \frac{\widehat{\overline{\rho u_l}}\,\widehat{\overline{\rho T}}}{\widehat{\overline{\rho}}}.$$
(5.15b)

Of course, mixed models for the SGS stress and heat flux can also be adapted to include a dynamic procedure.

It is also possible to reformulate the momentum and energy equations for the resolved field by introducing modified pressure and temperature variables (Lesieur, Métais, & Comte, 2005; Vreman, Geurts, & Kuerton, 1995a; Vreman, Geurts, & Kuerton, 1997). In what were termed macro-pressure and macro-temperature variables (Lesieur et al., 2005), given by

$$\overline{P}_m = \overline{p} + \frac{\overline{R}_{kk}}{3}$$
(5.16a)

$$\widetilde{T}_m = \widetilde{T} + \frac{\overline{R}_{kk}}{2},$$
(5.16b)

the resolved momentum and energy equations could be recast in terms of the deviatoric SGS stress $\overline{R}_{ij}^d$. The macro-pressure and macro-temperature are then related through an equation of state given by

$$\overline{P}_m = \overline{\rho}\mathcal{R}\widetilde{T}_m - \frac{3\gamma - 5}{6}\overline{R}_{kk},$$
(5.17)

which, to lowest order, can be rewritten as

$$\overline{P}_m \approx \overline{\rho}\mathcal{R}\widetilde{T}_m \left(1 - \frac{3\gamma - 5}{6}\gamma M_{sgs}^2\right).$$
(5.18)

Assuming that the second term is small in Eq. (5.18), it is clearly better than the previous assumption of $\gamma M_{sgs}^2 \ll 1$, and also suggests that the macro-pressure and macro-temperature formulation is better suited to neglecting the effect of the isotropic SGS stress contribution.

As Eqs. (5.5) and (5.8) showed, other correlations appear in the various energy equations that require SGS models. Their relative importance depends on the flow (free shear or wall-bounded), but in general their effect is of a higher-order than the heat flux contribution. The viscous dissipation $\bar{\varepsilon}_v$ and pressure dilatation in the internal energy equation require models. Vreman, Geurts, & Kuerton, 1995b in their study of a compressible mixing-layer proposed scale similarity models for these terms which could be written as

$$\bar{\varepsilon}_v = C_{\varepsilon v}\left(\widetilde{\tilde{\sigma}_{ij}\tilde{S}_{ji}} - \tilde{\tilde{\sigma}}_{ij}\tilde{\tilde{S}}_{ij}\right), \tag{5.19}$$

and

$$\overline{\Pi} = C_\pi\left(\widetilde{\tilde{p}\tilde{S}_{kk}} - \tilde{\tilde{p}}\tilde{\tilde{S}}_{kk}\right). \tag{5.20}$$

Additionally, alternative models to Eq. (5.19) were proposed (Vreman et al., 1995b) that were proportional to $\overline{\rho}\overline{R}_{kk}/\Delta$. These were also evaluated (Martin et al., 2000) in an *a priori* assessment using homogeneous, isotropic DNS. In both studies, the scale similarity model performed the best especially when adapted to the dynamic procedure for model constant determination.

Since many of the initial compressible LES studies utilized either the internal energy or enthalpy equations, the turbulent transport and viscous diffusion terms in Eq. (5.8) have received even less attention than the viscous dissipation and pressure–dilatation terms. A combined model for these two terms of the form $(\overline{R}_{ij} - \overline{\sigma}_{ij})\tilde{u}_i$ was proposed (Knight, Zhou, Okong'o, & Shukla, 1998). In the flow considered, isotropic, decaying turbulence, the viscous term was neglected and only the turbulent transport contribution was accounted for. A more elaborate dynamic scale-similarity closure using Germano's generalized central moments has been considered (Martin et al., 2000), as well as a dynamic scale-similarity model for the viscous diffusion term.

The usual suite of subgrid scale closures, such as the dynamic model, dynamic mixed model, and more recently, approximate deconvolution models, have been adapted to the compressible regime and evaluated along with various numerical schemes using compressible DNS databases (e.g. Grube, Taylor, & Martin, 2007; Kosović, Pullin, & Samtaney, 2002; von Kaenel, Adams, Kleiser, & Vos, 2002; Lenormand, Sagaut, Ta Phuoc, & Comte, 2000; Martin, 2005; Mathew, Foysi, & Friedrich, 2006). In contrast to these studies, Sun and Lu (2006) have not used density-weighted variables in forming the filtered compressible equations. They found in an *a priori* assessment of their models results that were in good agreement with data from DNS channel flow simulations. These compressible models are more directly based on adaptations of the incompressible forms, and are not necessarily formulated from fundamental compressible dynamics. It can be assumed that while the closures proposed for terms such as the SGS turbulent transport and viscous diffusion are rational proposals, there has not been sufficient validation to fully assess their accuracy or impact on the full equation set.

Another complicating factor in validating results occurs when large eddy simulations data is analyzed. In such cases, the statistical moments are extracted from the resolved flow field results which, by formulation of the governing equations, are density-weighted filtered variables. For example, the statistical moments for the turbulent stress tensor can be expressed in terms of fluctuating quantities given by (Stolz & Adams, 2003)

$$(\widetilde{u}_i)'' = \widetilde{u}_i - \frac{\mathfrak{E}\{\overline{\rho}\widetilde{u}_i\}}{\mathfrak{E}\{\overline{\rho}\}}, \tag{5.21}$$

where all the implied operations are filtered operations, and $\mathfrak{E}\{\cdot\}$ simply represents the expected value of the variable which in this case is some ensemble average or, if the flow is stationary, a long time average. This variable is easily contrasted with the more common relation used in averaged methods or extracted from physical experiments and given here with the same notation as

$$u_i'' = u_i - \frac{\mathfrak{E}\{\rho u_i\}}{\mathfrak{E}\{\rho\}}. \tag{5.22}$$

The extent to which the Eqs. (5.21) and (5.22) yield similar results depends on how well the fluctuating density and velocity fields, ρ and u_i, are reproduced by the filtered fields, $\overline{\rho}$ and $\widetilde{u}_i$. The assessments that have been carried out to date (e.g. Stolz & Adams, 2003; Stolz, 2005) for the supersonic Mach numbers investigated have shown agreement consistent with the corresponding direct simulation studies. This has also recently been considered for channel flow large eddy simulations (Brun, Boiarciuc, Haberkorn, & Comte, 2008) where only small percentage differences between the density-weighted and Reynolds-averaged quantities have been found.

Nevertheless, the various ambiguities and uncertainties discussed highlight the need for caution when assessing compressible simulation data. Just as in the case of experiments and experimental measurements, numerous factors can complicate any quantitative assessment of data.

The discussion in this section has shown that an important focus of early work in LES of compressible flows was dominated by an assessment of the SGS models and associated numerical discretization issues. Some studies (for example Spyropoulos & Blaisdell, 1998; Vreman et al., 1995b), however, focused on probing some physical features of the compressible mixing-layer and the supersonic boundary layer, respectively. While these were validated by comparison with DNS simulations, they were also instrumental in showing that reliable physics could be obtained with such methods. It still remains that resolved fields and the associated resolved field stresses do not fully represent the entire stress field. As such, comparisons to experiment and direct simulation studies need to made carefully.

It is premature to probe too deeply into the development of hybrid-models for compressible flows. It is not that such methods have not been used, since they have and on a variety of flow fields, but the level of compressibility in such flows and the ability of variable density extensions of incompressible forms to capture the essential physics leaves the question of a truly compressible model open. Nevertheless, possibly the first hybrid (RANS/LES) model proposed for compressible flows was put forth by Speziale (1998). The subgrid scale models required for closure of the filtered momentum and (internal) energy equations were simply related to the corresponding RANS closure models through an exponential function based on the (numerical) grid size and Kolmogorov length scale. The approach has been implemented into a flow simulation methodology (FSM) developed by Fasel and colleagues (see Fasel, von Terzi, & Sandberg, 2006). Other approaches, such as the popular (delayed) detached eddy simulation methodology, for example, have been straightforwardly applied to a variety of compressible flows (e.g. Deck, 2005a, 2005b). As cautioned, it remains to validate in many of these cases whether significant compressibility effects on the turbulence existed.

5.3 REYNOLDS-AVERAGED NAVIER–STOKES FORMULATION

The last two sections have discussed simulation methods where either all (in theory) the turbulence scales are resolved (DNS) or where only a portion of the turbulent scales are resolved (LES). In the former case, since there were no unresolved scales, no modeling was required, and in the latter case unresolved scales existed and required modeling. For the Reynolds-averaged approach (RANS), all the turbulent scales are unresolved and require modeling. As mentioned at the outset of this chapter, the reader is assumed to have a knowledge of RANS modeling for incompressible flows so that the material here is focused on issues related to closures in the compressible regime. The recent text by Hanjalić & Launder (2011) is ideally suited for those interested in getting a thorough background on the topic of modeling for incompressible flows. Recall in the derivation of the relevant equations in Section 3.3, utilization of Favre or density-weighted variables can produce a set of mean conservation equations that retain a form-invariance with the corresponding incompressible set. As also noted, however, this form-invariance does not carry over when fluctuations in viscosity or heat conductivity occur. Such fluctuations are usually regarded as higher-order effects and both viscosity and heat conductivity are replaced with their (Reynolds) mean value. In addition, in this case form-invariance does not necessarily imply that the Favre-mean and Favre statistical correlations, are the same as the corresponding Reynolds averaged variables. In fact, from the relations between the two variables derived in Sections (3.2) and (4.4), it is apparent that caution may be necessary in extending incompressible models to their compressible counterparts. However, this has not been the case and many closure models have simply used variable density extensions to the incompressible models to close the equations. Although the hierarchy of Reynolds-averaged turbulent closure models range from zero-equation (algebraic) to differential Reynolds stress models, only the higher-order differential models and their subsets will be discussed here. The reader is referred to the text by Pope (2000) and the linear and nonlinear eddy viscosity review by Gatski and Rumsey (2002) for additional details and references to earlier work. An evaluation of two-equation turbulence in the hypersonic regime by Roy and Blottner (2006) is also of interest.

It is important to recognize that unlike direct numerical simulations, and to a lesser extent large eddy simulation type methods, averaged methods such as RANS (and even hybrid RANS/LES) simply yield solutions consistent with the quality of models used in closing the various mean and turbulent transport equations. As such, results from RANS and RANS related models need thorough validation before being generally usable as a predictive tool. Results from numerical simulations such as DNS (and possibly LES) can be used for model calibration and as a basis for physical analysis in much the same way as laboratory experiments. An objective throughout this book has been to provide the reader with a knowledge of the various tools, both for physical measurements and numerical computations, applicable to the study of compressible flows. In the remainder of this chapter, some of the closure models necessary for accurate flow field predictions are discussed and developed. However, only a limited number of results from RANS computations are presented throughout since replication of observed phenomena can often be successful but not, necessarily, for the correct reasons, and it is not the intent, nor is it feasible here, to try to assess the innumerable set of RANS and RANS-type calculations that have been performed in compressible and supersonic flows.

5.3.1 Turbulent Stress and Stress Anisotropy

At the highest level of turbulent closure, the analysis of compressible, turbulent flow fields through either physical or numerical experiments, will require equations describing the transport of the turbulent stresses. As was seen in the development of the mean flow equations, the turbulent stresses are needed to close the equations, either directly or through an eddy viscosity relationship using the turbulent kinetic energy. Correspondingly, transport equations for the turbulent stresses require even higher-order correlations (the closure problem). In this section, the focus will be on both the formulation of the relevant turbulent correlation equations, and also on the various features unique to the compressible formulation.

5.3.1.1 Turbulent Stress and Kinetic Energy Transport Equations
The transport equations for the turbulent stresses in the compressible case are obtained in a manner analogous to the incompressible formulation; although, now the dependent variables are decomposed into Favre-mean and fluctuating

components and the equations are then Reynolds-averaged (e.g. Gatski, 1996). The equations for the second-moment of fluctuating velocity can be generated using Eq. (3.59), by averaging the product of u_i'' with the fluctuating momentum equation for u_j'', and summing with the average of the product of u_j'' with the fluctuating momentum equation for u_i''. This yields for the stress tensor $\overline{\rho}\tau_{ij}(=\overline{\rho u_i'' u_j''})$,

$$\overline{\rho}\frac{D\tau_{ij}}{Dt} = \frac{\partial(\overline{\rho}\tau_{ij})}{\partial t} + \frac{\partial}{\partial x_k}(\widetilde{u}_k \overline{\rho}\tau_{ij})$$
$$= P_{ij} + \Pi_{ij} - \epsilon_{ij} + M_{ij} + D_{ij}, \qquad (5.23)$$

where

$$P_{ij} = -\overline{\rho}\left[\tau_{ik}\frac{\partial \widetilde{u}_j}{\partial x_k} + \frac{\partial \widetilde{u}_i}{\partial x_k}\tau_{kj}\right]$$
$$= -\overline{\rho}\left[\tau_{ik}\left(\widetilde{S}_{kj} - \widetilde{W}_{kj}\right) + \left(\widetilde{S}_{ik} + \widetilde{W}_{ik}\right)\tau_{kj}\right], \qquad (5.24a)$$

$$\Pi_{ij} = \overline{p'\left(\frac{\partial u_i''}{\partial x_j} + \frac{\partial u_j''}{\partial x_i}\right)} = \overline{p'\left(\frac{\partial u_i'}{\partial x_j} + \frac{\partial u_j'}{\partial x_i}\right)}, \qquad (5.24b)$$

$$\epsilon_{ij} = \overline{\sigma_{ik}'\frac{\partial u_j''}{\partial x_k}} + \overline{\sigma_{jk}'\frac{\partial u_i''}{\partial x_k}} = \overline{\sigma_{ik}'\frac{\partial u_j'}{\partial x_k}} + \overline{\sigma_{jk}'\frac{\partial u_i'}{\partial x_k}}, \qquad (5.24c)$$

$$M_{ij} = \overline{u_i''}\left(\frac{\partial \overline{\sigma}_{jk}}{\partial x_k} - \frac{\partial \overline{p}}{\partial x_j}\right) + \overline{u_j''}\left(\frac{\partial \overline{\sigma}_{ik}}{\partial x_k} - \frac{\partial \overline{p}}{\partial x_i}\right)$$
$$= \frac{\overline{\rho' u_i'}}{\overline{\rho}}\left(\frac{\partial \overline{p}}{\partial x_j} - \frac{\partial \overline{\sigma}_{jk}}{\partial x_k}\right) + \frac{\overline{\rho' u_j'}}{\overline{\rho}}\left(\frac{\partial \overline{p}}{\partial x_i} - \frac{\partial \overline{\sigma}_{ik}}{\partial x_k}\right), \qquad (5.24d)$$

$$D_{ij} = -\frac{\partial}{\partial x_k}\left[\overline{\rho u_i'' u_j'' u_k''} + (\delta_{ik}\overline{p' u_j''} + \overline{p' u_i''}\delta_{jk}) - (\overline{\sigma_{ik}'' u_j''} + \overline{\sigma_{jk}'' u_i''})\right]$$
$$= -\frac{\partial}{\partial x_k}\left[\underbrace{\overline{\rho u_i'' u_j'' u_k''} + (\delta_{ik}\overline{p' u_j'} + \overline{p' u_i'}\delta_{jk})}_{\text{turbulent transport}} - \underbrace{(\overline{\sigma_{ik}' u_j'} + \overline{\sigma_{jk}' u_i'})}_{\text{viscous diffusion}}\right]. \qquad (5.24e)$$

As the equation shows, the transport of the turbulent stress is controlled by a balance among its production P_{ij}, redistribution Π_{ij}, destruction (energy dissipation) ϵ_{ij}, mass flux contribution M_{ij}, and transport and diffusion D_{ij}. With the exception of the production term, all these terms require modeling in general. Where possible, these higher-order correlations are also written in terms of their Reynolds-variable counterparts by explicitly factoring out the mean density variation. Although the equation retains a form invariance with its incompressible counterpart recall, that in the discussion of density-weighted variables, it was cautioned that the physical meaning relative to the Reynolds variable counterparts was changed. The fundamental source of this change was the assimilation of a mass flux contribution into the relationship between the Favre and Reynolds averaged variables (cf. Eq. (3.23)). In the closure of these higher-order correlations it is beneficial to rewrite them in terms of Reynolds variables so that adaptations from their incompressible counterparts are more apparent.

Although turbulent stress transport models are used in flow predictions, and the closure of higher-order correlations is still a research topic, many compressible flow field predictions are made using two equation closures, that is closures based on the compressible, turbulent kinetic energy and (often) the isotropic form of the destruction term or turbulent energy (or specific) dissipation rate, $\epsilon(\epsilon_{ii}/2)$.

The turbulent kinetic energy ($\overline{\rho}K = \overline{\rho}\tau_{ii}/2$) equation is simply formed by taking the trace of stress transport equation, Eq. (5.23)

$$\overline{\rho}\frac{DK}{Dt} = P + \Pi - \epsilon + M + D, \tag{5.25}$$

where

$$P = \frac{P_{ii}}{2} = -\overline{\rho}\tau_{ik}\frac{\partial \tilde{u}_i}{\partial x_k} = -\overline{\rho}\tau_{ik}\tilde{S}_{ki}, \tag{5.26a}$$

$$\Pi = \frac{\Pi_{ii}}{2} = \overline{p'\frac{\partial u'_i}{\partial x_i}}, \tag{5.26b}$$

$$\epsilon = \frac{\epsilon_{ii}}{2} = \overline{\sigma'_{ik}\frac{\partial u'_i}{\partial x_k}} = \overline{\sigma'_{ik}s'_{ki}} \tag{5.26c}$$

$$M = \frac{M_{ii}}{2} = \overline{u''_i\left(\frac{\partial \overline{\sigma}_{ik}}{\partial x_k} - \frac{\partial \overline{p}}{\partial x_i}\right)} \tag{5.26d}$$

$$D = \frac{D_{ii}}{2} = -\frac{\partial}{\partial x_k}\left[\frac{\overline{\rho u''_i u''_i u''_k}}{2} + \overline{p'u'_i}\delta_{ik} - \overline{\sigma'_{ik}u'_i}\right]. \tag{5.26e}$$

Once again, all the terms on the right hand side of Eq. (5.25), now including the production term, require closure. The production term requires a specification of the stress tensor $\overline{\rho}\tau_{ij}$ which is equivalent to specifying the turbulent stress anisotropy since the isotropic part (the turbulent kinetic energy K) is governed by the transport equation, Eq. (5.25). For the production term and the turbulent transport and viscous diffusion term closure is most often achieved through variable density extensions of the incompressible forms, and will not be discussed in detail. For the production term, the usual two-equation closure is to assume an isotropic eddy viscosity model for the stress tensor of the form

$$\overline{\rho}\tau_{ij} = 2\overline{\rho}K\frac{\delta_{ij}}{3} + 2\overline{\mu}_t\left(\tilde{S}_{ij} - \tilde{S}_{kk}\frac{\delta_{ij}}{3}\right), \tag{5.27}$$

where $\overline{\mu}_t = \overline{\rho} C_\mu K \tau$ with τ a turbulent time scale, and for the turbulent transport and viscous diffusion term,

$$D = \frac{\partial}{\partial x_k} \left[\left(\overline{\mu} + \frac{\overline{\mu}_t}{\sigma_K} \right) \frac{\partial K}{\partial x_k} \right], \tag{5.28}$$

where σ_K acts as a Prandtl number for the turbulent kinetic energy diffusion. The coefficient $C_\mu (\approx 0.09)$ is formally extracted from the proportionality assumption between the turbulent shear stress and turbulent kinetic energy assumed for the equilibrium layer of two-dimensional boundary layer flows. The remaining terms associated with the pressure–dilatation, dissipation rate and mass flux are influenced by flow compressibility and will be discussed separately.

5.3.1.2 Turbulent Stress Anisotropy Transport Equation
It is becoming more common in the analysis of turbulent flows to examine the anisotropy tensor formed from the turbulent correlations. The anisotropy tensor formed from the turbulent stress tensor $\overline{\rho} \tau_{ij}$ and the corresponding kinetic energy can be written as

$$\widetilde{b}_{ij} = \frac{\overline{\rho u_i'' u_j''}}{\overline{\rho u_i'' u_i''}} - \frac{\delta_{ij}}{3} = \frac{\overline{\rho} \tau_{ij}}{2 \overline{\rho} K} - \frac{\delta_{ij}}{3}, \tag{5.29}$$

where the mean density $\overline{\rho}$ has been retained in the numerator and denominator to highlight the similarity with the corresponding definition of $\overline{b}_{ij}$ where Reynolds averages are used. This is a traceless, symmetric tensor which provides a measure of the deviation from the normal stress isotropy. When Eqs. (5.23) and (5.25) are used, the exact transport equation for the turbulent stress anisotropy tensor $\widetilde{b}_{ij}$ is given by

$$\frac{D\widetilde{b}_{ij}}{Dt} - \frac{D}{K} \left[d_{ij}^{(t-\mu)} - \widetilde{b}_{ij} \right] = -\left[\frac{P}{\epsilon} + \frac{\Pi}{\epsilon} + \frac{M}{\epsilon} - 1 \right] \frac{\widetilde{b}_{ij}}{\tau} + \frac{d_{ij}^{(\epsilon)}}{\tau}$$
$$+ \frac{1}{2K} \left[\left(P_{ij} - \frac{\delta_{ij}}{3} P_{kk} \right) \right.$$
$$+ \left(\Pi_{ij} - \frac{\delta_{ij}}{3} \Pi_{kk} \right)$$
$$+ \left. \left(M_{ij} - \frac{\delta_{ij}}{3} M_{kk} \right) \right], \tag{5.30}$$

where $\tau = K/\epsilon$, a turbulent time scale, has been introduced as well as the additional anisotropy tensor for turbulent transport-diffusion term

$$d_{ij}^{(t-\mu)} = \frac{D_{ij}}{2D} - \frac{\delta_{ij}}{3}, \tag{5.31}$$

and the anisotropy tensor for turbulent energy dissipation rate term

$$d_{ij}^{(\epsilon)} = \frac{\epsilon_{ij}}{2\epsilon} - \frac{\delta_{ij}}{3}. \tag{5.32}$$

If the turbulent production term P_{ij}, Eq. (5.24a), is rewritten in terms of the stress tensor anisotropy $\widetilde{b}_{ij}$

$$P_{ij} = -2\overline{\rho}K\left[\frac{2}{3}\widetilde{S}_{ij} + (\widetilde{b}_{ik}\widetilde{S}_{kj} + \widetilde{S}_{ik}\widetilde{b}_{kj}) - (\widetilde{b}_{ik}\widetilde{W}_{kj} - \widetilde{W}_{ik}\widetilde{b}_{kj})\right], \tag{5.33}$$

then Eq. (5.30) can be written as

$$\begin{aligned}
\frac{D\widetilde{b}_{ij}}{Dt} - \frac{D}{K}\left[d_{ij}^{(t-\mu)} - \widetilde{b}_{ij}\right] &= -\left[\frac{P}{\epsilon} + \frac{\Pi}{\epsilon} + \frac{M}{\epsilon} - 1\right]\frac{\widetilde{b}_{ij}}{\tau} + \frac{d_{ij}^{(\epsilon)}}{\tau} \\
&\quad + \frac{1}{2K}\left[\left(\Pi_{ij} - \frac{\delta_{ij}}{3}\Pi_{kk}\right) + \left(M_{ij} - \frac{\delta_{ij}}{3}M_{kk}\right)\right] \\
&\quad - \frac{2}{3}\left(\widetilde{S}_{ij} - \frac{\delta_{ij}}{3}\widetilde{S}_{ii}\right) + \left(\widetilde{b}_{ik}\widetilde{W}_{kj} - \widetilde{W}_{ik}\widetilde{b}_{kj}\right) \\
&\quad - \left(\widetilde{b}_{ik}\widetilde{S}_{kj} + \widetilde{S}_{ik}\widetilde{b}_{kj} - \frac{2}{3}\left[\mathbf{b}{:}\widetilde{\mathbf{S}}\right]\delta_{ij}\right), \tag{5.34}
\end{aligned}$$

where the double-dot notation [:] has been introduced indicating a scalar product over both tensor indices.

A comparison of Eq. (5.34) with the incompressible equation for the anisotropy tensor formed from the Reynolds stress tensor (e.g. Gatski, 2004) shows that the two have a similar functional form. The explicit difference being the added compressibility effects of mean velocity dilatational, $\widetilde{S}_{ii}$, and pressure dilatation, Π, and mass flux variation, M_{ij}. However, an implicit and more fundamental difference is the fact that the unknown turbulent stress tensor is not necessarily equivalent to the incompressible Reynolds stress tensor. For example, in Section 3.2 the relationship between the second-moment correlation in Favre and Reynolds variables was given in Eq. (3.24), and when adapted to the velocity field can be written as

$$\widetilde{u_i''u_j''} = \overline{u_i'u_j'} + \frac{\overline{\rho'u_i'u_j'}}{\overline{\rho}} - \frac{\left(\overline{\rho'u_i'}\right)\left(\overline{\rho'u_j'}\right)}{\overline{\rho}^2}. \tag{5.35}$$

In incompressible flows, in order to have realizable turbulence, the anisotropy tensor $\overline{b}_{ij}$ formed from the Reynolds variables is delimited by realizability constraints (Lumley, 1978) (see also Simonsen & Krogstad, 2005) on its second and third invariants, II $(= -\overline{b}_{ij}\overline{b}_{ji}/2 = -\overline{b}_{ii}^2/2)$ and III $(=\overline{b}_{ij}\overline{b}_{jk}\overline{b}_{ki}/3 = \overline{b}_{ii}^3/3)$, respectively. (The reader should be cautioned that unfortunately the defining relations for II and III can differ amongst authors. Thus, quantitative

comparisons between results needs to be done carefully.) These constraints
are kinematically based and founded on the Cayley–Hamilton theorem applied
to second-order symmetric tensors. (By construction, the first invariant—the
trace—vanishes.) As Eq. (5.35) shows, a stress anisotropy tensor defined using
Reynolds variables $\overline{u_i' u_j'}$ could potentially yield a different mapping of the
invariants II and III than one defined using $\widetilde{u_i'' u_j''}$. However, even in flows
with shocks, the anisotropy maps for both $\overline{b}_{ij}$ and b_{ij} are essentially identical.
In Figure 5.2, simulation data from a shock/boundary layer interaction at
$M_\infty = 2.25$ is used to compare both the anisotropies computed from the Favre
(density-weighted) and Reynolds variables and the impact a shock structure
has on these anisotropy variables. As the figure shows, there is minimal
difference between the invariants calculated from the Reynolds variables and
density-weighted variables, even downstream of the shock impingement point.
However, the influence of the shock is clearly felt on the distribution within the
invariant triangle since the mapping from data closest to the shock is heavily
biased near one of the axisymmetric boundaries of the triangle. This boundary
represents a compressive effect on the turbulence structure consistent with the
influence of the shock (Simonsen & Krogstad, 2005). Farther downstream,
the figure shows that the invariant mapping relaxes back to near the upstream
level. In Figure 7.21 (see Section 7.2.2), simulation data for this same Mach
number case will be examined upstream and downstream of an impinging
shock. An alternative representation of the invariants will be discussed to
show how the different dynamic boundaries of the invariant map can be
highlighted.

This result is also substantiated by DNS results (Maeder et al., 2001) for
a supersonic boundary layer flow at $M_\infty = 3, 4.5$, and 6 that were briefly
presented in Section 3.5.2. A comparison was shown in Figure 3.4 between
the density-weighted and Reynolds variable shear stress (in this simulation the
z-axis is normal to the plate surface). Although the $M_\infty = 3$ case (solid line)
shows little difference, the higher Mach number cases, $M_\infty = 4.5$, and 6, do
begin to show a difference. (This numerical result is in contrast, however, with
some experimental results (Owen, 1990) where contributions from the triple
correlation in the shear stress relation $\overline{\rho' u' v'}$ was negligible (cf. Eq. (5.35))).

Figure 5.2 is also consistent with some results (Frohnapfel, Lammers,
Jovanović, & Durst, 2007) that examined the invariant mapping of the turbulent
stress anisotropy in supersonic channel flow where a decrease in skin-friction
(normalized with the incompressible value) was observed with increasing Mach
number. Figure 5.3 shows a comparison between the incompressible case
$Ma = 0.3$ and three supersonic channel flow cases using DNS data (Foysi,
Sarkar, & Friedrich, 2004). (Figure 5.3 should be compared with Figure 8
in Frohnapfel et al., 2007 who used the same data but defined the invariant
differently.) In Figure 5.2, the wall point clearly moves along the two-component
limit line away from the axisymmetric limit toward the one-component limit as

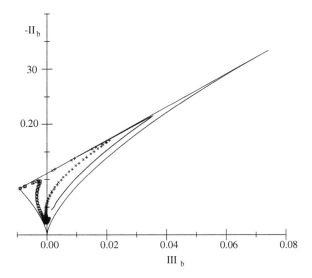

FIGURE 5.2 Invariant mapping of turbulent stress anisotropy in boundary layer flow with an impinging shock. Solid line (Favre and Reynolds): distribution 3δ upstream of shock impingement point (δ is boundary layer thickness in the undisturbed boundary layer at the location of the impingement point); $\circ$ (Favre), $\square$ (Reynolds), distribution at location 2δ downstream of impingement point; $+$ (Favre), $\times$ (Reynolds), distribution at location 10δ downstream of impingement point.

the distance downstream of the shock impingement point increases and local Mach number increases.

There has been very limited direct use of the differential transport equation for the anisotropy tensor and this has only been for incompressible flows (see Sjögren & Johansson, 2000). Equation (5.34) is, nevertheless, the starting point for the construction of explicit algebraic Reynolds stress models. When dynamic equilibrium conditions are applied to the left side of Eq. (5.34), an implicit system of algebraic equations for the stress tensor anisotropy results, and a solution to this system is assumed to be a polynomial tensor representation. These algebraic, polynomial representations are derived from a set of invariants formed from the mean flow functional dependencies. Such representations have been used extensively in incompressible modeling (Gatski, 2004, 2009; Deville & Gatski, 2012), and can be adapted to the compressible regime, although as is often the case, variable density extensions are common. Additional comments on the construction of such representations will be given in Section 5.3.5.

5.3.2 Turbulent Energy Dissipation Rate

The kinetic energy dissipation rate tensor that appears in the turbulent stress transport equation, Eq. (5.23), and defined in Eq. (5.24c), plays a crucial role in the overall dynamic balance of terms in both incompressible and compressible

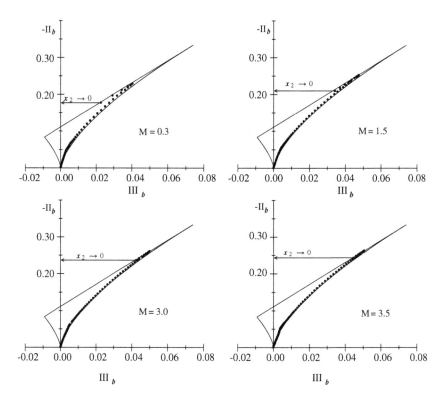

FIGURE 5.3　Anisotropy invariant mapping in supersonic channel flow (cf. Figure 8, Frohnapfel et al., 2007).

flows. When expressed in terms of the fluctuating strain rate and rotation rate tensors, Eq. (5.24c) can be written as

$$\epsilon_{ij} = \overline{\sigma'_{ik}s'_{kj}} + \overline{s'_{ik}\sigma'_{kj}} - \left(\overline{\sigma'_{ik}w'_{kj}} - \overline{w'_{ik}\sigma'_{kj}}\right), \qquad (5.36)$$

with expressions for the mean and fluctuating viscous stress tensors given in Eqs. (3.35) and (3.60), respectively. (Note that the definition in Reynolds variables is consistent with the derivations in Section 3.3 and provides a formal link with the incompressible forms.) It is possible to derive a transport equation for ϵ_{ij} using the relations given in Eq. (5.36), although by inspection, such an equation would be extremely complex with even higher-order correlations requiring closure. There have been some attempts in the incompressible regime to have the anisotropic effects accounted for (Hanjalič & Launder, 1976; Hallbäck, Groth, & Johansson, 1990; Oberlack, 1997; Speziale & Gatski, 1997), but none has been carried over to the compressible case. Thus, as might be suspected, any transport equation associated with the dissipation rate has been formed using the scalar form ϵ which then corresponds to the vanishing of the dissipation rate anisotropy $d_{ij}^{(\epsilon)}$,

$$\epsilon_{ij} = \frac{2}{3}\epsilon\delta_{ij}, \qquad d_{ij}^{(\epsilon)} = 0. \tag{5.37}$$

From Eqs. (5.36) and (3.60), the scalar dissipation rate can then be written as

$$\epsilon = \frac{\epsilon_{ii}}{2} = 2\overline{\mu}\,\overline{s_{ik}'s_{ki}'} - \frac{2}{3}\overline{\mu}\,\overline{s_{kk}'s_{ll}'}$$

$$+2\overline{\mu's_{ij}'}\,\overline{S}_{ij} - \frac{2}{3}\overline{\mu's_{kk}'}\,\overline{S}_{ll}$$

$$+2\overline{\mu's_{ik}'s_{ki}'} - \frac{2}{3}\overline{\mu's_{kk}'s_{ll}'}. \tag{5.38}$$

As was discussed in Section 3.3, contributions from the fluctuating viscosity are most often neglected in the development of the mean conservation equations; however, their role in defining the dissipation rate can be assessed more fully. In results from the cold-wall channel flow simulations (Huang et al., 1995), the term in Eq. (5.38) comprising the product of the fluctuating viscosity and fluctuating strain rate correlation, and the mean strain rate only made a significant contribution (from 6% to 16% depending on wall temperature) to the total dissipation rate in close proximity to the wall. In contrast, the term involving the third-order fluctuations was small throughout the flow.

In light of these results, model development focuses on the contributions from the term with the mean viscosity $\overline{\mu}$, and which can be further partitioned as

$$\epsilon \approx 2\overline{\mu}\,\overline{s_{ik}'s_{ki}'} - \frac{2}{3}\overline{\mu}\,\overline{s_{kk}'s_{ll}'}$$

$$= 2\overline{\mu}\,\overline{w_{ik}'w_{ik}'} + 2\overline{\mu}\,\overline{\frac{\partial u_i'}{\partial x_k}\frac{\partial u_k'}{\partial x_i}} - \frac{2}{3}\overline{\mu}\,\overline{s_{kk}'s_{ll}'}$$

$$= \overline{\mu}\,\overline{\omega_i'\omega_i'} + 2\overline{\mu}\,\overline{\frac{\partial u_i'}{\partial x_k}\frac{\partial u_k'}{\partial x_i}} - \frac{2}{3}\overline{\mu}\,\overline{s_{kk}'s_{ll}'}$$

$$= \underbrace{\overline{\mu\omega_i'\omega_i'}}_{\varepsilon} + \underbrace{2\overline{\mu}\frac{\partial}{\partial x_k}\left[\frac{\partial(\overline{u_k'u_l'})}{\partial x_l} - 2\left(\overline{u_k's_{ll}'}\right)\right]}_{\varepsilon_{inh}} + \underbrace{\frac{4}{3}\overline{\mu}\,\overline{s_{kk}'s_{ll}'}}_{\epsilon_d}, \tag{5.39}$$

where the vorticity vector ω_k' $(=e_{kji}w_{ij}'$ with e_{kij} the permutation tensor) has been introduced, and the relations,

$$\overline{s_{ik}'s_{ki}'} = \overline{w_{ik}'w_{ik}'} + \overline{\frac{\partial u_i'}{\partial x_k}\frac{\partial u_k'}{\partial x_i}}, \tag{5.40a}$$

$$\overline{\frac{\partial u_i'}{\partial x_k}\frac{\partial u_k'}{\partial x_i}} = \frac{\partial(\overline{u_k'u_i'})}{\partial x_i\partial x_k} - 2\frac{\partial}{\partial x_k}\left(\overline{u_k's_{ii}'}\right) + \overline{s_{kk}'s_{ii}'}, \tag{5.40b}$$

have been used. The contribution from ε is the solenoidal contribution since it is directly analogous to an incompressible counterpart, $\overline{\rho}\varepsilon_{inh}$ is an inhomogeneous contribution, and $\overline{\rho}\epsilon_d$ is the dilatation dissipation rate. The inhomogeneous contribution ε_{inh} consists of a term associated with the gradient of the fluctuating velocity second-moment and a term that includes the fluctuating dilatation. The term involving the velocity second-moment can be related to the turbulent stress field in Favre variables and can be assumed a known quantity. This turbulent stress gradient term should then dominate any contribution due to the fluctuating dilatation so that the inhomogeneous contribution need not be modeled separately. Since both $\overline{\rho}\varepsilon$ and $\overline{\rho}\epsilon_d$ are present in compressible homogeneous flows their relative importance can be assessed from numerical simulations.

5.3.2.1 Dilatation Dissipation Rate

Based on the observed phenomena of reduced growth rates in compressible mixing-layer flows that has been discussed earlier as an influential factor in compressible flow research, the effects of Mach number on the turbulent dissipation rate was the first higher-order statistical quantity investigated within the context of second-moment closures. In addition, recall that early simulations of isotropic decaying compressible turbulence had identified shock-type structures (shocklets) (Passot & Pouquet, 1987) associated with the small-scale turbulent structures (see Section 5.1.1). Using the direct simulation results for homogeneous turbulence, (Zeman, 1990, 1991a; Sarkar et al., 1991) models were proposed for the turbulent energy dissipation rate of the form $\epsilon = \varepsilon + \epsilon_d$ (cf. Eq. (5.39)), with ϵ_d proportional to the solenoidal dissipation rate ε through a proportionality factor that was a function of the turbulent Mach number M_t. These approaches, however, were quite dissimilar. Zeman (1990) derived an expression based on turbulent statistics in the vicinity of shocks (see Figure 5.4),

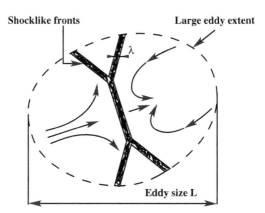

FIGURE 5.4 Sketch of eddy shocklet and normal shock relations (cf. Figure 5.1). From Zeman (1990) with permission.

and motivated by the observance of shocklets by the simulations of Passot and Pouquet (1987) (cf. Figure 5.1). Sarkar et al. (1991) exploited the concepts developed from linear theory (see Section 3.4) where the vorticity, acoustic, and entropy modes were decoupled. The acoustic mode could then be effectively isolated and an asymptotic analysis followed. Adhering to the same general functional form as the Sarkar et al. (1991) model, an alternative form was proposed (Wilcox, 1992) that was primarily applicable to the $K - \omega$ formulation where, in the discussion here, ω is the specific dissipation rate. These three forms can be summarized and written as

$$\epsilon = \varepsilon + \epsilon_d$$
$$= \varepsilon + \varepsilon F(M_t), \qquad (5.41)$$

with

$$F(M_t) = \begin{cases} M_t^2 & \text{Sarkar et al. (1991)} \\ \dfrac{3}{4}\left[1 - e^{-\frac{1}{2}(\gamma+1)[(M_t - M_{t0})/0.6]^2}\right] & \text{Zeman (1990)} \\ \mathcal{H}(M_t - M_{t0}) & \\ \dfrac{3}{2}\left[M_t^2 - M_{t0}^2\right]\mathcal{H}(M_t - M_{t0}) & \text{Wilcox (1992),} \end{cases} \qquad (5.42)$$

where $M_{t0} = 0.25$ for the Wilcox model, and $M_{t0} = 0.10\sqrt{2/(\gamma + 1)}$ in the Zeman model for free shear flows. (For boundary layers, Wilcox, (1992) noted that the Zeman model then has $M_{t0} = 0.25\sqrt{2/(\gamma + 1)}$ and the factor 0.6 increases to 0.66.) Blaisdell et al. (1993) attempted to evaluate the Sarkar and Zeman models and found that as the effects of initial conditions vanished, the dilatation dissipation ϵ_d had a behavior consistent with an M_t^2 variation. This behavior has also been more recently confirmed (Pirozzoli & Grasso, 2004) in simulations of isotropic decaying turbulence. Although the functional dependency on M_t derived by Zeman does not appear to be correct, he did correctly observe that instantaneous values of turbulent Mach number can indeed be supersonic and sufficiently large values of $F(M_t)$ can occur. This is consistent with the experimental observation of shocklet dynamics which was briefly given in Section 1.4.

This form of the partitioning allowed for the direct extension of the incompressible form of the solenoidal dissipation rate transport equation, and brought the calculations into agreement with the experimental results available. The performance of these models using both the $K - \varepsilon$ and $K - \omega$ two-equation linear eddy viscosity models in predicting the growth rate of compressible mixing-layer has been examined (Barone, Oberkampf, & Blottner, 2006). The growth rate was parameterized by a compressibility factor Φ defined as the compressible growth rate normalized by the incompressible growth rate at the

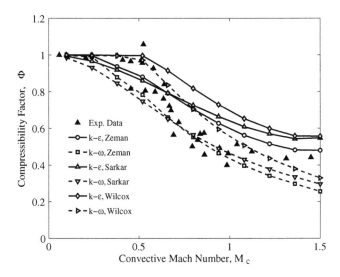

FIGURE 5.5 Comparison of calculated and measured normalized spreading rate for a compressible mixing-layer. From Barone et al. (2006) with permission.

same velocity and density ratios, that is

$$\Phi = \left(\frac{d\delta}{dx}\right)_c \left(\frac{d\delta}{dx}\right)_i^{-1}, \tag{5.43}$$

where δ is a measure of the mixing-layer thickness. As will be discussed in Section 6.2, it is a function of the convective Mach number $M_c = (U_1 - U_2)/(c_1 + c_2)$, where U_1, c_1 and U_2, c_2 are the respective free-stream velocities and speeds of sound. (These parameters will be more fully discussed in Section 6.2 but are introduced here to show the impact the dilatation models have on reproducing key phenomenological features of the flow.) Figure 5.5 illustrates this by showing the significant influence this modification to the solenoidal dissipation has on the predicted spreading rate. Similar results were also obtained (Sarkar & Lakshmanan, 1991) using a differential Reynolds stress transport model. Such results originally suggested that the increase in compressible dissipation rate and the resultant decrease in the turbulence was the underlying dynamic reason for the reduction in spreading rate. Many of the ideas about the role of the dilatation dissipation rate were originally founded in the simulations of homogeneous flows. However, the introduction of mean shear and the subsequent turbulence inhomogeneity and anisotropy brought into question many of the earlier conclusions about the role of ϵ_d. As will be discussed in Section 5.3.3, the dynamics of the pressure fluctuations lie at the base of this observed phenomena of reduced mixing-layer spreading rate.

5.3.2.2 Solenoidal Dissipation Rate Transport Equation

In incompressible flows, the often used definition for the solenoidal dissipation rate is $2\nu\overline{s'_{ik}s'_{ki}}$ which can be approximated by its familiar isotropic form $\nu\overline{(\partial u'_i/\partial x_k)^2}$. The transport equation for these dissipation rate quantities can then be derived and has been the starting point for developing closures for the exact transport equation for incompressible flows (e.g. Rodi & Mansour, 1993; Nagano & Shimada, 1995). Alternately in incompressible flows, the solenoidal dissipation rate equation has been approximated by the enstrophy equation $\overline{\omega'_i\omega'_i}$, and as has been shown in the last section, the compressible solenoidal dissipation rate equation can also be suitably approximated by the enstrophy, and expressed in terms of Reynolds fluctuation variables. For example, Hassan and colleagues (e.g. Robinson, Harris, & Hassan, 1995; Robinson & Hassan, 1998) have adopted the enstrophy equation as the scale equation in their compressible two-equation formulations; although, in their compressible form they utilized the Favre mean and fluctuating variables. Sinha and Candler (2003) have provided a useful comparison of the incompressible and compressible enstrophy equations (in Reynolds variables) as well as the corresponding solenoidal dissipation rate equation. They have also examined the budget of terms in a high speed boundary layer flow to further assess the relative importance of terms. More recently, the compressible enstrophy equation has been examined (Kreuzinger, Friedrich, & Gatski, 2006) with a slightly different partitioning of the terms originating from the viscous term in the momentum equation. Their *a priori* assessment involved both channel flow and mixing-layer flow.

A general form for the solenoidal dissipation rate transport equation can be written as (cf. Sinha & Candler, 2003; Kreuzinger et al., 2006)

$$\frac{D\varepsilon}{Dt} = P_\varepsilon^1 + P_\varepsilon^2 + P_\varepsilon^3 - \Upsilon + T_\varepsilon + D_\varepsilon + T_\varepsilon^c + B_\varepsilon + F_\varepsilon + \frac{\varepsilon}{\overline{\mu}}\frac{D\overline{\mu}}{Dt}, \quad (5.44)$$

where

$$\begin{aligned}
P_\varepsilon^1 &= 8\overline{\mu}\left(\overline{w'_{ij}w'_{jk}}\,\overline{S}_{ki} + \overline{w'_{ij}s'_{jk}}\,\overline{W}_{ki}\right)\\
&= 2\overline{\mu}\left[\left(\overline{\omega'_i\omega'_j}\,\overline{S}_{ij} + \overline{\omega'_i s'_{ij}}\,\overline{\Omega}_j\right) - \left(\overline{\omega'_i\omega'_i}\,\overline{S}_{jj} + \overline{\Omega}_i\overline{\omega'_i s'_{jj}}\right)\right]
\end{aligned} \quad (5.45a)$$

is the turbulent production of dissipation associated with the mean velocity gradients,

$$P_\varepsilon^2 = -2\overline{\mu}\,\overline{(u'_k\omega'_i)}\,\frac{\partial\overline{\Omega}_i}{\partial x_k} \quad (5.45b)$$

is the production due to the mean vorticity gradient,

$$P_\varepsilon^3 = 2\overline{\mu}\left(\overline{\omega'_i\omega'_j s'_{ji}} - \overline{\omega'_i\omega'_i s'_{jj}}\right) \quad (5.45c)$$

is the production due to vortex stretching by fluctuating velocity gradient field, and the viscous destruction term is

$$\Upsilon = 2\overline{\mu}\overline{\left[\frac{\partial}{\partial x_k}\left(\frac{\omega_r'}{\rho}\right)\right]\left(e_{rji}\frac{\partial \sigma_{ik}}{\partial x_j}\right)}. \tag{5.46}$$

The turbulent transport and viscous diffusion terms are given, respectively, by

$$T_\varepsilon = -\overline{\mu}\frac{\partial \overline{(u_k'\omega_i'\omega_i')}}{\partial x_k} \tag{5.47a}$$

and

$$D_\varepsilon = 2\overline{\mu}\frac{\partial}{\partial x_k}\overline{\left[\frac{\omega_r'}{\rho}\left(e_{rji}\frac{\partial \sigma_{ik}}{\partial x_j}\right)\right]}. \tag{5.47b}$$

The remaining terms constitute the higher-order contributions of compressibility or density and viscosity gradients. These include the compressible turbulent transport,

$$T_\varepsilon^c = \overline{\mu(\omega_i'\omega_i's_{kk}')}, \tag{5.48a}$$

the baroclinic term B_ε due to the force exerted by the pressure gradient,

$$B_\varepsilon = -2\overline{\mu}\overline{\left(\frac{\omega_r'}{\rho^2}\right)\left[e_{rij}\frac{\partial \rho}{\partial x_j}\frac{\partial p}{\partial x_i}\right]}, \tag{5.48b}$$

and a force term F_ε resulting from the viscous stress gradient,

$$F_\varepsilon = 2\overline{\mu}\overline{\left(\frac{\omega_r'}{\rho^2}\right)\left[e_{rji}\frac{\partial \sigma_{ik}'}{\partial x_k}\frac{\partial \rho}{\partial x_j}\right]}. \tag{5.48c}$$

Both B_ε and F_ε are contributions due to forces on a volume element that are normal to the density gradient. Finally, the last term on the right of Eq. (5.44) requires no modeling and is simply the variation of the mean viscosity.

It is clear that Eq. (5.44) will present a formidable obstacle to any attempts at developing a rigorous closed transport equation. However, with this construction of the solenoidal dissipation rate equation, it is logical to construct a model equation in a manner analogous to that used for incompressible flows (Hanjalić & Launder, 2011). Many such models exist in the incompressible literature, and suffice it to say that their structure consists of source, sink and diffusion terms (see Vallet, 2008, for a recent example of dissipation rate closure). The following is just an example of some of the closure issues and strategies involved.

Recall that it was emphasized at the outset that the construction of the solenoidal dissipation rate equation was based on Reynolds variables since it has been recognized that Favre variable decomposition of viscous terms was not the most concise partitioning. However, the mean momentum and energy equations are in terms of Favre variables and the turbulent correlations are in terms of Favre variables. This is not surprising since the closure problem arises in the nonlinear advection terms in the mean equations and these are optimally partitioned using density-weighted variables. This requires the solenoidal dissipation rate equation, Eq. (5.44), to be closed using density-weighted variables. The source term P_ε^1 in Eq. (5.45a) can be approximated in the form

$$P_\varepsilon^1 = -C_\varepsilon^1 \tau_{ij} \left(\frac{\widetilde{S}_{ji}}{\tau_s} \right) - C_\varepsilon^{1'} \varepsilon \widetilde{S}_{jj}, \tag{5.49}$$

where $\tau_s (= K/\varepsilon)$ is a turbulent time scale based on the solenoidal dissipation rate, and C_ε^1 and $C_\varepsilon^{1'}$ are closure coefficients. The first term is a straightforward extension of an incompressible model, and the second term is necessary to properly account for dilatation effects resulting from bulk compressions and expansions. Omission of the mean dilatation term causes the model to incorrectly predict the decrease of the integral length scale for isotropic expansion and the increase for isotropic compression (see for a summary Lele, 1994). An exact model for the mean dilatation term in Eq. (5.49) has been developed (Coleman & Mansour, 1991a) for rapid spherical (isotropic) compressions of incompressible turbulence, and applied (Coleman & Mansour, 1991b) to the case of compressible turbulence. In the latter case, pressure–dilatation effects need to be considered as well. In the former case, an exact solution to the evolution equations resulting from rapid distortion theory (RDT) was used as a modeling guide. In such flows, the enstrophy equation is a balance between the (dilatation) production and the viscosity variation and a comparison with the RDT equation yielded a value for the closure coefficient $C_\varepsilon^{1'}$. This procedure (Coleman & Mansour, 1991a) yielded an expression for $C_\varepsilon^{1'}$ given by $[1 + 3n(\gamma - 1) - 2C_\varepsilon^1]/3$, where n is the exponent in the viscosity law.

The other source term P_ε^2 is often neglected in many two equation models, but as has been shown for incompressible flow (Rodi & Mansour, 1993) and for compressible flow (Kreuzinger et al., 2006), it is of comparable size to the combination of the other source terms and the turbulent diffusion term. The form proposed for incompressible (Rodi & Mansour, 1993) has been generalized and adapted to the compressible case as

$$P_\varepsilon^2 = -\bar{\mu} \tau_s \frac{\partial}{\partial x_n} \left(\widetilde{S}_{in} - \frac{1}{2} e_{ijn} \widetilde{\Omega}_j \right) \left[C_\varepsilon^2 (\bar{\rho} K) \frac{\partial}{\partial x_k} \left(\widetilde{S}_{ik} - \frac{1}{2} e_{ipk} \widetilde{\Omega}_p \right) \right.$$
$$\left. + C_\varepsilon^{2'} \left(\widetilde{S}_{ik} - \frac{1}{2} e_{iqk} \widetilde{\Omega}_q \right) \frac{\partial (\bar{\rho} K)}{\partial x_k} \right], \tag{5.50}$$

where C_ε^2 and $C_\varepsilon^{2'}$ are closure coefficients. The sink term in Eq. (5.44) is represented by the combination of the vortex stretching term P_ε^3 and the destruction term Υ and the terms P_ε^3 and Υ are assumed to have the same asymptotic behavior and combine to form a sink term approximated as

$$P_\varepsilon^3 - \Upsilon = -C_\varepsilon^3 \frac{\varepsilon}{\tau_s}. \tag{5.51}$$

The turbulent transport and viscous diffusion terms can be combined, as in the incompressible case, for a model of the form

$$D_\varepsilon = \frac{1}{\overline{\rho}} \frac{\partial}{\partial x_k} \left[\left(\overline{\mu} \delta_{kj} + \tau_s C_\varepsilon \overline{\rho} \tau_{kj} \right) \frac{\partial \varepsilon}{\partial x_j} \right]. \tag{5.52}$$

For eddy viscosity models, a simpler form of closure for D_ε can be used, and is given by

$$D_\varepsilon = \frac{1}{\overline{\rho}} \frac{\partial}{\partial x_k} \left[\left(\overline{\mu} + \frac{\overline{\mu}_t}{\sigma_\varepsilon} \right) \frac{\partial \varepsilon}{\partial x_k} \right], \tag{5.53}$$

where $\overline{\mu}_t (= C_\mu \overline{\rho} K \tau_s)$ is the turbulent eddy viscosity with C_μ a closure constant and $\sigma_\varepsilon \approx 1.3$ is a dissipation rate Prandtl number. The closure models given in Eqs. (5.49), (5.50), (5.51), and (5.52) or (5.53) that have been adapted from their incompressible forms for the production, destruction and transport-diffusion terms respectively, can be introduced into the solenoidal dissipation rate equation, Eq. (5.44). The reader is referred to the books by Pope (2000) and Launder and Sandham (2000) and related references for a state-of-the-art overview of incompressible modeling.

The majority of incompressible models for the solenoidal dissipation rate do not include any contribution from Eq. (5.50), and are only composed of the modeled terms Eqs. (5.49), (5.51), and (5.52) or (5.53) with closure coefficients given by values $C_\varepsilon^1 \approx 1.44, C_\varepsilon^3 \approx 1.92$, and $C_\varepsilon \approx 0.15$, or $C_\mu \approx 0.09$, depending on the calibration. It is clear that if these are the only terms included in the solenoidal dissipation rate equation then it is essentially a variable density extension of the incompressible form (Vallet, 2008). In Section 6.4, the predictive capabilities of such a model are evaluated for the case of a zero pressure gradient boundary layer flow. Specifically, the self-consistency of the closure coefficients and models in predicting the correct equilibrium log-layer is assessed.

What remains is the closure of terms attributed to compressibility effects and given by Eqs. (5.48a), (5.48b), and (5.48c). An *a priori* assessment of the compressibility terms $T_\varepsilon^c, B_\varepsilon$, and F_ε in channel, boundary layer and mixing-layer flows has been performed (Kreuzinger et al., 2006). In the channel flow (cold wall) and boundary layer flow (adiabatic wall), the peak amplitude of these terms was less than 10% of the production/destruction terms. There was, however, an increase with Mach number in the range examined ($1.5 < M < 3.2$) although the relative contribution remained small. In the mixing-layer case, the compressible turbulent transport, T_ε^c, and the force from the viscous stress gradient, F_ε, are a negligible effect in cases with and without density differences across the layer. However, in the case with a density difference, it was found (Kreuzinger et al., 2006) that the baroclinic term increased to a level comparable with the production term. It should be cautioned that in the absence of shear, the baroclinic term has not been found (e.g. Pirozzoli & Grasso, 2004) to affect the dynamic balance in the enstrophy (solenoidal dissipation rate) equation.

As a first step in gaining a little more insight into the baroclinic term, Eq. (5.48b) can be rewritten as

$$
B_\varepsilon \approx -\frac{2\overline{\mu}}{\overline{\rho}^2}\left[e_{rij}\overline{\left(\omega_r'\frac{\partial\rho}{\partial x_j}\frac{\partial p}{\partial x_i}\right)} \right.
$$
$$
\left. + e_{rij}\overline{\left(\omega_r'\frac{\partial\rho}{\partial x_j}\frac{\partial p}{\partial x_i}\right)}\sum_{n=1}^{\infty}(-1)^n\left(2\frac{\rho'}{\overline{\rho}}+\frac{\rho'^2}{\overline{\rho}^2}\right)^n \right], \qquad (5.54)
$$

and to the lowest-order can be approximated by the first term (Kreuzinger et al., 2006). This term has been partitioned (Krishnamurty & Shyy, 1997) into three parts given by,

$$
B_\varepsilon \approx -\frac{2\overline{\mu}}{\overline{\rho}^2}\left(\overline{e_{rij}\omega_r'\frac{\partial\rho}{\partial x_j}\frac{\partial p}{\partial x_i}}\right)
$$
$$
= -\frac{2\overline{\mu}}{\overline{\rho}^2}\left[\left(\overline{e_{rij}\omega_r'\frac{\partial p'}{\partial x_i}\frac{\partial\overline{\rho}}{\partial x_j}}\right) + \left(\overline{e_{rij}\omega_r'\frac{\partial\rho'}{\partial x_j}\frac{\partial\overline{p}}{\partial x_i}}\right) \right.
$$
$$
\left. + \left(\overline{e_{rij}\omega_r'\frac{\partial\rho'}{\partial x_j}\frac{\partial p'}{\partial x_i}}\right)\right], \qquad (5.55)
$$

and concluded from an order of magnitude analysis that the second term on the right in Eq. (5.55) is the dominate one. These same contributions have been analyzed (Aupoix, 2004) and it was concluded for compressible mixing-layers that the first term dominates in light of the absence of pressure gradient effects in such flows. The third term involving the higher-order correlation was neglected in both studies. Contrary to both of these studies, however, it was found from an *a priori* analysis of DNS mixing-layer data that the higher-order triple-correlation term dominated (Kreuzinger et al., 2006).

It is interesting that even with the same starting point for the baroclinic term, there has been a rather diverse set of proposed models. For example, Krishnamurty and Shyy (1997) retained the mean pressure gradient term in Eq. (5.55) and proposed a model that included the mass flux. In contrast, Aupoix (2004) argued that in mixing-layers the mean pressure is constant so that the baroclinic term is proportional to the mean density gradient, and coupled this with the assumption of isentropic flow so that the fluctuating pressure-gradient term could be related to the mean velocity gradient. Unfortunately, the proposed model was only specified for two-dimensional flow. A third example is the proposal made by Kreuzinger et al. (2006) who developed a model for the triple-correlation term, and estimated the fluctuating pressure-gradient term from spectral analysis of free shear flows as well as isotropic turbulence, and the fluctuating density gradient term from mixing-length ideas. These three proposed models can be summarized as

$$
B_\varepsilon \propto \begin{cases} \tau_s^{-1} \left(\dfrac{\overline{\rho' u_j'}}{\overline{\rho}} \right) \dfrac{\partial \overline{p}}{\partial x_j} & \text{Krishnamurty and Shyy (1997)} \\[2ex] K^{3/2} \left(\dfrac{\partial \widetilde{u}}{\partial y} \right) \dfrac{\partial \overline{\rho}}{\partial y} \quad (2D) & \text{Aupoix (2004)} \\[2ex] \tau_s^{-1} K^{3/2} \left| \dfrac{1}{\overline{\rho}} \dfrac{\partial \overline{\rho}}{\partial x_j} \right| & \text{Kreuzinger et al. (2006).} \end{cases}
$$

$$(5.56)$$

It is apparent from this discussion that while variable density extensions to incompressible dissipation rate models suffice for many of the unknown correlations in the solenoidal dissipation rate equation, models for terms directly associated with compressibility have yet to be finalized. Although in the preceding paragraphs, a method of analysis and assessment has been presented that extensively utilizes DNS data to obtain rational closures for these models, additional simulation studies with broader parameter ranges need to be assessed.

Most RANS applications utilizing the solenoidal dissipation rate (as well as other scale variables) were without the terms directly due to compressibility effects. The remaining terms need to properly account for any variable density effects and yet be able to produce solutions capable of yielding correct mean field variables. Huang, Bradshaw, and Coakley (1994) (see also Huang, Bradshaw, & Coakley, 1992) recognized this problem for compressible wall-bounded flows and provided a rational modification to the (incompressible, solenoidal) dissipation rate equation (as well as other potential scale equations such as ω). Subsequently, this formulation was extended (Catris & Aupoix, 2000a) and generalized to a variety of two-equation models. Nevertheless, a brief outline of the analysis is given here and is necessarily constrained to two-dimensional equilibrium boundary layer flows. A more complete discussion of such flows is given in Section 6.4.

In the logarithmic region of the equilibrium boundary layer, the turbulent shear stress is assumed equal to the wall shear stress τ_w. A Boussinesq relation for the eddy viscosity, coupled with a mixing-length formula for closure, yields for the turbulent eddy viscosity (see Eq. (6.26))

$$
\overline{\mu}_t = \overline{\rho} C_\mu \frac{K^2}{\varepsilon} = \sqrt{\overline{\rho} \, \overline{\rho}_w} u_\tau \ell, \tag{5.57}
$$

with $C_\mu \approx 0.09$ from the fact that the Reynolds stress anisotropy is constant in the equilibrium log-layer, u_τ is the friction velocity, and the mixing-length ℓ given by κy where κ is the von Kármán constant and y is the distance from the wall. Eq. (5.57) now provides a relationship between the dissipation rate ε and K. Unfortunately, a consequence of this relationship is that the turbulent kinetic energy equation (cf. Eq. (5.25)) is no longer balanced in the near-wall region where the advection terms can be neglected, and that it is not possible to predict a log-law slope of $1/\kappa$ from the usual modeled equations. Since there is a unique relationship between κ and the other coefficients, a compatibility relation (Huang et al., 1992, 1994) can be extracted between the relevant parameters

$$\frac{\sqrt{C_\mu}\sigma_\phi}{\ell^2\kappa^2}\left(C_2 - C_1\right) = 1 + \frac{1}{\ell^2}\left[d_1\frac{y}{\overline{\rho}}\frac{\partial\overline{\rho}}{\partial y} + d_2\frac{y^2}{\overline{\rho}}\frac{\partial^2\overline{\rho}}{\partial y^2} + d_3\left(\frac{y}{\overline{\rho}}\frac{\partial\overline{\rho}}{\partial y}\right)^2\right],$$

(5.58)

where the variable ϕ can be defined as $\overline{\rho}^n K^m \varepsilon^\ell$ ($\phi = \varepsilon, n = m = 0$ and $\ell = 1$), and the d_i are functions of the n, m, ℓ, σ_K, and σ_ϕ. For $\phi = \varepsilon$ in incompressible flows, Eq. (5.58) reduces to the well known relation

$$\frac{\sqrt{C_\mu}\sigma_\varepsilon}{\kappa^2}\left(C_\varepsilon^3 - C_\varepsilon^1\right) = 1,$$

(5.59)

with $C_1 = C_\varepsilon^1$ and $C_2 = C_\varepsilon^3$. In general, the coefficients C_1 and C_2 are related to the usual solenoidal dissipation rate coefficients by $C_1 = \ell C_\varepsilon^1 + m$ and $C_2 = \ell C_\varepsilon^3 + m$. The effects of pressure gradients on the law of the wall have been investigated (Huang et al., 1995) and found to have minimal impact on the velocity law, but a strong effect on the temperature law. A variety of prediction models were evaluated and important trends in model applications were presented. Once again, however, the length scale equation had a significant impact on prediction accuracy.

These ideas were extended (Catris & Aupoix, 2000a) by first proposing an alternate functional form for the diffusive flux in the turbulent kinetic energy equation so that (cf. Eq. (5.28))

$$D = \frac{\partial}{\partial x_k}\left[\frac{\overline{\mu}_t}{\overline{\rho}\sigma_K}\frac{\partial(\overline{\rho}K)}{\partial x_k}\right],$$

(5.60)

where the viscous term contribution has been neglected. A redefined length scale flux in the transport equation for ϕ was then given by (cf. Eq. (5.53))

$$D_\phi = \frac{\partial}{\partial x_k}\left[\frac{\overline{\mu}_t}{\sqrt{\overline{\rho}}\sigma_\phi}\frac{\partial(\sqrt{\overline{\rho}}\phi)}{\partial x_k}\right],$$

(5.61)

so that for the solenoidal dissipation rate $\phi = \varepsilon$.

These last two sections have provided a fundamental framework from which transport equations and closures for the unknown energy dissipation rate can be extracted. Obviously, many variations of these fundamentals can be conceived and rationalized so that a viable energy dissipation rate equation can be obtained and utilized in conjunction with a Reynolds stress or kinetic energy closure. Most notable of these alternatives is the specific dissipation rate ω which has been extensively validated and used for enumerable flow problems. Other scale variables can also be used that are directly derivable from the more fundamental quantities K and ϵ (or ε). An early example of an alternative closure, formulated during the period when dilatation dissipation effects were deemed important simply used (El Baz, & Launder, 1993) a solenoidal dissipation rate equation that sensitized the destruction term to compressibility effects through a factor containing the turbulent Mach number. The closure was justified by arguing it was a more consistent alteration to the dissipation rate equation within a single-scale framework in which the dissipation equation provides a model for the energy transfer to the small scales.

5.3.3 Velocity–Pressure Gradient Correlation

Although the velocity–pressure gradient correlation does not explicitly appear in the transport equations associated with the Reynolds stress tensor (kinetic energy and stress anisotropy), Eq. (5.23), it follows directly from the construction of the transport equation from the fluctuating momentum equation. For RANS modeling in incompressible flows, the partitioning of the velocity–pressure gradient correlation most often used is simply,

$$-\overline{u_i' \frac{\partial p'}{\partial x_j}} - \overline{u_j' \frac{\partial p'}{\partial x_i}} = -\frac{\partial}{\partial x_k}\left(\delta_{ik}\overline{p'u_j'} + \overline{p'u_i'}\delta_{jk}\right) + \overline{\rho}\Pi_{ij}, \qquad (5.62)$$

with $\overline{\rho}\Pi_{ij}$ given in Eq. (5.24b). In the compressible case, however, the velocity gradient tensor is not traceless so that $\overline{\rho}\Pi_{ij}$ can be rewritten as

$$\overline{\rho}\Pi_{ij} = \frac{2}{3}\overline{p's_{kk}'}\delta_{ij} + 2\left(\overline{p's_{ij}'} - \frac{\delta_{ij}}{3}\overline{p's_{kk}'}\right)$$

$$= \frac{2}{3}\overline{\rho}\Pi\delta_{ij} + \overline{\rho}\Pi_{ij}^d, \qquad (5.63)$$

where $\overline{\rho}\Pi$ is the pressure dilatation given in Eq. (5.26b) and $\overline{\rho}\Pi_{ij}^d$ is the deviatoric part of the pressure–strain rate correlation. In this form, the suggested role of the pressure–dilatation parallels that of the dilatation dissipation (although the underlying physics are only weakly connected), and the deviatoric pressure–strain rate correlation $\overline{\rho}\Pi_{ij}^d$ makes no contribution to the turbulent kinetic energy budget (as its incompressible counterpart). Although such form-similarities give a certain measure of confidence in simple adaptation of incompressible models to the compressible case, it is important to recall the inherent differences in the pressure field between incompressible and compressible flows.

5.3.3.1 Pressure–Dilatation

As mentioned in Section 5.1.1, in addition to the dilatation dissipation, early modeling studies focused on the pressure–dilatation term since dilatation effects were initially and generally viewed as an important contributor to the flow dynamics. As with the dilatation dissipation, the starting point for developing a pressure–dilatation model was the direct simulation of compressible isotropic turbulence. Sarkar et al. (1991) and Zeman (1991a) used direct simulation data and theoretical analysis of compressible decaying turbulence to assess the role of pressure–dilatation and proposed closures. The simulation data used by each, though different, showed that after an initial transient, the pressure–dilatation relaxed to some acoustic equilibrium value (zero for Sarkar et al., 1991, but nonzero for Zeman, 1991a). It was then concluded (Sarkar et al., 1991) that the term could be neglected in such flows or, if necessary, assimilated into the dilatation dissipation model.

Further studies (Zeman, 1991a; Zeman & Coleman, 1993) analyzed the pressure–variance equation as the equation basis for model development. In the absence of molecular effects, the pressure–variance equation could be written as (cf. Eq. (3.66))

$$
\frac{D\overline{p'^2}}{Dt} \approx -2\overline{p'u'_j}\frac{\partial \overline{p}}{\partial x_j} - 2\gamma \overline{p'^2}\widetilde{S}_{jj} - 2\gamma \overline{p}\,\overline{p's'_{jj}}
$$

$$
= -2\overline{c^2}\,\overline{p'u'_j}\frac{\partial \overline{\rho}}{\partial x_j} - 2\gamma \overline{p'^2}\widetilde{S}_{jj} - 2\overline{c^2}\overline{\rho}^2\Pi, \tag{5.64}
$$

where terms of $\mathcal{O}(\rho'/\rho)$ and higher have been neglected, and an isentropic speed of sound, $\overline{c^2} = \gamma \overline{p}/\overline{\rho}$, has been used. In the case of decaying isotropic turbulence, Eq. (5.64) simplifies so that the only term remaining is the pressure–dilatation. A relaxation model was proposed (Zeman, 1991a) based on a decay constant associated with the acoustic propagation velocity of the large scales. Other effects could be added, such as rapid compression (spherical and one-dimensional) (Durbin & Zeman, 1992; Zeman & Coleman, 1993), and this would necessitate the inclusion of the mean dilatation term $\overline{p'^2}\widetilde{S}_{jj}$ in Eq. (5.64). Although model development from such benchmark flows is not uncommon, a comparison with Eq. (3.66) shows that such flows severely truncate the form of the equation and, more importantly, potentially important physical behavior.

In contrast to this approach, a Poisson equation for the fluctuating pressure field derived from the mass and momentum conservation equations, and applicable to homogeneous flows was analyzed (Sarkar, 1992). It resulted in a partitioning of the fluctuating pressure field into an incompressible part that was associated with the mean density field and a compressible part associated with the fluctuating density field (see also Erlebacher, Hussaini, Kreiss, & Sarkar, 1990). DNS data for homogeneous shear flow was utilized to show that only the contribution from the incompressible pressure field p'^I to the pressure–dilatation correlation was significant. This justified the approach of further partitioning the pressure field into slow and rapid parts that both contributed to the pressure–dilatation model. Although some of the early studies on compressible turbulence utilized this partitioning of the fluctuating pressure fields into incompressible and compressible parts, it was shown (Blaisdell & Sarkar, 1993) that the two fields were reasonably correlated in the case of homogeneous shear. This is shown in Figure 5.6 where the various combinations of pressure component correlations are presented. As the figure shows, the

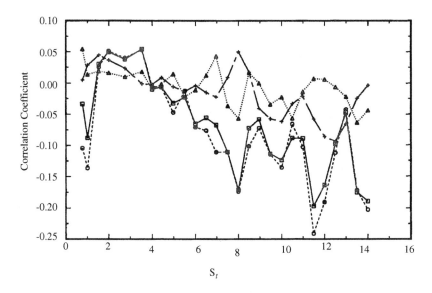

FIGURE 5.6 Temporal evolution of correlation coefficients for the fluctuating pressure field components in homogeneous shear flow (S, the shear rate): p'^I, incompressible pressure, p'^r, rapid part of incompressible pressure, p'^s, slow part of incompressible pressure, and p'^C, compressible pressure. Correlation coefficient $\overline{p'^I\,p'^C}$, □; correlation coefficient $\overline{p'^r\,p'^C}$, o; correlation coefficient $\overline{p'^s\,p'^C}$, △; correlation coefficient $\overline{p'^r\,p'^s}$, +. From Blaisdell and Sarkar (1993) with permission.

correlation coefficient associated with $\overline{p'^I\,p'^C}$ has a minimum value of almost -0.20. Such a strong coupling brought into question the decomposition of the pressure field into these two components. The figure also shows the correlation coefficient associated with $\overline{p'^r\,p'^s}$ which has values ranging between 0.05 and -0.09. This rather weak coupling does suggest, however, that a decomposition into rapid and slow parts is rational and potentially useful.

Nevertheless, both Zeman (1991a) and Sarkar (1992) used their analyses to formulate the following models for Π,

$$\overline{\rho}\Pi = \overline{p's'_{kk}} = \begin{cases} \alpha_{Zd}\left(\dfrac{\overline{p'^2}}{p_e^2} - 1\right)\overline{\rho}\varepsilon M_t + \alpha_{Zc}\dfrac{\overline{p'^2}}{\gamma\overline{p}}\widetilde{S}_{kk} & \text{Zeman (1991a)} \\[4mm] \alpha_{Sd}\overline{\rho}\varepsilon M_t^2 + \alpha_{Sc}\gamma\,\overline{p}\widetilde{S}_{kk}M_t^4 & \text{Sarkar (1992),} \end{cases}$$
$$\tag{5.65}$$

where α_{Zd}, α_{Zc} and α_{Sd}, α_{Sc} are closure coefficients, and p_e^2 is the pressure–variance equilibrium value. Since p_e^2 is proportional to $\overline{p}^2 M_t^4$, the second term of the Zeman model is consistent with the second term of the Sarkar model. Other terms can be added to the forms given in Eq. (5.65) to account for anisotropic effects (such as rapid one-dimensional compressions) as well as flow inhomogeneities (Sarkar, 1992; Zeman, 1991b; Zeman & Coleman, 1993). Taulbee and VanOsdol (1991) proposed a model for the combination $\overline{\rho}\Pi - \overline{\rho}\epsilon$. The model was dependent on both the density variance and mass flux so that modeled transport equations for the density variance and mass flux were included in the governing equation set. The full set of equations was then rather extensive and, owing to the uncertainty in closing the differential equations, required an extensive validation process to be generally applicable.

The cases of asymmetric (Durbin & Zeman, 1992) and axial (Jacquin, Cambon & Blin, 1993; Cambon et al., 1993) compressions were studied using RDT. In these studies, and for $M \ll 1$ (Durbin & Zeman, 1992) and for (initial) $M \gg 1$ (Jacquin, Cambon & Blin, 1993), the solenoidal fluctuating pressure contribution to the pressure–dilatation term dominated. For spherical compressions, the solenoidal pressure fluctuations are zero; whereas, the dilatational pressure fluctuations are unaltered by an increase in the dimensionality of the turbulence. Since such flows were amenable to theoretical analysis through RDT, it was possible to gain insight into the effects of a shock (rapid compressions) on the turbulence.

5.3.3.2 Pressure–Strain Rate

Although the studies just discussed were useful in understanding compressible flow dynamics, they still were not able to uniquely identify the causes for the decreased levels of turbulence in the important case of high speed mixing-layers. Sarkar (1995) (homogeneous shear flow) and Vreman et al. (1996) (mixing-layer) further focused on this question and were able to identify the reduction of pressure fluctuations as the source of the reduced mixing-layer (and uniform shear) growth rates. The gradient Mach number M_g was also identified (Sarkar, 1995) as an important parameter in assessing the relative compressibility influence on the flow. These reduced pressure fluctuations directly affect the pressure–strain rate correlation, reduce the transverse velocity intensity and transverse length scale (Freund et al., 2000a) which in turn reduces the dominant shear stress component. Vreman et al. (1996) showed how the corresponding reduction in turbulent kinetic energy production directly affected the mixing-layer growth rate. Later studies of the annular mixing layer (Freund et al., 2000a) and planar mixing-layer (Pantano & Sarkar, 2002) also found a reduction in pressure fluctuations and pressure–strain rate terms.

A more general analysis than the thermodynamically based analysis of the pressure–variance and the subsequent pressure–dilatation models discussed can be performed. The starting point is the fluctuating pressure Poisson equation given in Eq. (3.63). With this equation, the fluctuating density term $D^2\rho'/Dt^2$ can be taken either as a source term or under assumed isentropic conditions $(Dp/Dt = c^2 D\rho/Dt)$ used to recast the fluctuating pressure equation into a convective wave equation written as

$$\left[\frac{1}{c^2} \frac{D^2}{Dt^2} - \frac{\partial^2}{\partial x_i \partial x_i} \right] p' = 2 \frac{\partial \tilde{u}_i}{\partial x_j} \frac{\partial (\rho u_j'')}{\partial x_i} + \rho' \frac{\partial \tilde{u}_i}{\partial x_j} \frac{\partial \tilde{u}_j}{\partial x_i}$$

$$+ \frac{\partial^2}{\partial x_i \partial x_j} \left(\rho u_i'' u_j'' - \widetilde{\rho u_i'' u_j''} \right), \qquad (5.66)$$

where the operator D^2/Dt^2 is defined in Eq. (3.64), and mean dilatation and molecular effects are neglected so that the speed of sound, c, is assumed constant, and the flow is assumed to be homogeneous in mean shear. It is important to recognize that a fundamental difference between compressible and incompressible flow lies in the equations governing the pressure. In incompressible flow, the speed of sound is taken to be infinite and fluctuating pressure is governed by a Poisson equation (cf. Eq. (3.63) with constant density), and pressure fluctuations at any given point are determined by the mean and fluctuating velocity everywhere else in the flow. In compressible flow, the fluctuating pressure is, in general, governed by the convective wave equation, Eq. (5.66), and fluctuations propagate at the speed of sound relative to the medium. Papamoschou (1993) and Papamoschou and Lele (1993) initially studied the consequence of the wave operator on pressure fluctuations in a supersonic shear layer using ray theory (Papamoschou, 1993) and using DNS of an induced vortex disturbance (Papamoschou & Lele, 1993) to delineate dominant communication zones of influence. In the absence of solid boundaries, for example, the fluctuating pressure at a point P in the compressible shear layer is influenced only by the (backward) sonic cone from P. Within this sonic zone are points whose signal can reach P by propagating at the speed of sound. This scenario is consistent with the sonic eddy discussion in Section 5.1.2. An example of the anisotropic propagation characteristics introduced by the mean shear on the convective wave operator is sketched in Figure 5.7.

The right side of Eq. (3.63) (or of the simplified form Eq. (5.66)), presents a formidable challenge for any analysis when considered in its entirety, and the general solution of Eq. (5.66) then requires a knowledge of the associated Green's function. In order to effectively parameterize the problem, it is useful to introduce a set of dimensional scaling factors that can be used to highlight the

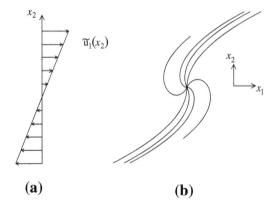

(a)   **(b)**

FIGURE 5.7 Sketch of: (a) homogeneous shear mean velocity profile; and (b) ray trajectories in homogeneous shear flow ($M_g = 1$). Rays emanating from source at initial angular spacings of $\pi/4$. From Thacker et al. (2007) with permission.

dominant dynamics. Consider, for example, a high Reynolds number flow in the absence of solid boundaries so the viscous contribution can be neglected. Let the fluctuating fields be scaled by a characteristic length scale l_e (the transverse integral scale of the turbulence), and velocity scale u_e with corresponding time scale l_e/u_e, and the thermodynamic variables scaled with the characteristic density and pressure scales ρ_e and $\rho_e u_e^2$, respectively. In addition, let the mean velocity field be scaled by a characteristic velocity difference across the length scale l_e which in a flow with characteristic mean velocity gradient S is Sl_e/c. With these scalings, Eq. (5.66) can then be rewritten as

$$\left\{ M_t^2 \left[\frac{\partial}{\partial t} + \tilde{u}_j \left(\frac{M_g}{M_t} \right) \frac{\partial}{\partial x_j} \right]^2 - \frac{\partial^2}{\partial x_j \partial x_j} \right\} p'$$

$$= 2 \left(\frac{M_g}{M_t} \right) \left\{ \left[\frac{\partial \tilde{u}_i}{\partial x_j} \frac{\partial (\rho u_j'')}{\partial x_i} \right] + \left(\frac{M_g}{M_t} \right) \rho' \left[\frac{\partial \tilde{u}_i}{\partial x_j} \frac{\partial \tilde{u}_j}{\partial x_i} \right] \right\}$$

$$+ \frac{\partial^2}{\partial x_i \partial x_j} \left[\overline{\rho} (u_i'' u_j'' - \widetilde{u_i'' u_j''}) + \rho' u_i'' u_j'' \right], \tag{5.67}$$

where the turbulent Mach number M_t is u_e/c and the gradient Mach number M_g is Sl_e/c. (Note that the notation used in Eq. (5.67) is the same as in Eq. (5.66) in order to avoid a cumbersome introduction of new variable definitions.) As is commonly done in the incompressible case, the pressure field can be decomposed into rapid and slow parts (e.g. Pope, 2000; Durbin & Pettersson-Reif, 2001), where the rapid part is associated with terms involving the mean velocity gradient and the slow part involves terms with turbulence–turbulence interactions. As discussed previously, in the compressible case this partitioning can also be employed to aid in the modeling of correlations involving the fluctuating pressure as well as its decomposition into incompressible and compressible parts (Blaisdell & Sarkar, 1993; Sarkar, 1992; Thacker, Sarkar, & Gatski, 2007). The last term on the right in Eq. (5.67) is clearly associated with the slow part, and the remaining terms on the right are associated with the rapid part. For small turbulent Mach number ($M_t \ll 1$), the dynamics in both the rapid and slow cases are governed by a Poisson equation. The relative influence of shear is then further delimited by the ratio M_g/M_t. For cases where the ratio M_g/M_t is also small (such as in wall-bounded flows), only the first term on the right remains.

With a knowledge of the pressure fluctuations, it is possible to formally construct the correlations involving the fluctuating pressure field, such as the pressure–strain rate correlation, defined in Eq. (5.24b). The shear-free Green's function was used to study the behavior of the pressure–strain rate correlation in the vicinity of the center of a free shear layer (Pantano & Sarkar, 2002), and the effects of homogeneous shear on the behavior of the pressure–strain rate correlation was later studied (Thacker et al., 2007). An outline of the procedure to follow is analogous to the incompressible case (Kim, 1989).

The case of a homogeneous mean shear is considered. The solution for the fluctuating pressure field is formally written in terms of a Green's function

$$p'(\mathbf{x},t) = \int_0^t dt' \int d^3\mathbf{x}' G(\mathbf{x} - \mathbf{x}', t - t') f(\mathbf{x}',t'), \qquad (5.68a)$$

or in wave-vector space by

$$\hat{p}(\mathbf{k},t) = \int_0^t dt' \widehat{G}(\mathbf{k}, t - t') \hat{f}(\mathbf{k},t'), \qquad (5.68b)$$

where $f(\mathbf{x},t)$ (or $\hat{f}(\mathbf{k},t)$) represents the source term on the right side of Eq. (5.67). This Green's function is defined by the Fourier transform

$$G(\mathbf{x},t) = \frac{1}{(2\pi)^3} \int d^3\mathbf{k} e^{\iota\varkappa(\mathbf{k},t)\cdot\mathbf{x}} \widetilde{G}(\mathbf{k},t), \qquad (5.69)$$

where $\varkappa(\mathbf{k},t)$ is a moving wave-vector and $\widetilde{G}(\mathbf{k},t)$ is the solution of

$$\left(M_t^2 \frac{\partial^2}{\partial t^2} + \varkappa^2 \right) \widetilde{G}(\mathbf{k},t) = \delta(t), \qquad (5.70)$$

with $\varkappa^2 = \varkappa \cdot \varkappa$ and $\varkappa$ given by (Thacker et al., 2007)

$$\varkappa(\mathbf{k},t) = \mathbf{k} e^{-(M_g/M_t)\lambda t}, \qquad (5.71a)$$

or in component form as

$$\left[\varkappa_1, \varkappa_2, \varkappa_3 \right] = \left[k_1, k_2 - \frac{M_g}{M_t} k_1 t, k_3 \right]. \qquad (5.71b)$$

(Note that the matrix exponential is represented in terms of a low-order series expansion.) The type of ray trajectories corresponding to this moving wave-vector pattern was shown previously in Figure 5.7. In the limiting case of shear-free flow ($M_g = 0, \varkappa_i \varkappa_i = k_i k_i$) (Pantano & Sarkar, 2002) the Green's function $\widetilde{G}_{sf}(\mathbf{k},t)$ is given by

$$\widetilde{G}_{sf}(k,t) = \frac{\mathcal{H}(t)}{M_t k} \sin(kt/M_t), \qquad (5.72)$$

where $k^2 = k_i k_i$, and $\mathcal{H}(t)$ is the Heaviside function. For the shear flow case, the moving Green's function solution to Eq. (5.70), $\widetilde{G}(\mathbf{k},t)$, can be obtained (see for details Thacker et al., 2007).

The rapid part of the pressure–strain rate correlation can be represented in wave-vector space by the spectral function,

$$\begin{aligned} \Psi_{ij}(\mathbf{k},t) &= \frac{1}{(2\pi)^3} \int_{\mathscr{V}} \Pi_{ij}(\mathbf{r},t) e^{-\iota\mathbf{k}\cdot\mathbf{r}} d^3\mathbf{r} \\ &= -\frac{\iota(2\pi)^3}{\mathscr{V}} \langle \hat{p}(\mathbf{k},t) \left(k_j \hat{u}_i(-\mathbf{k},t) + k_i \hat{u}_j(-\mathbf{k},t) \right) \rangle, \end{aligned} \qquad (5.73)$$

where $\langle \cdot \rangle$ represents an ensemble mean, and $\Pi_{ij}(\mathbf{r} = \mathbf{0}, t)$ is the single-point pressure–strain rate correlation given in Eq. (5.24b). With the pressure mode $\hat{p}(\mathbf{k}, t)$ determined from Eq. (5.68b), the pressure–strain rate spectrum $\Psi_{ij}(\mathbf{k}, t)$ for this homogeneous shear example is then given by

$$
\begin{aligned}
\Psi_{ij}(\mathbf{k}, t) = {} & 2\langle \rho \rangle \lambda_{lm} \left(\frac{M_g}{M_t} \right) \left[\frac{(2\pi)^3}{\mathcal{V}} \right] \int_0^t dt' \widehat{G}(\mathbf{k}, t - t') \\
& \times \left[k_l k_j \langle \hat{u}_m(\mathbf{k}, t') \hat{u}_i(-\mathbf{k}, t) \rangle + k_l k_i \langle \hat{u}_m(\mathbf{k}, t') \hat{u}_j(-\mathbf{k}, t) \rangle \right],
\end{aligned}
\tag{5.74}
$$

with the ratio M_g / M_t assumed small, and the term $\langle \rho' u_i'' \rangle /$ vanishing (Blaisdell et al., 1993) for the homogeneous turbulence. The function $\widehat{G}(\mathbf{k}, t)$ defined by Eq. (5.68b) is the fixed wave-vector Green's function required for the specification of the pressure field and is directly related to $\widetilde{G}(\mathbf{k}, t)$ (see Thacker et al., 2007).

Analogous to the incompressible case, an energy spectrum tensor, E_{ij}, can be introduced,

$$
\langle \hat{u}_i(\mathbf{k}', t') \hat{u}_j(\mathbf{k}, t) \rangle = e^{-(t-t')/\tau_I} \delta^3(\mathbf{k}' + \mathbf{k}) E_{ij}(\mathbf{k}, t),
\tag{5.75}
$$

where a temporal decorrelation function (τ_I an isotropic decorrelation time) is necessarily introduced (Pantano & Sarkar, 2002; Thacker et al., 2007) to account for the $t, t' (<t)$ two time dependency introduced at the outset in the Green's function (see Eq. (5.68a)). In the incompressible case, the decorrelation time $\tau_I \to \infty$ since the speed of sound is infinite, and the decorrelation function becomes unity.

If it is now assumed that a statistically steady (stationary) turbulence is reached after a long time, the rapid pressure–strain rate spectrum tensor $\Psi_{ij}(\mathbf{k}, t)$ can be rewritten as

$$
\Psi_{ij}(\mathbf{k}) = 2\overline{\rho} \left(\frac{M_g}{M_t} \right) \mathcal{G}(\mathbf{k}; \tau_I) \lambda_{lm} k_l \left[E_{mi}(\mathbf{k}) k_j + E_{mj}(\mathbf{k}) k_i \right],
\tag{5.76a}
$$

in the limit as $t \to \infty$, with

$$
\mathcal{G}(\mathbf{k}; \tau_I) = \int_0^\infty \widehat{G}(\mathbf{k}, t - t') e^{-(t-t')/\tau_I} dt'.
\tag{5.76b}
$$

The rapid part of the single-point ($\mathbf{r} = 0$) pressure–strain rate correlation Π_{ij} can be obtained directly from the spectrum tensor $\Psi_{ij}(\mathbf{k})$ by integrating over all wave-vectors yielding

$$
\begin{aligned}
\Pi_{ij} = {} & 2\overline{\rho} \left(\frac{M_g}{M_t} \right) \lambda_{ln} \int_0^\infty dk \int_0^\pi \sin\theta \, d\theta \int_0^{2\pi} d\phi \, R_c(k, \theta, \phi; \tau_I) \\
& \times k_l \left[E_{ni}(k, \theta, \phi) k_j + E_{nj}(k, \theta, \phi) k_i \right]
\end{aligned}
$$

$$= 2\bar{\rho} \left(\frac{M_g}{M_t} \right) R(k,\theta,\phi; \tau_I)\lambda_{ln} \left(M_{inlj} + M_{jnli} \right), \tag{5.77}$$

with

$$R_c(k,\theta,\phi; \tau_I) = k^2 \mathcal{G}(k,\theta,\phi; \tau_I). \tag{5.78}$$

Aside from the scaling factor M_g/M_t and the compressible, anisotropy factor $R_c(k,\theta,\phi; \tau_I)$, the functional form in Eq. (5.77) is the same as in the incompressible case (cf. Shih, Reynolds, & Mansour, 1990). For incompressible flows, the factor $R_c(k,\theta,\phi) = 1$, and for shear-free compressible flows (Pantano & Sarkar, 2002) it is given by

$$R_c(k,\theta,\phi; \tau_I) = \left(1 + \frac{M_t^2}{\tau_I^2 k^2} \right)^{-1}, \tag{5.79}$$

which is isotropic but does exhibit a decaying amplitude behavior with wavenumber. The angular and radial (k_i) dependency shown in Eq. (5.76b) then highlights both the directional anisotropy and wavenumber amplitude variation introduced in the presence of mean shear.

In order to complete the representation of the pressure–strain rate correlation, a suitable representation for the energy spectrum tensor E_{ij} is needed. For incompressible flows, a four-term representation (Shih et al., 1990) can be written in the form

$$E_{ij}(\mathbf{k}; \mathbf{b}) = \sum_{n=1}^{4} \psi_n \Phi_{ij}^{(n)}, \tag{5.80a}$$

with

$$\begin{aligned} \Phi_{ij}^{(1)} &= \delta_{ij} & \Phi_{ij}^{(3)} &= b_{ij} \\ \Phi_{ij}^{(2)} &= \frac{k_i k_j}{k^2} & \Phi_{ij}^{(4)} &= \frac{b_{ir} k_r k_j + b_{jr} k_r k_i}{k^2}, \end{aligned} \tag{5.80b}$$

where the expansion coefficients ψ_n can, in general, be functions of the invariants formed from the bases $\Phi_{ij}^{(n)}$. In the incompressible case, these expansion coefficients can be associated with the isotropic energy spectral density with any spectral anisotropies dependent on those associated with the basis tensors.

Unlike the incompressible case, but as suggested by the ray trajectories in Figure 5.7, spatial anisotropies are introduced into the signal propagation associated with the fluctuating pressure field. This anisotropy is directly related to the Green's function of the convective wave equation, Eq. (5.67), and is dependent on both the gradient and turbulent Mach numbers through $R(k,\theta,\phi; \tau_I)$. Even though the tensor polynomial representation given in Eq. (5.80a) still applies, the corresponding expansion coefficients required in Eq. (5.80a) are more complex and this is reflected in the need for the specification of an anisotropic energy spectral density in addition to the isotropic spectral energy density

(Thacker et al., 2007). If properly specified, however, such a representation could be used in the formulation of a pressure–strain rate model as can be done in the incompressible case (see Shih et al., 1990). To date, this has not been done and the analytical challenges may preclude a unique, well-posed form in physical space.

Cambon and Jacquin (2006) have also recently examined the issue of the stabilizing effect of compressibility on the turbulence dynamics. This directly relates to the pressure fields studies since the ultimate effect is the potential reduction of the turbulence energetics in influencing the flow (such as the mixing-layer growth rate). Their analysis suggests that the alteration of the pressure field is more importantly felt through a degradation of the nonlinearity of the system in contrast to the anisotropic wave propagation as suggested by the analysis here. Future developments will hopefully lead to a truly compressible (rather than a variable density extended) pressure–strain rate model.

5.3.4 Scalar Fluxes and Variances

The remaining terms to complete closure of the full turbulent stress and stress anisotropy transport equations involve the fluctuations of the thermodynamic variables density (mass flux), temperature (heat flux), and pressure (pressure–velocity correlation). As such, any consistent closure strategy should account for the relations and approximations relating these variables. For example, assuming an equation of state for an ideal gas, and the fact that $\overline{p'u_i''} = \overline{p'u_i'}$ results in a consistency relation given by,

$$\frac{\overline{p'u_i'}}{\overline{p}} = \frac{\overline{u_i''T''}}{\widetilde{T}} + \frac{\overline{\rho'u_i'}}{\overline{\rho}}. \tag{5.81}$$

Recall that for homogeneous turbulence (see Section 5.1), it was pointed out for $M \ll 1$, the acoustic mode is decoupled from the vortical mode (Durbin & Zeman, 1992), and for $M \gg 1$ (Jacquin, Cambon, & Blin, 1993) both the acoustic and turbulent time scales are large compared to the mean distortion so that pressure-velocity correlations can possibly be neglected. In addition, the mass-flux term vanishes, so that in the absence of any mean temperature gradients, the heat flux term also vanishes.

Since the pressure–velocity correlation term need only be considered for second-moment closure modeling, it is often neglected or assimilated into a model for the full turbulent transport term (the $d_{ij}^{(t-\mu)}$ term in Eq. (5.34)). If modeling were required, it could be constructed in a manner analogous to the pressure–strain rate correlation; although, some account would need to be taken of the contribution to the slow part of the fluctuating pressure $\left[\overline{\rho}(u_i''u_j'' - \widetilde{u_i''u_j''}) + \rho'u_i''u_j''\right]_{ij}$ in Eq. (5.67). Transport equations for the heat and mass flux can be derived from the fluctuation equations given in Section 3.4 with subsequent closure. An excellent analysis, discussion and review of scalar flux modeling, including both the heat flux and mass flux correlations, is given by Chassaing (2001).

5.3.4.1 Heat Flux and Temperature Variance

Unlike the incompressible case, where except for buoyancy effects, the mean momentum equation is decoupled from the mean energy equation with the velocity field independent of the temperature field, a direct coupling exists in the compressible case that requires a more careful examination of the turbulent heat flux modeling.

The transport equation for the heat flux vector $\widetilde{u_i''T''}$ for a perfect gas, can be extracted from the mass conservation, and fluctuating velocity and temperature equations given in Eqs. (3.57), (3.59) and (3.61) (or (3.62)). The resulting form is then

$$\bar{\rho}\frac{D\widetilde{u_i''T''}}{Dt} = \bar{\rho}P_{Ti}^1 + \bar{\rho}P_{Ti}^2 + \bar{\rho}\Phi_{Ti} - \bar{\rho}\varepsilon_{Ti} + \bar{\rho}D_{Ti} + \bar{\rho}C_{Ti}, \qquad (5.82)$$

where

$$\bar{\rho}P_{Ti}^1 = -\bar{\rho}\tau_{ik}\frac{\partial\widetilde{T}}{\partial x_k} - \bar{\rho}\widetilde{u_k''T''}\frac{\partial\widetilde{u}_i}{\partial x_k}$$
$$-\left(\frac{\gamma}{c_p}\right)\left[\widetilde{pu_i''\widetilde{S}_{kk}} - \widetilde{u_i''\sigma_{jk}'\widetilde{S}_{kj}}\right] \qquad (5.83a)$$

is the production due to the mean temperature and velocity gradients,

$$\bar{\rho}P_{Ti}^2 = \left(\frac{\gamma}{c_p}\right)\left[\overline{u_i''\sigma_{jk}'s_{kj}''} - \overline{u_i''p's_{kk}''}\right] \qquad (5.83b)$$

is the production due to the fluctuating strain rate field,

$$\bar{\rho}\Phi_{Ti} = -\overline{T''\frac{\partial p}{\partial x_i}} \qquad (5.83c)$$

is the "pressure-scrambling" term,

$$\bar{\rho}\varepsilon_{Ti} = \left(\frac{\gamma}{c_p}\right)\left[\bar{k}_T\overline{\frac{\partial u_i''}{\partial x_k}\frac{\partial T'}{\partial x_k}} + k_T'\overline{\frac{\partial u_i''}{\partial x_k}\frac{\partial\overline{T}}{\partial x_k}} + k_T'\overline{\frac{\partial u_i''}{\partial x_k}\frac{\partial T'}{\partial x_k}}\right]$$
$$+ \left[\overline{\sigma_{ik}'\frac{\partial T''}{\partial x_k}}\right] \qquad (5.83d)$$

is the turbulent viscous-thermal dissipation term,

$$\overline{\rho}D_{Ti} = \left(\frac{\gamma}{c_p}\right)\frac{\partial}{\partial x_k}\left[\overline{k_T u_i''\frac{\partial T'}{\partial x_k}} + \overline{k_T' u_i''\frac{\partial \overline{T}}{\partial x_k}} + \overline{k_T' u_i''\frac{\partial T'}{\partial x_k}}\right]$$
$$-\frac{\partial}{\partial x_k}\left[\overline{\rho u_i''T''u_k''} - \overline{\sigma_{ik}'T''}\right] \tag{5.83e}$$

is the turbulent and viscous-thermal transport contribution, and

$$\overline{\rho}C_{Ti} = \overline{T''\frac{\partial \overline{\sigma}_{ik}}{\partial x_k}} + \left(\frac{\gamma}{c_p}\right)\left[\overline{u_i''\sigma_{jk}}\tilde{S}_{kj} + \overline{u_i''s_{kj}''}\left(\overline{\sigma}_{kj} - \overline{p}\delta_{kj}\right)\right] \tag{5.83f}$$

are compressibility terms. As is readily seen, additional unknown higher-order correlations appear that need to be parameterized. For the most part in the compressible case, heat transfer models for terms such as the pressure-scrambling and viscous-thermal dissipation, $\Phi_{Ti} - \varepsilon_{Ti}$, and the viscous-thermal transport D_{Ti}, have been variable density extensions of their incompressible counterparts. In addition, in engineering calculations gradient transport models for the heat flux are most prevalent and these are, in turn, also variable density extensions of their incompressible counterparts. As a matter of practice, the heat flux equation itself is rarely if ever solved in conjunction with the mean momentum and turbulent velocity closures in a RANS closure. Explicit polynomial representations of the heat flux vector are most often used and even at this level, the majority of representations used are based on a one- or two-term isotropic or anisotropic eddy diffusivity model. In the isotropic case the diffusivity field is proportional to the turbulent eddy viscosity coupled with a turbulent Prandtl number, and in the anisotropic case, the eddy viscosity is replaced with some factor associated with the Reynolds stress tensor. In both instances, the dependency is proportional to the mean temperature gradient.

The turbulent Prandtl number in the isotropic representation for the heat flux vector has often been assumed fixed ($Pr_t = 0.9$). As is well-known in wall-bounded flows, this assumption is not generally valid across the entire boundary layer, since in close proximity to the wall, the molecular Prandtl number should be the relevant parameter. It can be desirable to then replace Pr_t with a mixed Prandtl number, Pr_m, that takes into account both the molecular and turbulent diffusivities (see Eq. (6.43)). This will be further discussed in Section 6.4.4. In addition to a parameter such as Pr_t (or Pr_m), the turbulent time scale ratio between the thermal and velocity fields is often introduced and provides a further quantification of the coupling between these fields (see Section 6.4.4).

For the anisotropic representation of the heat flux vector, the development is analogous to the turbulent Reynolds stress, where it was optimal to pose the governing equations in terms of a scaled variable, that is, the anisotropy tensor b_{ij}. A similar procedure can be followed for the heat flux vector in order to identify a proper normalization. The approach commonly followed was proposed by Shih and Lumley (1993) (see also; Shih, 1996) and has been used in previous algebraic representations of scaled heat flux models by Abe, Kondoh, and Nagano (1996) and Wikström, Wallin, and Johansson (2000). The starting point is the formation of an equation for the corresponding correlation coefficient, that is

$$\zeta_i = \frac{\overline{u_i'' T''}}{\sqrt{K K_T}}. \tag{5.84}$$

This transport equation for ζ_i would then be related to the corresponding heat flux derivative by

$$\frac{D\zeta_i}{Dt} = \frac{1}{\sqrt{K K_T}} \frac{D\overline{u_i'' T''}}{Dt} - \frac{\zeta_i}{2} \left[\frac{1}{K} \frac{DK}{Dt} + \frac{1}{K_T} \frac{DK_T}{Dt} \right]. \tag{5.85}$$

In order to complete the system of equations, a (closed) equation is needed for the temperature variance K_T.

Vector algebraic representations can be extracted from the transport equation derived from Eq. (5.85) through the imposition of equilibrium conditions on the variable ζ_i. These equilibrium conditions reduce the differential transport equations to a set of implicit algebraic equations for the vector ζ_i. Once extracted, however, an explicit representation for ζ_i in terms of the mean temperature gradients can be obtained.

The temperature variance equation can be derived directly from Eq. (3.61) and can be written as (cf. Favre, 1976, Eq. (3.121) B)

$$\overline{\rho} \frac{D K_T}{Dt} = -\overline{\rho u_k'' T''} \frac{\partial \tilde{T}}{\partial x_k} - \frac{\partial}{\partial x_k} \left(\frac{\overline{\rho u_k'' T''^2}}{2} \right) + \gamma \overline{T''} \frac{\partial}{\partial x_k} \left(\frac{\overline{k_T}}{c_p} \frac{\partial \overline{T}}{\partial x_k} \right)$$
$$+ \overline{\rho} D_T - \overline{\rho} \varepsilon_T + \overline{\rho} C_T, \tag{5.86}$$

where the first three terms on the right are contributions due to thermal production, turbulent velocity transport, and mean thermal conduction, respectively, (constant specific heats are assumed throughout) with

$$\overline{\rho} D_T = \left(\frac{\gamma}{c_p} \right) \frac{\partial}{\partial x_j} \left[\overline{k_T T''} \frac{\partial T'}{\partial x_j} + \overline{k_T' T''} \frac{\partial \overline{T}}{\partial x_j} + \overline{k_T' T''} \frac{\partial T'}{\partial x_j} \right], \tag{5.87a}$$

thermal diffusion,

$$\overline{\rho} \varepsilon_T = \left(\frac{\gamma}{c_p} \right) \left[\overline{k_T} \frac{\overline{\partial T'' \partial T'}}{\partial x_j \partial x_j} + \overline{k_T' \frac{\partial T'' \partial \overline{T}}{\partial x_j \partial x_j}} + \overline{k_T' \frac{\partial T'' \partial T'}{\partial x_j \partial x_j}} \right], \tag{5.87b}$$

thermal dissipation rate,

$$\overline{\rho} C_T = -\left(\frac{\gamma}{c_p} \right) \left[\overline{T'' \left(p \frac{\partial u_k}{\partial x_k} - \sigma_{ik} \frac{\partial u_i}{\partial x_k} \right)} \right], \tag{5.87c}$$

and contributions due to pressure–dilatation and viscous dissipation, respectively. As with the turbulent kinetic energy equation, the dissipation of temperature variance ε_T can, in general, be determined from a modeled transport equation. For the incompressible case, an assessment of a variety of ε_T transport equations along with related references is given in Zhao, So, and Gatski (2001) and Nagano (2002). It should be recognized that in Eq. (5.87c) instantaneous variables are used, and these need to be fully partitioned into their respective mean and fluctuating parts, either through a Reynolds or Favre decomposition, and including the dynamic viscosity μ (see Tamano & Morinishi, 2006).

5.3.4.2 Mass Flux and Density Variance

There have been attempts in the past to model the $\overline{\rho' u_i'}$ correlation but none of the models have been extensively tested in compressible flows (see for a brief historical survey Rubesin, 1990), and no dominant form has emerged. Zeman and Coleman (1993), in their studies on compressed turbulence concluded that a significant contributor to the energetic balance in the presence of shocks was an acceleration term associated with the product of the mass flux and mean pressure gradient, and subsequently proposed a transport equation that represented the evolution of the mass flux through a relaxation term driven by both mean density and mean velocity gradients. It is worth noting that this conclusion about the importance of the mass flux term is contrasted by the results shown in Figure 5.2 for the invariants of the turbulent stress anisotropy tensor. Such differences are not uncommon in analyzing compressible flows, and is an indicator of the varying impact compressibility has on the turbulence itself. In this case, the difference may lie in the fact that at the Mach number of the results shown in Figure 5.2 the difference between Favre and Reynolds stresses was minimal implying a rather limited role for the mass flux variable. However at higher Mach numbers, Figure 3.4 suggests that mass flux effects may play a more important dynamic role.

The mass flux $\overline{\rho' u_i'}$ and average fluctuating velocity $\overline{u_i''}$, which often appear in the governing compressible transport equations, are related through the defining relation

$$\overline{\rho' u_i'} = \overline{\rho' u_i''} = \overline{\rho}\, \widetilde{u_i'} = -\overline{\rho}\, \overline{u_i''}. \tag{5.88}$$

It is then, of course, possible to formulate a transport equation for the mass flux using a combination of the fluctuating density equation and the equation for the Reynolds fluctuating variable, or using the Favre fluctuating momentum equation directly and then the mean density equation. The latter alternative is the most direct from the prior presentation of fluctuating transport equations in Section 3.4. The transport equation for the mass flux is then obtained by averaging the fluctuating momentum equation, Eq. (3.59), and applying the mean density equation in Eq. (3.30a). This yields

$$\frac{\partial(\overline{\rho'u_i'})}{\partial t} + \frac{\partial}{\partial x_k}\left(\tilde{u}_k\overline{\rho'u_i'}\right) = P_{\rho i}^1 + P_{\rho i}^2 + \Phi_{\rho i} - \varepsilon_{\rho i} + D_{\rho i}, \qquad (5.89)$$

where

$$P_{\rho i}^1 = \overline{\rho'u_k'}\frac{\partial\tilde{u}_i}{\partial x_k} - \tau_{ik}\frac{\partial\overline{\rho}}{\partial x_k} + \left(\frac{\overline{\rho'^2}}{\overline{\rho}^2}\right)\left(\frac{\partial\overline{p}}{\partial x_i} - \frac{\partial\overline{\sigma}_{ik}}{\partial x_k}\right) \qquad (5.90a)$$

is the production due to the mean field gradients,

$$P_{\rho i}^2 = -\overline{\rho}\overline{u_i''s_{kk}''} \qquad (5.90b)$$

is the production due to the fluctuating dilatation,

$$\Phi_{\rho i} = \frac{\overline{p'\frac{\partial\rho'}{\partial x_i}}}{\overline{\rho}} \qquad (5.90c)$$

is the pressure-scrambling term,

$$\varepsilon_{\rho i} = \frac{\overline{\sigma_{ik}'\frac{\partial\rho'}{\partial x_k}}}{\overline{\rho}} \qquad (5.90d)$$

is the fluctuating density dissipation rate, and

$$D_{\rho i} = -\overline{\rho}\frac{\partial}{\partial x_k}\left(\frac{\overline{\rho'u_k''u_i''}}{\overline{\rho}}\right) - \frac{1}{\overline{\rho}}\frac{\partial}{\partial x_k}\left[\overline{p'\rho'}\delta_{ik} - \overline{\rho'\sigma_{ik}'}\right] \qquad (5.90e)$$

is the turbulent transport term. Note that the factor $1/\rho$ appearing with pressure gradient terms has been expanded in the (truncated) series $1/\rho \approx (1/\overline{\rho})$ $(1 - \rho'/\overline{\rho} + \rho'^2/\overline{\rho}^2)$ so that its dynamic role in the equations could be explicitly shown (cf. Grégoire, Souffland, & Gauthier, 2005). In addition, the fluctuating viscous stress that appears in the above terms needs to be expanded carefully and should, in general, include fluctuations to the dynamic viscosity just as the fluctuations in the thermal conductivity were included in the thermal dissipation rate. There are a variety of alternate forms this equation can take (e.g. Taulbee & VanOsdol, 1991; Yoshizawa, 2003); however, the form presented in Eq. (5.89) minimizes the higher-order correlations that need to be closed. A transport equation for the average fluctuating velocity $\overline{u_i''}$ was also formulated (Taulbee & VanOsdol, 1991), and provided a closure model for the higher-order correlations. As in the equation here, the density variance appears in the transport equation, so that an additional modeled transport equation for the density variance is needed, which is easily derived from Eq. (3.58) and given by

$$\frac{\partial K_\rho}{\partial t} + \frac{\partial}{\partial x_k}\left(\tilde{u}_k K_\rho\right) = -\overline{\rho'u_k'}\frac{\partial\overline{\rho}}{\partial x_k} - 2K_\rho\tilde{S}_{kk}$$

$$-\frac{1}{2}\frac{\partial\overline{\rho'^2u_k''}}{\partial x_k} - \frac{\overline{\rho'^2s_{kk}''}}{2} + \overline{\rho}\,\overline{\rho's_{kk}''}, \qquad (5.91)$$

where $K_\rho = \overline{\rho'^2}/2$ has been defined. The first two terms on the right are production terms, the third is a diffusion term and the remaining two are compressible destruction terms. Using low Reynolds number and Mach number direct simulations, an *a priori* assessment of the various terms in the density variance budget was performed (Gerolymos, Sénéchal, & Vallet, 2007; Gerolymos, Sénéchal, & Vallet, 2008). Not surprisingly, little variation in density variance was found over the two Mach numbers examined (0.3 and 1.5) (Gerolymos et al., 2007). The corresponding budgets (Gerolymos et al., 2008) showed the usual dominant balance between production and destruction over most of the flow with diffusion effects contributing in the sublayer region.

Rubesin (1990) also considered the average fluctuating velocity due to its presence in the turbulent kinetic energy equation and possible role in the dynamics. However, rather than starting from a transport equation, the starting point was an assumption that the density and pressure fluctuations were related through a polytropic gas law. The fluctuating enthalpy was then related to the average fluctuating velocity through assumptions consistent with the SRA discussed in Section 3.5. The result was a gradient diffusion model with a tensor diffusivity coefficient related to the turbulent stress field. A low-order truncation of the transport equation for the average fluctuating velocity without the pressure and viscous forces has been considered (Adumitroaie, Ristorcelli, & Taulbee, 1999; Ristorcelli, 1993). A weak equilibrium condition was applied of the form $D\widetilde{b}_{ij}/Dt = 0$ and an algebraic representation obtained.

In addition to the review of Adumitroaie et al. (1999), Yoshizawa (2003) also considered the mass flux term, but in the context of turbulent combustion and the issue of counter-gradient diffusion of heat. While combustion is outside of the topical scope here, his analysis is worth noting in that it focused on the derived variable $\overline{\rho u_i''}/\rho$ and its transport equation.

Lastly, from a straightforward analysis involving only the equation of state (Chassaing, 2001), it was found that the sign of the density correlation with fluctuating quantities, such as temperature, may depend only on the corresponding mean field. In Figure 5.8, a sketch of different scenarios for co- and counter-gradient diffusion is shown. Such a figure suggests no simple diffusion scheme applicable for such flows. Within the context of a second-moment closure, a modeled transport equation for the one-dimensional mixing-flow induced by a Richtmyer–Meshkov instability can be solved (Grégoire et al., 2005). Although these are noteworthy examples, and clearly indicate the potential of successfully using such mass flux modeling, many studies simply neglect the effect of this term—even when shocks are present and the term may have a significant impact.

5.3.5 Other Closure Issues

It is clear from the previous discussion of closure models, that implementing a solution based on an averaged or RANS based method significantly increases the

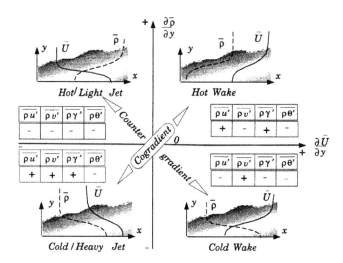

FIGURE 5.8 Signs of density fluctuation correlations in free shear flows. From Chassaing (2001).

modeling burden relative to a filtered approach such as LES. In addition, even with the significant increase in computer power, there is a continued reluctance to solve additional transport equations for the higher-order tensor and vector correlations, such as the Reynolds stresses and the scalar fluxes. Thus, the two-equation and even one-equation formulations are still often used (e.g. Roy & Blottner, 2006). As the interested reader will find, enumerable modeling issues arise in the development of closure models and it is not the intent here to delve into these. In the remainder of this section, however, two topics of relevance will be briefly focused on. These two topics are the correlation model representations and wall-proximity effects.

5.3.5.1 Polynomial Representations for Second-Moments and Scalar Fluxes

Although model development using polynomial representations has focused, almost exclusively, on the incompressible regime and has only been applied to the compressible regime through variable density extensions, one should recall that for the most part the lower order (isotropic eddy viscosity) models are usually altered to replicate an observed phenomena rather than to replicate the underlying physics. As such the models do indeed "work," but identifying their range of applicability can be difficult.

In order to both increase the appeal of higher-order closures and to fulfill the need to account for anisotropic effects, explicit algebraic Reynolds stress (EASM or EARSM), and heat flux (EAHFM) models or representations have been developed. (Though not discussed, mass flux models can also be developed in a manner analogous to the heat flux models, but this has not been investigated extensively.) This is in addition to the utilization of representations for the pressure–strain rate correlation or even for the spectral energy density tensor mentioned in Section 5.3.3.2.

The tensor and vector representations, that are often used in forming models of the Reynolds stress (anisotropy) tensor and vector scalar (heat or mass) fluxes, range from the popular gradient transport models associated with a linear isotropic turbulent eddy viscosity and/or eddy diffusivity to those associated with nonlinear anisotropic forms. Both the eddy viscosity and eddy diffusivity are modeled through a proportionality with a turbulent velocity scale and turbulent velocity and thermal time scales.

The predictive capability of such models can be potentially augmented by going to higher-order representations that effectively include more dynamics. This could mean the inclusion of anisotropic effects into the eddy viscosity and eddy diffusivity coefficients through a dependency on either a turbulent stress tensor or flux vector. The procedure is to extract tensor and vector representations with expansion coefficients that are directly related to the original set of governing equations. In this context, the tensor or vector bases used in the representations for the incompressible regime can also be used in the compressible regime; however, it will be the associated expansion coefficients that are altered between the two regimes.

This procedure can be divided into three tasks. The first task involves the imposition of suitable "equilibrium conditions" on the relevant governing transport equations for the turbulent stresses and the scalar fluxes which then yields a system of implicit algebraic equations. As was the case for the turbulent Reynolds stress, it was optimal to pose the governing equations in terms of a scaled variable, that is, the anisotropy tensor b_{ij}. A similar procedure can be followed for the heat flux (and mass flux) vectors. The scaled variable for the heat flux vector was given in Eq. (5.85) and was based on proposals by Shih and Lumley (1993) and Shih (1996) and has been used in previous algebraic representations of scaled heat flux models (Abe et al., 1996; Wikström et al., 2000). Since these equations involve tensor and vector quantities associated with the mean flow field, the second task is to construct tensor and vector bases from the mean variables. A general procedure can be established where a polynomial expansion, founded on these tensor and vector dependencies, can be constructed (Gatski, 2009). The third and final task in the construction of explicit representations is the determination of the respective expansion coefficients needed in the polynomial representation. An approach that can be used, for example, is a Galerkin-based projection methodology. The projection method allows for the use of incomplete sets of basis tensors and vectors, and since the projection method produces scalar products of the associated tensors and vectors the method is independent of the coordinate frame. Thus, a system of equivalent scalar equations is constructed.

5.3.5.2 Wall Proximity Effects

Irrespective of whether a differential or algebraic representation approach is used for closure, it is necessary to account for the presence of any solid boundaries in developing closure models. The usual calibration procedures for closure models are almost always based on equilibrium turbulence dynamics. As just discussed in the context of the polynomial algebraic representations, this equilibrium is not necessarily an energetic equilibrium (e.g. $DK/Dt = 0$), but can also be a structural equilibrium that requires a state variable, such as the turbulent stress anisotropy, or scaled heat flux vector, to remain constant over some region of the flow. With solid boundaries, both the pressure field and imposition of the no-slip condition at the wall, necessitate modifications to these equilibrium forms.

In the compressible case, this has generally been done through the wall function approach that specifies the functional behavior of the mean and turbulent variables in proximity of the solid boundary, or through the introduction of damping function factors so that the modeling correlations have the correct asymptotic behavior in close proximity to the wall (Vallet, 2008). The compressible wall function approach has been investigated thoroughly in a number of early studies (Huang & Coakley, 1993; Huang et al., 1994; So, Zhang, Gatski, & Speziale, 1994; Viegas, Rubesin, & Horstman, 1985) and have been summarized in Gatski (1996). Although there have been some wall function formulations for three-dimensional flows with imposed pressure gradients (Shih, Povinelli, Liu, Potapczuk, & Lumley, 1999; Shih, Povinelli & Liu, 2003), this has been in the incompressible regime, so that their extension to the compressible regime remains a topic of future work.

Considerably less research has concentrated on the development of compressible near-wall models than on the incompressible counterparts. In addition, the various compressible correlations discussed previously, such as the pressure–strain rate correlation, and the heat and mass flux, are generally not developed for near-wall applications and have often been neglected in the near-wall models. Both a two-equation and second-moment near-wall closure model have been developed (Zhang, So, Gatski, & Speziale, 1993; see also Zhang, So, Speziale, & Lai, 1993) based on a variable density extension of an incompressible model (Lai & So, 1990), and evaluated against experimental data on flat-plate boundary-layer flows with adiabatic and cold wall conditions up to a free-stream Mach number of 10 (Zhang, So, Gatski, & Speziale, 1993) as well as on compressible ramp flows (Morrison, Gatski, Sommer, Zhang, & So, 1993). Two-equation compressible models have also been derived (for additional details on these formulations see Gatski, 1996). Since the local Mach number in the vicinity of the solid boundary will be subsonic and, in the absence of shocks or changes in geometry, imposed flow distortions will be absent, incompressible based near-wall corrections have been implemented (Vallet, 2008).

Chapter 6

Compressible Shear Layers

Turbulent shear flows have traditionally been classified into two main categories: free shear layers and wall-bounded flows. For free shear flows, jets, mixing-layers, and wakes can be viewed as the building blocks of more complex flow configurations of engineering interest. These flows, being far from boundaries, are relatively insensitive to low-Reynolds number effects as mentioned at the end of the last chapter, but are often dominated by large scale events and global or convective instabilities. As the Mach number increases, both the large scale structure and the (smaller scale) turbulence field become increasingly influenced by the compressibility conditions. Ultimately, these two characterizing features, large scale structure and fluctuation statistics, are strongly modified by the compressibility effect, as evidenced by the strong dependence on Mach number of both the spreading rate and turbulent stresses and fluxes. Physical and numerical experiments have, therefore, focused on determining and analyzing both the spreading rate and turbulence statistics in the study of this class of flows.

As for wall-bounded flows, irrespective of the speed regime, boundary layers are sensitive to low-Reynolds number effects due to the presence of the no-slip wall and wall-blocking (pressure field) effects. In the simple case of a zero pressure-gradient planar flow, the influence of compressibility and thermal wall conditions can have a significant impact on the skin-friction parameter, which is of engineering importance. In general, a key variable that can impact the characteristics of the boundary layer is the temperature distribution at the wall. The heat transfer in the compressible (particularly supersonic) regime can be very large and can impose severe engineering constraints on designs. These effects are quite variable and depend on the wall condition (engineering application)—adiabatic or heated/cooled wall. While the effects just discussed are intrinsic to the (high speed) fluid flow, an important influence on wall-bounded flows is the distortion effect of shocks. This topic will be deferred

Compressibility, Turbulence and High Speed Flow. http://dx.doi.org/10.1016/B978-0-12-397027-5.00006-X
169

until Chapter 7, and the "simpler" no shock supersonic boundary layer will be the focus here.

It is apparent, then, that the free shear and wall-bounded flows, with their underlying physics being so different, need to be considered separately in order to best characterize their compressibility effects from both physical and numerical experiments. In Sections 6.1, 6.2, 6.3 the free shear flows of jets, mixing-layers and wakes will be discussed, and then followed by a discussion of the supersonic boundary layer flow in Section 6.4. In each of these flow cases, the primary area of study has differed depending on the flow physics as well as the characterizing features of each flow. In each section, these primary areas of study will be the focus of discussion.

6.1 JETS

Supersonic jets are geometrically "simple" flows of practical interest and are encountered in many industrial applications. In the aeronautics and space industries, supersonic jets are an integral part of internal (engine) and external (aerodynamic) flows, rocket applications (both missiles and booster rockets), and compressible flow control, where supersonic microjets have been used for separation control (Alvi, Shih, Elavarasan, Garg, & Krothapalli, 2003). In this latter case, the characteristics of these micro-jet flows themselves, distinct from the flow being controlled, is an interesting (and new) area of study that will be discussed further in Chapter 8. The Mach number of the (propulsive) jets of civilian aircraft at cruise conditions is approximately 0.8 and can reach values up to 1.2, and in future designs of transport aircraft with high nozzle pressure ratios, the jet Mach number may reach values up to 1.7–1.8 (Ennix, 1993). For military aircraft, or self-propelled missiles, larger Mach numbers can be often encountered. It is in these Mach number ranges that most studies have been focused.

As with subsonic jets, supersonic jets impact the environment in areas such as pollutant dispersion, mixing, and noise, and in some instances, these added compressibility effects can have a negative impact. In pollutant dispersion, for example, a longer potential core region (to be discussed shortly) corresponds to less dilution of the high temperature gases in propulsive jets, and in sound generation, an increase in sound emission during aircraft take-off that affects the airport surroundings. Another area is in booster rockets where safety and reliability constraints surface, and in military applications, the infra-red signature (IRS) is also affected by the reduced dilution of the high temperature gases. Of course, as shown in Chapter 1 other more diverse industrial applications exist such as in the case of sootblowers, oxygen injection in electric arc furnaces, and airblast injectors. These industrial examples can also have an environmental impact through influences on manufacturing fuel consumption and emissions.

The origin of all jets can be traced back to the mixing-layers that develop from the trailing edge of the nozzle. The (virtual) origin of the developing jet itself begins in a transition region that immediately follows the end of the potential core. Downstream of this region, a fully developed and, in general, self-similar state is eventually reached. In the case of a plane jet, the potential core exists from the trailing edge of the nozzle to the confluence of the two (planar) mixing-layers, and in the case of an axisymmetric jet from the trailing edge of the nozzle to the collapse of the annular mixing-layer. In both cases, the spreading rates of the mixing-layers will influence the length of the potential core. It is not surprising that studies have shown that plane mixing-layers, annular mixing-layers, and jets can often be assumed to exhibit very comparable behavior when dealing with mean or turbulent flow characteristics.

Since an important identifying characteristic of supersonic mixing-layers is the decrease in spreading rate (relevant to equivalent subsonic ones) with Mach number (see Section 6.2.2), an immediate consequence for supersonic jets, both planar and annular, is that the potential core length L_c (normalized with respect to the nozzle diameter) is increased. An empirical formula for L_c can be deduced (Lau, Morris, & Fisher, 1979) from a variety of experiments and is of the form

$$L_c \simeq 4.2 + 1.1 M_j^2, \tag{6.1}$$

where M_j is the exit Mach number of the jet. This relation has only been established for isothermal, fully expanded jets in still air; however, it should be emphasized that the length of the potential core can strongly depend on the external pressure conditions. A major difference between supersonic and subsonic jets is due to the fact that the exit pressure of subsonic jets is naturally balanced with the ambient; whereas, for supersonic flows, the exit pressure can be adjusted independently of the external conditions. Thus, a supersonic jet can then be either under-expanded, fully-expanded or over-expanded. Only the fully-expanded condition is free of expansion and shock systems. In contrast, under-expanded jets create sets of Mach disks that strongly influence the mixing-layer behavior and consequently the length of the core. In addition, the turbulent structure is changed with the appearance of different azimuthal modes and the appearance of three-dimensional structures (Lesieur, Métais, & Comte, 2005).

The dynamical behavior of jets is best viewed by considering the far and near regions of the jet separately. In the far jet region, the natural frequency is associated with a global instability mode, generally termed a column mode, and associated with a Strouhal number based on the nozzle exit diameter and velocity, $St_g = f_g D / U_j$. In addition to the global instability mode, in the near jet region between the nozzle exit and the end of the potential core, the annular shear layer contains higher frequency instability modes, namely shear layers modes. The Strouhal number associated with a shear layer mode is then dependent on the local mixing-layer thickness that evolves from the exit downstream to the location where the mixing-layers merge at the end of the potential core, and is defined as $St_e = f_e \delta_\omega / U_c$ (see Section 4.1.2). In this near jet region St_e varies little, so that the evolution of δ_ω (increasing with distance from the jet exit) leads to a decrease of f_e. Thus, in supersonic jets this potential core region allows for a double-frequency partitioning between the temporal shear layer and column modes, that is the so-called preferred modes, and parameterized by Strouhal numbers (0.2–0.25) that are only weakly dependent on Mach number for moderate supersonic flows. Although the focus here is on the temporal instabilities associated with far and near jet regions axisymmetric jets are also sensitive to azimuthal (spatial) instability modes (e.g. Glauser & George, 1992; Jung, Gamard, & George, 2004; Samimy, Kim, Kastner, Adamovich, & Utkin, 2007b). These extra azimuthal modes have to be considered for flow control and will be discussed further in Chapter 8.

Even for perfectly expanded jets, in some experiments the potential core length can be influenced by the proximity of the tunnel wall in (physical and numerical) experiments. In (axisymmetric) jets discharging in an infinite medium (still air), the axis of the mixing-layers in the core region is located at a constant distance from the (geometric) center of the jet. This characteristic symmetry can be modified if, for example, a secondary flow is created by a suction of the outer flow. This can be the case in some industrial systems when supersonic jet ejectors are used for suction, or in some experiments where flight effects have to be introduced. In the latter case, the wind tunnel has a limited lateral extent, allowing for the suction of ambient air. External Mach numbers on the order of 0.1– 0.3 can then be obtained by adjusting the size of the test section and the driving pressure of the supersonic jet without any additional blowing system. In this case, however, the entrainment of ambient fluid at the inner supersonic side and at the subsonic side of the jet in the core region are not the same as in the case of jets in still air. The spreading rate of the mixing-layer, for a given convective Mach number, is unchanged. However, the length of the potential core can be dramatically changed by the "ejector effect." This is illustrated by the results from a variety of experiments (Bellaud, 1999) and shown in Figure 6.1 ($L_c = x_c / D$ in the figure), where the convective Mach number is given by $M_c = (U_1 - U_2)/(c_1 + c_2)$ (Papamoschou, 1989). (In order to conform with the usual notation adopted in discussing free shear flows, the velocity variables $U_1 = \overline{u}_1(y = y_{max})$ and $U_2 = \overline{u}_1(y = y_{min})$ are introduced.) The "ejector effect" on the potential core length L_c (relative to still air behavior) can be seen when the external Mach number is increased to a range of 0.1– 0.2 by limiting the test section installed around the jet (Bellaud, 1999, jet diameter was 50 mm and the test section was 500×500 mm^2). While the values of L_c for jets in still air exhibit some scatter, they generally follow the behavior given in Eq. (6.1), but in contrast with the "ejector effect" the results show a large discrepancy with Eq. (6.1), that is a large increase of the core length. In this case, the axis of the mixing-layer is displaced farther from the axis of the jet. The spreading rate is unaffected by the external wall proximity,

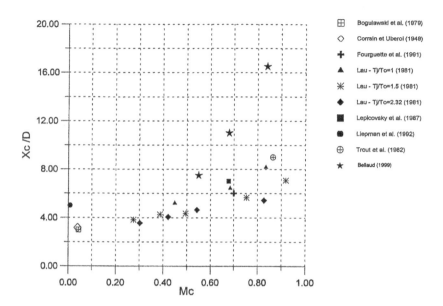

FIGURE 6.1 Variation of potential core length with convective Mach number for supersonic jets. From Bellaud, (1999).

but the distance where the axisymmetric annular mixing-layers merge is larger. This increase in potential core length is easily explained by the entrainment of ambient air which is, in this case, of finite total mass flux. In order to balance the internal and external entrainment, a displacement of the entire mixing-layer is necessary, with an adjustment of the static pressure. When considering the whole jet flow, whether in physical or numerical experiments, it is important to account for the (finite) size of the control sections (computational domains). Dimotakis (1986) addressed this issue and provided an example based on a two-dimensional mixing-layer that can be used to obtain a quantitative estimate of the ratio of induced fluid fluxes from each of the co-flowing free streams. Such a model for estimating the flow deviation is shown schematically in Figure 6.2.

A continuing focus of jet flow research is in the study of jet noise. Although the questions related to the origin of the noise emitted by supersonic jets have not been definitively answered, it is generally acknowledged that the sources are located between the exit plane of the nozzle and the end of the potential core (see Tam, 1991). The power of the noise emitted by turbulent jets is strongly affected by the Mach number in this regime. The power is known to be proportional to U_j^7 or U_j^8 for subsonic jets depending, among other parameters, on the direction of the observation and the Reynolds number (where U_j is the jet exit velocity). For example, a power law of $U_j^{7.5}$ has been obtained by large eddy simulation (LES) and has also been confirmed by several experiments (Bogey & Bailly, 2006); however a more precise scaling yields an acoustic energy proportional

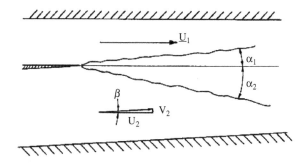

FIGURE 6.2 Schematic of the flow deviation due to asymmetric entrainment. From Dimotakis (1986) with permission.

to $U^3 M^5$, independent of possible (contaminating) noise sources from shocks or from feedback (screech noise). In contrast, in the supersonic case there is a consensus that the noise is only proportional to M^3 (Goldstein, 1976).

Most of the compressibility effects occur in the early stages of the jet development, mainly in the potential core region or immediately downstream (Tam, 1991). It is generally acknowledged that in this region the noise generation is linked to large scale structures and turbulent energy levels—both quantities being affected by compressibility effects. Also in this region, high mean shear is present and the turbulent energy levels are high. If the driving velocity, that is the difference between the jet exit velocity and the outer velocity, is constant, then most of the important dynamics for turbulence production and energy spectra development occur in this region. Downstream of the confluence of the mixing-layers, the flow dynamics evolves more gradually, with a decrease of the driving velocity (the axis velocity). Since the primary noise sources are located between the nozzle exit and the end of the potential core, the precise knowledge and prediction of mixing-layers is then a key element in supersonic jet flow analysis. This is yet another reason why only a few supersonic flow studies have been devoted to jets downstream of the potential core, while most of the attention has been on the study of mixing-layers.

6.2 MIXING-LAYERS

Compressible mixing-layers contribute significantly to the overall dynamics of many practical aerodynamic flows including, as just discussed, the developing jet flow that is dominated by the annular mixing-layer surrounding the potential core region and, as will be discussed next, wake base flows. The study of mixing-layers in isolation, as a canonical flow, is an important tool in the understanding of these more complex flow

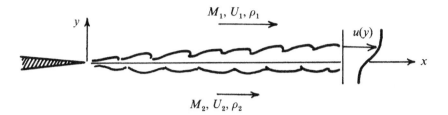

FIGURE 6.3 Schematic representation of a plane mixing-layer. From Papamoschou and Roshko (1988a) with permission.

fields. A planar mixing-layer, shown in Figure 6.3, can be generated by two parallel flows that merge after the trailing edge of a splitter plate, with corresponding mean velocities U_1 and U_2 $(U_1 > U_2)$, and Mach numbers M_1 and M_2. In each of the flows, the speeds of sound and densities can differ, and are given by c_1 and c_2, and ρ_1 and ρ_2, respectively. (As in the previous section on jets, the velocity variables $U_1 = \overline{u}_1(y = y_{max})$ and $U_2 = \overline{u}_1(y = y_{min})$ are introduced.) The discussion here will primarily focus on the planar mixing-layer; although, relevant results from the annular mixing-layer will also be presented where necessary.

6.2.1 Flow Structure

As the flow fields in Figure 6.3 merge, a Kelvin–Helmholtz (KH) instability generates the structure shown in Figure 6.4. The figure shows the effect of varying Reynolds number with the flow, in case (a) having Reynolds number four times the one in case (e). Thus, increasing the Reynolds number corresponds to an increase in the presence of small scale turbulent structures; however, the initial large scale structures resulting from the KH instability persist even at the highest Reynolds numbers.

These large-scale structures are still present in supersonic flows. Such observations of large scale turbulent structures suggest that a relevant measure of the Mach number of the flow should be based on the relative motion of the large scale structures to the free stream flow. For mixing-layers where $U_2 = 0$, the Mach number parameterization is clearly in terms of U_1; however, for $U_2 \neq 0$ the optimal parameterization becomes less obvious; although the question has been investigated extensively (Bogdanoff, 1983; Coles, 1981; Dimotakis, 1984; Papamoschou & Roshko, 1988a, 1988b). A convective Mach number parameterization (Papamoschou & Roshko, 1988a, 1988b), based on a stability analysis of a temporally evolving vortex sheet, has been proposed. If it is assumed that the flow structures are being convected at a (relatively) constant velocity, then in a coordinate frame translating at this convection velocity U_c,

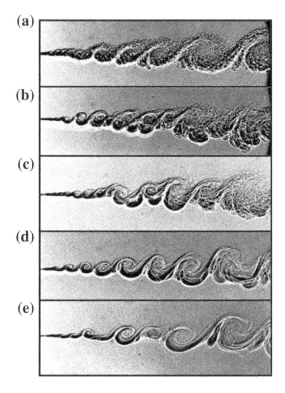

FIGURE 6.4 Effect of Reynolds number on the structure of subsonic plane mixing-layers. From (a), high Reynolds number, to (e), low Reynolds number. Shadowgraph visualizations with different densities. From Brown and Roshko (1974) with permission.

the corresponding convective Mach numbers relative to each stream are

$$M_{c1} = \frac{U_1 - U_c}{c_1} \quad \text{and} \quad M_{c2} = \frac{U_c - U_2}{c_2}. \tag{6.2}$$

As Figure 6.5 illustrates, in this convected frame of reference a saddle point is present between two adjacent eddies. If, in this frame of reference an isentropic, steady flow is assumed and on either side of the stagnation point, the total, or stagnation, pressures are assumed equal, then a mechanical equilibrium of the saddle point leads to the relation

$$\left[1 + \frac{(\gamma_1 - 1)}{2} M_{c_1}^2 \right]^{\frac{\gamma_1}{(\gamma_1 - 1)}} = \left[1 + \frac{(\gamma_2 - 1)}{2} M_{c_2}^2 \right]^{\frac{\gamma_2}{(\gamma_2 - 1)}}, \tag{6.3}$$

where γ_1 and γ_2 are the ratios of specific heats for the two streams of fluid. Using the definition of the two convective Mach numbers, and the added assumption

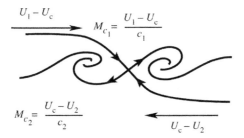

FIGURE 6.5 Sketch of large scale structures in convective reference frame. From Papamoschou and Roshko (1988a) with permission.

of $\gamma_1 = \gamma_2$, expressions for U_c and M_c can be derived and are given by

$$U_c = \frac{c_2 U_1 + c_1 U_2}{c_1 + c_2} \tag{6.4}$$

$$M_c = \frac{U_1 - U_2}{c_1 + c_2} = \frac{\Delta U}{c_1 + c_2}. \tag{6.5}$$

Mach number effects on these large-scale structures are shown in the tomographic visualizations of Figure 6.6 covering a convective Mach number range from 0.525 to 1. (Such visualizations can be viewed as instantaneous images of the flow structure.)

In the $M_c = 0.525$ case, traces of two-dimensional large scale structures appear clearly, and are almost as sharp as in the subsonic case (Brown & Roshko, 1974); however, when the convective Mach number is increased up to 0.64, the flow appears less organized. In particular, the upper edge of the shear layer is essentially flat and less disturbed by the large scale structures than in the $M_c = 0.525$ case. The decrease in spreading rate, or mixing efficiency, with increasing M_c is also visible in these pictures. When the convective Mach number is increased up to $M_c = 1$, the visualization shows a more uniform mixing zone and a more homogeneous internal flow, and a sharp boundary with the outer flow.

The proper orthogonal decomposition (POD) (see Section 4.3) is a tool that is often used for an "objective" analysis of the large-scale organization in flows; however, it is not that easily adapted to supersonic flows. Among the few examples available, is the application of the POD to the longitudinal mass flux fluctuations measured by hot-wire rakes (Debisschop, Sapin, Delville, & Bonnet, 1993). Figure 6.7 shows a comparison between a subsonic and supersonic $M_c = 0.64$ mixing-layer in transverse planes (the same $M_c = 0.64$ flow as presented in Figure 6.6). It is also possible to apply the POD to data from numerical experiments such as a compressible mixing-layer direct numerical simulation (DNS) (Yang & Fu, 2008). Figure 6.8 shows some results obtained at different transverse planes. As can be seen from the above

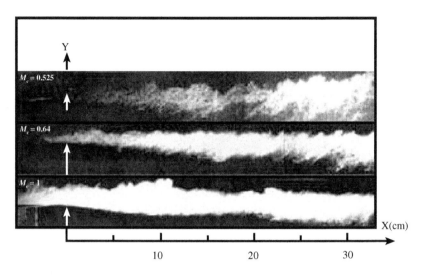

FIGURE 6.6 Tomographic visualizations of three kinds of compressible plane mixing-layers at $M_c = 0.525, 0.64$, and 1. From Bonnet and Debisschop (1992) and Chambres (1997) with permission.

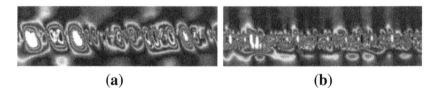

(a)   **(b)**

FIGURE 6.7 Reconstruction of the iso-values of the longitudinal mass flux fluctuations from the first POD mode obtained from hot-wire anemometry in turbulent mixing-layers: (a) incompressible, $M = 0.05$; (b) supersonic, $M_c = 0.64$. The vertical scale is the transverse direction, the horizontal is the time (arbitrary units). From Debisschop et al. (1993) with permission.

illustrations, the large-scale structures are less organized when the convective Mach number increases. The pioneering spanwise visualizations obtained with filtered Rayleigh scattering (Section 4.2.2) showed that these structures are highly three-dimensional (Elliott, Samimy, & Arnette, 1992). This structural evolution has been observed in physical experiments (e.g. Bonnet, Debisschop, & Chambres, 1993; Clemens & Mungal, 1995) as well as in numerical simulations (e.g. Comte, Fouillet, & Lesieur, 1992; Sandham & Reynolds, 1991). The trend towards three-dimensionality has also been analyzed using (spatial) linear stability theory (Sandham & Reynolds, 1990), where it was found that the maximum amplification was proportional to the growth rate of the developing layer, and that oblique waves played a dominate role in the

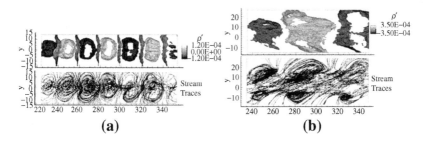

FIGURE 6.8 Fluctuating density (top) and streamlines (bottom) of first POD mode in supersonic mixing-layers: (a) $M_c = 0.4$; (b) $M_c = 1.2$. The vertical scale is the transverse direction, the horizontal is the time (arbitrary units). From Yang and Fu (2008) with permission.

mixing-layer above convective Mach numbers of 0.6. The DNS simulations (e.g. Freund, Lele, & Moin, 2000a; Pantano & Sarkar, 2002; Vreman, Sandham, & Luo, 1996) also reproduced the growth rate reduction of mixing-layers indicating that the mean flow instabilities, large scale structures, and shear layer spreading rate are coupled by the same underlying flow physics. This is a characterizing feature of supersonic flow physics that is not simply related to mean density effects. The persistence of large scale structures over a wide range of Mach and Reynolds numbers is an important feature of mixing-layers, and needs to be taken into account when developing models of such flows.

Since the defining relation for the convective Mach number involves a measure of relative velocity across an embedded structure, it can be formally linked (Sarkar, 1995) to the previous discussion and definitions of the deformation or gradient Mach number introduced in Section 5.1. The convective Mach number is a parameter characterizing the flow; whereas, the deformation or gradient Mach number varies locally within a particular flow and can vary from flow field to flow field. The question that naturally arises is whether the two are correlated. Both the convective Mach number and the gradient or deformation Mach number are ratios that include the mean velocity difference across either the shear layer (M_c) or a large scale eddy (M_g). Figure 6.9 shows the variation of gradient Mach number M_g with convective Mach number M_c in the center of an annular mixing-layer. After a gradual increase, the gradient Mach number appears to asymptote to a value near 2 for values of M_c greater than 1. Although this is but one example, it does show a correlation at some speed range interval, and also suggests that other mixing-layers and parameter ranges may exhibit a similar behavior.

The turbulent structure can also be investigated through a measure of the stress anisotropy. As was shown previously in Section 5.3, the construction of the anisotropy tensor defined in Eq. (5.29) and coupled with the associated invariant map, provides a rational and quantitative measure of the turbulent anisotropy (see also Figure 5.3). As a practical matter, many of the earlier

Compressibility, Turbulence and High Speed Flow

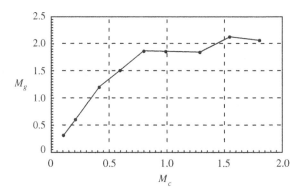

FIGURE 6.9 Gradient Mach number evolution versus convective Mach number at the center of an annular mixing-layer. From Barre et al. (2002) with permission.

experimental studies simply were not able to measure all the components of velocity necessary in obtaining all the components of the anisotropy tensor b_{ij}. Only a few attempts have been made to measure the three components of the velocity (Bellaud, 1999; Chambres, Barre, & Bonnet, 1997; Gruber, Messersmith, & Dutton, 1993). For this reason, anisotropy measures have in most cases been based on deviations from the quasi-isotropic relationship $\sqrt{\overline{u'^2}} \sim \sqrt{\overline{v'^2}} \sim \sqrt{\overline{w'^2}}$. Ratios between the various turbulent intensity components can be formed and a maximum identified, with the deviations from unity providing a measure of anisotropy. It has not been possible using such a ratio criteria to conclusively determine the behavior of the turbulence anisotropy across the flow. Since the third component is, in general, unknown from experiments, a b_{12} component is sometimes evaluated using the ratio of the shear stress to the longitudinal component (rather than twice the kinetic energy), that is, $b_{12} \approx \overline{u'v'}/\overline{u'^2}$. There is only a slight decrease of this b_{12} anisotropy measure when the Mach number increases (see discussion in Smits & Dussauge, 2006, their Figure 6.16), coupled with a large scatter in the data for both experiments and computations. While several studies (Bradshaw, Ferriss, & Johnson, 1964; Bellaud, 1999; Chambres, 1997; Debisschop, 1993; Denis, Delville, Garem, Barre, & Bonnet, 1998; Elliott & Samimy, 1990; Mistral, 1993) have found that the turbulent anisotropy expressed in terms of longitudinal versus transverse velocity fluctuations was relatively constant over a broad convective Mach number range ($0.2 < M_c < 0.9$) (see also Barre, Bonnet, Gatski, Sandham, 2002), other studies (Goebel & Dutton, 1991; and Gruber et al., 1993) have actually shown a growth with M_c. Figure 6.10 shows experimental results obtained for $\sqrt{\overline{u'^2}}/\sqrt{\overline{v'^2}}\Big|_{\text{max}}$. The reason for this growth in $\sqrt{\overline{u'^2}}/\sqrt{\overline{v'^2}}\Big|_{\text{max}}$ is that there is a relatively constant value in $\sqrt{\overline{u'^2}}/\Delta U$

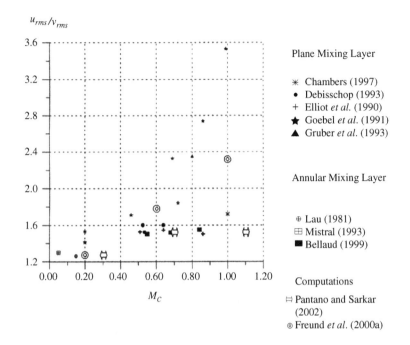

FIGURE 6.10 *rms* velocity ratio evolution as a function of M_c for plane and annular mixing-layers. From experiments and DNS.

but a decrease in $\sqrt{\overline{v'^2}}/\Delta U$ with M_c (Goebel & Dutton, 1991). Some direct numerical simulations (Freund et al., 2000a; Pantano & Sarkar, 2002) were also characterized by these two different trends. The Freund et al. (2000a) results showed a large departure from incompressible values that was consistent with the second category of experimental results. The Pantano and Sarkar (2002) set of results, however, showed a small increase with an increase in convective Mach number as observed in most of the experiments. In any case, these contradictions in both the experimental and simulation results makes it difficult to correctly identify the effects of compressibility and develop reliable predictive models.

Many of the physical experiments have preceded the numerical experiments and the quantities measured were often dictated by measurement capabilities. For example, the anisotropy measures just discussed focused on ratios of the normal turbulent stresses that can, of course, provide a measure of anisotropy. However, the anisotropy tensor b_{ij}, provides a more rational measure that can be used to objectively and quantitatively measure the degree of anisotropy. If accurate, both physically and quantitatively, models are to be developed for closure of prediction schemes, the relative assessment of such quantities needs to be made for both physical and numerical experiments that cover a broad parameter range.

Even though there are numerous physical and numerical experiments on planar and annular mixing-layers, many fundamental questions remain unanswered. In addition, there has not been any focused effort at collating the results from the numerous studies into a self-consistent dataset. Physical experiments are difficult and the parameter space is large, so studies that may appear similar can differ in some important details, and numerical experiments (simulations), are constrained by limitations on computational resources and are sometimes outside the parameter range of the physical experiments.

6.2.2 Spreading Rate

When the Reynolds number is sufficiently large, the mixing-layer reaches an asymptotic downstream state where self-similarity of the (properly) normalized mean and turbulent fields is observed. For the longitudinal mean velocity profile, it is well approximated for a subsonic flow by either an error function or a hyperbolic tangent function as sketched in Figure 6.11. From this profile, one can define the thickness of the shear layer. In fact, it is possible to define this thickness in several different ways that, unfortunately, are not completely equivalent. For example, the following measures can be used: vorticity thickness (based on maximum shear and probably the most popular), the velocity thickness (defined by the locations with a given ratio of the external velocities), the total pressure-based thickness (or Pitot thickness, often used for compressible flows between two gases), visual thickness (based on flow visualizations or schlieren pictures), momentum thickness, etc. Although all differ in detail (Bonnet, Moser, & Rodi, 1998), these thicknesses are related by simple relationships (e.g. see Pui & Gartshore, 1979), provided there is a unique analytical description of the velocity profile. Nevertheless, these different choices for the definition of the mixing-layer thickness may be a possible source of discrepancy between the different experiments.

Some slight departure, however, from the incompressible velocity distributions (in nondimensional form) can be observed when the Mach number increases. This has no influence on the vorticity thickness (based on the maximum shear), but can have a large influence on the other definitions, particularly if the thickness is based on a local criteria such as either the velocity or Pitot thicknesses, and even from an integral momentum thickness (Li, Chen, & Fu, 2003).

6.2.2.1 Subsonic Spreading Rate
When a side is at rest, such as for a subsonic regime, the conventional velocity thickness δ_u corresponds to a distance between the locations where the local velocity reaches 10% and 90% of the external velocity. The spreading rate in this

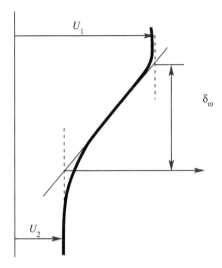

FIGURE 6.11 Schematic of self-similar mixing-layer longitudinal mean velocity profile.

case is related to a spreading parameter σ_0, that has been defined in such a way that the use of the similarity variable $\sigma y/x$ corresponds to mean velocity profiles closest to an error function (Rodi, 1975), and in this case $d\delta_u/dx \propto \sigma_0^{-1}$. Some scatter exists in the results (even in incompressible, subsonic flows); however, a value of $\sigma_0 \simeq 11$ has been accepted for more than six decades (cf. Liepmann & Laufer, 1947; Rodi, 1975).

When dealing with turbulent flows, the most straightforward measure of the spreading, is the evolution of the vorticity thickness given by

$$\delta_\omega = \Delta U / \left(\frac{\partial U}{\partial y} \right)_{max}, \tag{6.6}$$

where $\Delta U = U_1 - U_2$ is the velocity difference between the external flows, and $(\partial U/\partial y)_{max}$ is the maximum mean velocity gradient of the planar flow. When the self-similar state is reached, the plane mixing-layer spreads linearly with a constant value of the spreading rate $d\delta_\omega/dx$. It should be cautioned that the existence of an asymptotic state is not assured, but may be dependent on the initial (upstream) conditions that generate the mixing-layer. Some (upstream) conditions can influence the flow far downstream (George & Davidson, 2004) at least in subsonic mixing-layers (the equivalent has not been proven for supersonic mixing-layers, however). It should be cautioned that the determination of the convection velocities from experiments is not straightforward. It has been demonstrated from hot-wire measurements (Barre, Dupont, & Dussauge, 1997; Dupont, Muscat, & Dussauge, 1999) that the velocities obtained from an isentropic estimation, Eq. (6.4), can be observed, but that different velocities, closer to the external values can be measured. Thurow, Jiang, Kim, Lempert, and Samimy (2008) revisited the experimental methods and showed a strong dependency of the results on the measurement method itself when non-intrusive measurements are performed. From an experimental standpoint, this is still an open issue.

In subsonic flow, if the mixing-layer is formed between two streams with the same gas at the same temperature, the spreading rate is a function of the velocity ratio $r_u = U_2/U_1$, the subscripts 1 and 2 corresponding to the high and low external speeds respectively. In this case, the Sabin–Abramovich rule then applies (Abramovich, 1963; Sabin, 1965) and

$$\delta'_\omega = \frac{d\delta_\omega}{dx} = \left(\frac{\sqrt{\pi}}{\sigma_0}\right)\left(\frac{1 - r_u}{1 + r_u}\right) \simeq 0.16\left(\frac{1 - r_u}{1 + r_u}\right). \tag{6.7}$$

When two different gases are merging at a trailing edge, the spreading rate can be computed with a semi-empirical relationship (Brown and Roshko, 1974) that is a function of the density and the velocity ratio of the external flows,

$$\left(\frac{d\delta_\omega}{dx}\right)_i = C_\delta \frac{(1 - r_u)\left(1 + \sqrt{\rho_2/\rho_1}\right)}{2\left(1 + r_u\sqrt{\rho_2/\rho_1}\right)}, \tag{6.8}$$

where C_δ (≈ 0.18) is an (incompressible) growth rate constant. In order to take into account possible spatial asymmetries, an additional term can be introduced (Dimotakis, 1986) into the right side of Eq. (6.8),

$$\left(\frac{d\delta_\omega}{dx}\right)_i = \frac{C_\delta}{2}\left(\frac{1 - r_u}{1 + r_u\sqrt{\rho_2/\rho_1}}\right)$$
$$\times \left\{1 + \sqrt{\rho_2/\rho_1} - \frac{1 - \sqrt{\rho_2/\rho_1}}{1 + 2.9[(1 + r_u)/(1 - r_u)]}\right\}. \tag{6.9}$$

Although this form is not often used, it does introduce an adjustment of the incompressible value on the order of 10% in some extreme cases.

In the assessment of supersonic spreading rates to be discussed next, the basis for comparison will be the corresponding subsonic spreading rate. In order to compare the supersonic and equivalent subsonic case, it is necessary to determine the values of the incompressible reference growth rate with one side at rest, that is, determine the value of C_δ in Eq. (6.8), or (6.9), that will be used as a reference. Barone, Oberkampf, and Blottner (2006) has made a compilation of the values of C_δ that can be deduced from the value $\sigma_0 = 11$. Three different mixing-layer thicknesses were considered: (i) a velocity thickness δ_u that is based on the locations where $(\bar{u}_1(y) - U_2)$ is $0.10\Delta U$ and $0.90\Delta U$ (see Figure 6.11); (ii) an "energy" thickness δ_e (Bradshaw, 1981) that is based on the locations where the (mean) "energy" $(\bar{u}_1(y) - U_2)^2$ is $0.10(\Delta U)^2$ and $0.90(\Delta U)^2$; (iii) the vorticity thickness δ_ω, Eq. (6.6), which is presently the one most used. Coupled with these three thickness choices is the complicating factor that two similarity laws (error function or hyperbolic tangent function) can potentially be used to fit the experimental data. Cross-examining the three definitions and the two similarity functions, different values for C_δ can be obtained (see Table 6.1).

6.2.2.2 Supersonic Spreading Rate

For supersonic mixing-layers, Eq. (6.8) does not correctly describe the spreading rate. This is illustrated in Figure 6.12, where measured spreading rates in supersonic backward facing steps are plotted versus the corresponding density ratio (the free stream Mach number is also given). This distribution of measured δ'_ω values is compared to Eq. (6.8) for $r_u = 0$ at different density ratios corresponding to the experimental configuration tested. There is a ratio of about three between the measured values and the ones deduced from Eq. (6.8). The problem with both the Brown and Roshko (1974) and Dimotakis (1986) relationships is that the observed decrease in spreading rate is not entirely due to the density ratio variation, and that the additional compressibility effects need to be taken into account. This was basis for the compressibility corrections discussed in Section 5.3.

In order to empirically represent these compressibility effects, a new functional form for the mixing-layer spreading rate (Papamoschou & Roshko, 1988a) has been introduced and is given by

$$\delta'_\omega = \left(\frac{\mathrm{d}\delta_\omega}{\mathrm{d}x}\right)_c = \left(\frac{\mathrm{d}\delta_\omega}{\mathrm{d}x}\right)_i \Phi(M_c), \tag{6.10}$$

where $\Phi(M_c)$ is a compressibility function (cf. Eq. (5.4.3)) and as shown, is the ratio between the observed spreading rate of a compressible mixing-layer and the corresponding incompressible value at the same velocity and density ratios. Results from several experimental studies have supported this functional relationship and are often compared with the "Langley curve" that was initially compiled by Birch and Eggers (1972) (see also Papamoschou & Roshko, 1986).

Because of its historical significance, the original "Langley curve" (Birch & Eggers, 1972) is shown in Figure 6.13. There is a curve fitting of selected data that accurately represents the shear layer spreading rate. The value of $\sigma_0 = 11$ is the incompressible limit. Since Birch and Eggers (1972) stated, "It is therefore

TABLE 6.1 Evolution of C_δ for several definitions of mixing-layer thickness and similarity laws. From Barone et al. (2006).

	δ'_u	δ'_e	δ'_ω
erf	0.165	0.136	0.161
tanh	0.2	0.168	0.182

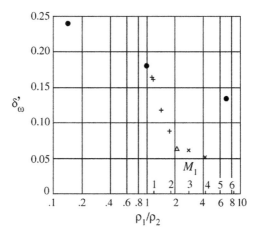

FIGURE 6.12 Effect of density ratio on spreading rate with $U_2 = 0$: •, values for incompressible flow, and other symbols for compressible mixing-layers. From Brown and Roshko (1974) with permission.

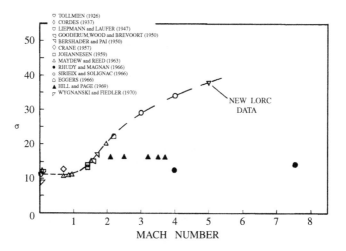

FIGURE 6.13 Variation of spreading rate with Mach number $U_2 = 0$: •, values for incompressible flow, and other symbols for compressible mixing-layers. From Birch and Eggers, (1972).

suggested that on the basis of the data available at present, the faired line shown in Figure 6 best represents the variation of σ with Mach number for developed shear layers," it suggests that they were not anticipating the importance that would be attached to the "faired line." Bradshaw (1981) later replotted the curve (see for example Figure 6.15) in terms of more generally accepted variables, that is, by plotting $d\delta/dx = 1.3/\sigma$ (the velocity thickness δ_u is used as the

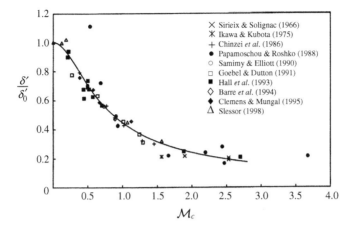

FIGURE 6.14 Growth rate from different experimental mixing-layer studies selected by Slessor et al. (2000) and plotted in terms of $\mathcal{M}_c$: ———, curve fit from Eq. (6.13). From Slessor et al. (2000) with permission.

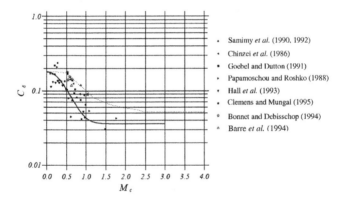

FIGURE 6.15 Growth rate from different experimental mixing-layer studies selected by Aupoix and Bézard (2006): ———, Dimotakis law, Eq. (6.11); ......, "Langley curve" after Bradshaw (1981). From Aupoix and Bézard (2006) with permission.

reference) and, when normalized by the incompressible limit, became the now accepted "Langley curve." The majority of contemporary studies have been summarized in Lele (1994) and Smits and Dussauge (2006), and in addition, some of the data have also become available in digital form (Bonnet, Moser & Rodi, 1998).

While the selection of the correct compressibility function $\Phi(M_c)$ is important in obtaining the correct compressible spreading rate, it is important to recognize that estimation of the incompressible spreading has some

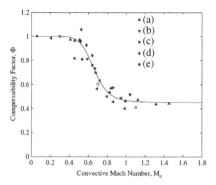

FIGURE 6.16 Growth rate from different experimental mixing-layer studies selected by Barone et al. (2006): (a) Bogdanoff (1983) and Papamoschou and Roshko (1988a); (b) Chinzei, Nasuya, Komuro, Murakami, and Kudou (1986); (c) Samimy and Elliott (1990); Elliott and Samimy (1990); (d) Goebel and Dutton (1991), Dutton, Burr, Goebel, and Messersmith (1990), Gruber et al. (1993) and (e) Debisschop and Bonnet (1993), Debisschop, Chambres, and Bonnet (1994) and Barre, Braud, Chambres & Bonnet (1997); □, DNS Pantano and Sarkar (2002); -----, curve fit. From Barone et al. (2006) with permission.

uncertainties as well. Recall that in Table 6.1, the choice of the subsonic similarity profile can introduce more than 10% difference in mixing-layer thickness depending on the thickness definition used. In order to avoid, or minimize, this source of uncertainty, alternative methods have been proposed.

Slessor, Zhuang, and Dimotakis (2000) attempted to better collapse the available data by allowing for different growth rate constants C_δ for each experiment. Using the compressibility function $\Phi(M_c)$ (Dimotakis, 1991),

$$\Phi(M_c) = 0.2 + 0.8 \exp(-3M_c^2), \qquad (6.11)$$

the incompressible relationship in Eq. (6.9), and the measured spreading rate from each experiment, a growth constant C_δ could be determined for each experiment. Although some improvement was found (see Table 1 of Slessor et al., 2000), the scatter among the growth constants of the experiments remained large.

Almost all of the assessments of compressibility effects are dependent on the convective Mach number M_c. The Dimotakis law that was defined in Eq. (6.11) is often compared with the experimental data and represents, in general, a satisfactory estimation of the compressibility effect (see for example Figure 6.15). However, the convective Mach number may under-estimate the compressibility effects, since the speed of sound ratio is inversely proportional to the density ratio, small density ratios lead to large speed of sound ratios so that two distinct effects can occur (Slessor et al., 2000). This suggests that an alternative compressibility function can be introduced that better accounts for these features. Instead of the convective Mach number M_c, an alternative

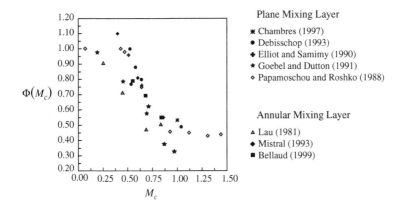

FIGURE 6.17 Normalized growth rate versus convective Mach number for different experimental mixing-layer studies.

compressibility parameter, $\mathcal{M}_c$, is introduced

$$\mathcal{M}_c = \max_{i=1,2} \left[\frac{\sqrt{\gamma_i - 1}}{c_i} \right] (U_1 - U_2), \qquad (6.12)$$

where γ_i and c_i are the ratios of specific heats and speeds of sound, respectively. Using this parameter, a new compressibility function was proposed (Slessor et al., 2000),

$$\Phi(\mathcal{M}_c) \approx 1/\sqrt{1 + 4\mathcal{M}_c^2}. \qquad (6.13)$$

As Figure 6.14 shows this alternative compressibility parameter provides a better collapse of the data. From the available experimental data, it can be shown that either M_c or $\mathcal{M}_c$ can adequately represent the effects of compressibility (Aupoix & Bézard, 2006).

Aupoix (2004) and Aupoix and Bézard (2006) re-evaluated most of the available experimental results by introducing, in place of the subsonic "growth rate constant" C_δ and the compressibility function $\Phi(M_c)$ (Eqs. (6.8) and (6.10)), a unique parameter that is the combination of the two, that is, $C_\delta(M_c) = C_\delta \Phi(M_c)$. They then computed for each experiment the corresponding value of $C_\delta(M_c)$. Although this value is still subject to the different uncertainties listed above, they can be better identified by plotting the results in a log-linear manner in order to translate the uncertainties to simple vertical shifts (see Figure 6.15). Unfortunately, when compared with the format used in Figure 6.17, the improvement in the scatter is not that large.

In another method (Barone et al., 2006), a validation metric that quantifies the accuracy of a model relative to an experimental data set was introduced. This process yielded a selected set of experiments that met the validation metric criteria and from which the compressibility function $\Phi(M_c)$ could be determined. From this set of experiments a curve fit of the data was obtained and is shown in Figure 6.16. Even with this different data selection method, the experiments shown in Figure 6.16 exhibit significant scatter. It is interesting to note, however, that the DNS results for a temporally developing mixing-layer (Pantano & Sarkar, 2002) are in agreement with the physical experiments.

It is clear that the identification of a universal spreading rate law is still not fully resolved due to wide-ranging origins of discrepancies. Only multiple replication of experiments will be able to extract reliable trends and realistic laws that can be expressed in term of a suitably defined Mach number. Although alternatives have been sought, the convective Mach number M_c is still the more popular and representative parameter, especially for mixing-layers created by the same gas.

6.2.2.3 Fluctuating Pressure Field and Prediction

In the last section on spreading rate, the discussion focused on the empirical parameterization of both the key functional parameter (a measure of the convective Mach number, e.g. $M_c, \mathcal{M}_c$), and the actual functional dependency on this parameter, that is the compressibility function Φ. In this section, some statistical characteristics of supersonic mixing-layers will be further explored in order to gain an understanding of the interactive dynamics occurring within the layers that lead to the effects represented by compressibility functions such as Φ.

Most of the available measurements of mean and turbulent quantities have been carried out using either hot-wires (Barre, Quine, & Dussauge, 1994) or laser-Doppler velocimetry (LDV) (Chambres, 1997; Debisschop, 1993; Elliott & Samimy, 1990; Goebel & Dutton, 1991; Gruber et al., 1993) in both plane mixing-layers and supersonic jet configurations (Bellaud, 1999; Lau, 1981). Figure 6.17 once again shows the reduction in the normalized spreading rate $\Phi(M_c)$ with the convective Mach number for a selection of these experiments. From these results, however, it can be seen that annular mixing-layers (jets) have, if perfectly expanded, the same spreading rate behavior as the plane mixing-layer. Between $M_c = 0.5$ and $M_c = 1$ there is a reduction by more than a factor of two, and above $M_c = 1$ the normalized spreading rate is relatively constant. When compared with Figure 6.9, the behavior is similar to the trend shown for the evolution of the gradient Mach number.

A relatively large scatter is observed in the figure, and it should be noted that discrepancies between different experiments are also observed even in less complex situations, namely in the incompressible regime where several contributing factors have been identified (Brown & Roshko, 1974; Browand & Latigo, 1979; Hussain & Zedan, 1978a, 1978b; Oster & Wygnanski, 1982): (i) turbulence in free streams, (ii) free stream oscillations, (iii) residual pressure gradients, (iv) vibration of splitter plate, (v) state of the boundary layers at the splitter plate, (vi) curvature and angle of confluence, (vii) aspect ratio, (viii) test section length, and (ix) Reynolds number. For supersonic flows, not only one, but several, of these factors can be responsible for the discrepancies between the different experiments, but it is difficult to quantify all these effects experimentally. Some progress can be made, however, from computations. As an example, Aupoix (2004) attempted to estimate the influence of the free stream turbulence (i) on a Reynolds-averaged Navier–Stokes (RANS) computation. With the model used, a quasi-linear increase of the spreading rate was found; however, this behavior was observed for very high turbulence levels, with the trend being less clear for moderate (realistic) external levels. In contrast, for the pressure drop (iii), Barre, Braud, Chambres, and Bonnet (1997) showed that even high pressure non-equilibrium at the trailing edge does not strongly influence flow compressibility effects on the spreading rate. In supersonic experimental studies, there are some common features encountered in the boundary layers along the splitter plate. In such experiments, most of the data has been obtained with fully turbulent boundary layers on the supersonic side; however, the state of the subsonic incoming boundary layer could not be as well defined. In Barre et al. (1994), the influence of the boundary layer on the low velocity side was investigated, and it was found that when the boundary layer thickness was small compared to the thickness on the supersonic side, the effect was negligible. Unfortunately, supersonic experiments introduce a practical limitation that needs to be considered. This is the limited size of most of the wind tunnel test sections in supersonic flows, where aspect ratios (vii) lying between 4 and 10 are not uncommon. A related parasitic effect (x), specific to supersonic tests lies in the wall proximity—an effect being particularly important in several small scale wind tunnels that can be susceptible to "blockage effects" (as suggested by Smits and Dussauge, (2006)). This effect is somewhat difficult to quantify, but the thickness of the mixing-layer has to be small enough when compared with the size of the test section of the wind tunnel.

The turbulent Reynolds stress statistics from several different planar and annular mixing-layer experimental studies over a wide convective Mach number range have been compared (Barre et al., 2002). Although there were significant differences between the various studies, several of the experimental studies confirmed the behavior of decreasing fluctuations with increasing M_c accompanied by a reduction in the spreading rate.

Direct numerical simulations (Freund et al., 2000a; Pantano & Sarkar, 2002; Sarkar, 1995; Vreman et al., 1996) have also confirmed that the spreading rate decreases when the convective Mach number increases, and as shown in Section 5.3.2 has been the basis for RANS model development and validation (see Figure 5.5). However, as cautioned in Chapter 5, the early simulation studies of homogeneous turbulence, that were partially motivated by this observed effect, did not correctly identify the underlying dynamic mechanism responsible for the spreading rate reduction. For mixing-layers, Vreman et al. (1996) showed

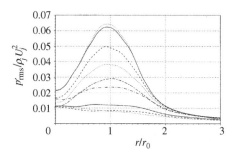

FIGURE 6.18 Normalized pressure fluctuations for different convective Mach numbers: $\cdots\cdots$, $M_c = 0.2$; ——, $M_c = 0.21$; — — —, $M_c = 0.41$; ------, $M_c = 0.59$; — · —, $M_c = 0.79$; — · — $M_c = 0.99$; ——, $M_c = 1.29$; — — — $M_c = 1.54$; - - - - - -, $M_c = 1.80$. From Freund et al. (2000a) with permission.

that compressible dissipation (and overall dilatation effects) remained negligible even at higher Mach numbers. According to their simulations, the difference in the level and distribution of the pressure fluctuations were central to the reduction in the turbulent activity. With an increase in Mach number, there was a reduction in the pressure fluctuations, an observation that was confirmed in simulations (Freund et al., 2000a; Pantano & Sarkar, 2002). Figure 6.18 (Freund et al., 2000a) shows the normalized (with centerline density and mean velocity) *rms* pressure fluctuation profiles from a DNS of an annular mixing-layer for nine convective Mach numbers M_c ranging from 0.2 to 1.8. It is clear from these results that a strong decrease in the maximum *rms* levels occurs with increasing Mach number. Vreman et al. (1996) used models of compressible vortices to confirm the origin of the reduction of pressure fluctuations. With this mechanism the reduction of growth rate is not due to extra explicit compressibility terms, but due to an implicit modification of existing contributions to the overall energetic balance—such as the pressure–strain rate correlation.

From a computational standpoint, it is easy to see from the turbulent kinetic energy equation, Eq. (5.25) for example, that the overall net production is achieved through a sum of production by the mean velocity gradient, pressure–dilatation, production by viscous stress gradient, and dissipation rate (destruction). At this level of prediction methodology, it is an accurate determination of the net effect rather than an accurate determination of the individual components that is relevant. Although, the simulations (Freund et al., 2000a; Pantano & Sarkar, 2002; Vreman et al., 1996) showed that dilatation effects were not the primary cause of the spreading rate reduction, but rather reduced pressure fluctuation levels that ultimately affected the turbulence production mechanisms, it is possible to replicate the correct overall energetic balance through an (incorrect) enhanced turbulent energy dissipation rate.

The determination of the terms in the energetic balance is delicate in incompressible flows, and much more difficult in compressible, high velocity

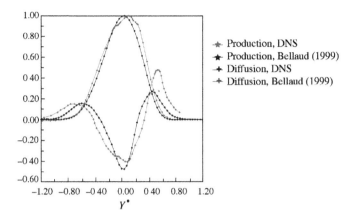

FIGURE 6.19   Comparison between experimental (Bellaud, 1999) and DNS (Freund et al., 2000a) turbulent kinetic energy budgets across annular mixing-layer at $M_c = 0.85$.

flows. On the experimental side, several limitations remain even today. For example, it is not possible at this time to obtain reliable experimental data for some important terms, such as the pressure–strain rate correlation; however, by using the Reynolds analogies and related assumptions, it is possible to establish budget estimates of turbulent kinetic energy in supersonic mixing-layers (Bellaud, 1999; Chambres et al., 1997). Such results can be useful in validating closure models and comparing the components of the Reynolds stress tensor with the ones obtained in incompressible flows. Figure 6.19 shows the production and diffusion terms in the kinetic energy budgets of an annular mixing-layer obtained from experiments (Bellaud, 1999) and DNS (Freund et al., 2000a) at $M_c = 0.85$. While these budgets are not completely equivalent, a qualitative agreement is clearly shown. The most important feature revealed by both these budgets is the asymmetry of the diffusion term between the subsonic and the supersonic edge of the flow. At present, this type of compressibility effect has no satisfactory explanation. Nevertheless, with this type of agreement one can assume that these budgets across the layer are a good approximation of the actual physical behavior that must be modeled.

While the constraints imposed on physical experiments are numerous and severely curtail their ability to completely map all the dynamics, a set of constraints are inherent in numerical simulations (direct or large eddy simulations), particularly those involving spatially developing mixing-layers. Such spatial simulations are less common than the temporal simulations due to the significantly increased computational requirements. In addition, as with all spatial simulations (and computations), effects of inflow conditions can influence the solution far downstream making spreading rate estimates susceptible to error. In addition, if trailing (or leading) edges of solid surfaces are embedded within the computational domain grid resolution issues can

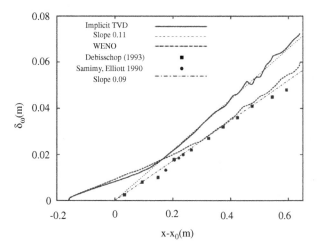

FIGURE 6.20 Longitudinal evolution of the vorticity thickness for different numerical schemes. From Doris, Tenaud, and Ta-Phuoc (2000) with permission.

predominate as well as algorithmic issues associated with the need for high accuracy in such highly resolved regions. In Figure 6.20, an example of this algorithmic sensitivity is shown, where the solutions from an implicit total variation diminishing (TVD) scheme and a weighted essentially non-oscillatory (WENO) scheme with both having overall second-order spatial accuracy are compared. The results show the overly dissipative behavior of the TVD scheme and its impact on the mixing-layer vorticity thickness.

While the first models developed to account for reduced turbulent activity in compressible flows were predicated on dilatation effects, and accurately represented results from numerical simulations of homogeneous flows, they were ultimately found to incorrectly represent the physics of the real inhomogeneous flows. Unfortunately, two decades later such models continue to be used since they are easily implemented and are able to reproduce results that match the observed phenomena of reduced spreading rate. Even though the exact quantitative extent of the reduced mixing-layer spreading rate remains an open question, computational validation studies, such as those often employed in RANS modeling, are often using closure models that don't correctly represent the flow physics.

6.3 WAKES

Wakes from obstacles traveling at supersonic speeds can be classified into two main categories: (i) base flows, and (ii) flat plate, or shaped body, wakes. The first category corresponds to wakes from a variety of objects such as rockets, powered missiles, and re-entry capsules in external aerodynamics, and flame holders in internal flows. As will be discussed next, these flows are quite complex with large separated zones containing supersonic mixing-layers and a recirculation region. The drag resulting from the base flow can often reach 35% (Sandberg & Fasel, 2006) and even 50% (Sahu, Nietubicz, & Steger, 1985) of the total drag for vehicles flying at supersonic speeds. In addition, the presence of supersonic mixing-layers naturally introduces dynamical features associated with them, such as noise generation. The second category corresponds to wakes from slender bodies, such as wings or thin bodies with thin trailing edges (that is, without separation at the trailing edge). In this category, the flat plate is the simple generic configuration. In such a configuration, there is no recirculation region and less turbulent energy is generated, with no significant drag increase or noise emissions. This flow is then less important for practical applications, but does contain many interesting features suitable for fundamental study.

6.3.1 Base Flows

Figure 6.21 provides a qualitative sketch of the important and complex dynamic regions comprising the supersonic base flow. The key characteristic of this flow is the separation that occurs at the trailing edge. Depending on the geometry, this separation creates either an annular or planar mixing-layer which is bounded on one side by a recirculation region. Due to the deviation of the streaklines, an expansion wave system develops from the trailing edge. The mixing-layer converges towards a stagnation point in a high pressure region where the wake itself starts. The change in the direction of the streaklines in the vicinity

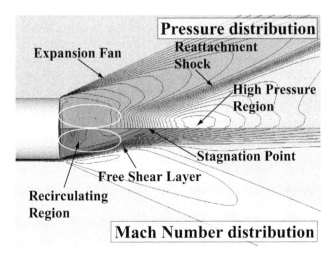

FIGURE 6.21 Schematic representation of a bluff body wake. From Kawai and Fujii (2005) with permission.

of the stagnation point creates recompression waves. The complex dynamic interactions are inherently unsteady with high turbulent energy levels, as well as associated absolute and convective instabilities.

Most of the different available experimental techniques have been applied to the study of base flows. The pioneering experimental work on turbulent base flows was performed by Gaviglio, Dussauge, Debiève, and Favre (1977), who made very detailed measurements of such flows using hot-wire anemometry. They were able to obtain mean flow variables and turbulent correlations, but also quantities such as the turbulent kinetic energy production term. The validity of the strong Reynolds analogy (SRA) (see Section 3.5) in different regions of the flow was investigated (Gaviglio et al., 1977) and confirmed to hold in all regions including the initial boundary layers, mixing-layer zone, compression region and, finally the wake itself. Later, experiments have been performed using LDV (Amatucci, Dutton, Kuntz, & Addy, 2003) and planar velocimetry (Scarano & van Oudheusden, 2003) in two-dimensional configurations and LDV in an axisymmetric configuration (Herrin & Dutton, 1994). Three-dimensional flows using Mie scattering have also been investigated (Kastengren & Dutton, 2004). All these studies have contributed to a knowledge of the large scale flow structures as well as an extensive database of mean and turbulent quantities over a Reynolds number range (based on diameter or height) between 0.8 and 3.3×10^6.

Analogous to the case of the supersonic jet flow, the mixing-layer plays a pivotal dynamic role in the development of the base wake flow. With the mixing-layers, the two main characterizing features are the flow structure and spreading rate. The presence of large scale structures in the near wake inside the mixing-layers is similar to the structure of the mixing-layers discussed in Section 6.2. In this region, however, the turbulent activity is intense (Amatucci et al., 2003; Scarano & van Oudheusden, 2003, see also Petrie, Samimy, & Addy, 1986), with observed values about 20% of the external velocity (lower values had been observed by Gaviglio et al. (1977) probably due to some limitations in the measurement system). These values are larger than those reported for the simple mixing-layers for which typical values of the order of 16% are observed (see Figure 6.22). A possible origin of these higher values can be the unsteadiness of the layers, since the stagnation point is not fixed as is the case for a backward facing step. At the end of this mixing region, the stagnation region is an unstable zone with a large excursion of the reattachment location, and two peaks of high turbulent activity. The wake develops downstream with the usual decrease in turbulent activity but with a complex flow structure. When the tunnel size is small, the reattachment shocks shown in Figure 6.21 can reflect at the walls and then interact downstream with the wake itself. This induces an increase of the turbulence and of the spreading rate (Nakagawa & Dahm, 2006).

These physical experiments have been supplemented by direct numerical simulations of axisymmetric transitional flows (Sandberg & Fasel, 2006). A detailed topology of the azimuthal mode has been obtained but is limited

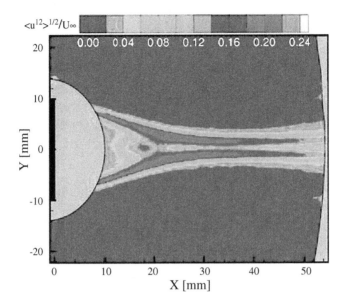

FIGURE 6.22 Turbulence intensity of streamwise velocity component in a supersonic ($M = 2$) bluff body wake. From Scarano and van Oudheusden (2003) with permission.

to a Reynolds number of 10^5, which is about an order of magnitude lower than the corresponding experiments. Nevertheless, these simulation results are in good agreement with the preferred mode experimental results. It is conjectured that both absolutely unstable global modes (linked to the presence of the recirculation region) and convective ones (entrained by the shear layer) are present.

Wakes of cylinders are often used for generating grid turbulence in subsonic flows. At higher speeds, the grid itself can be used as a crude nozzle generating the supersonic flow (Jacquin, Blin, & Geffroy, 1993), and in the transonic regime, grids have been used to produce weakly compressible turbulence. The turbulence generated with grids made from cylinders has characteristics close to subsonic isotropic turbulence (Briassulis, Agui, & Andreopoulos, 2001), with little effect on the power laws. A marked effect, however, is observed on the virtual origin that increases when the Mach number (here transonic) increases. The mechanism responsible for this is related to the compressibility effects on the spreading rate of the mixing-layers that develop on the rods (see Figure 6.23), and described in detail in Section 6.2.2.

It is clear from the discussion here that supersonic base flows (either two-dimensional or axisymmetric) are a difficult flow case for both experiments and computations. Their inherent composition of boundary layers, mixing-layers (with associated reattachment) and subsequent wake lead to both high (small-scale) turbulence intensity and large-scale unsteadiness. Since direct

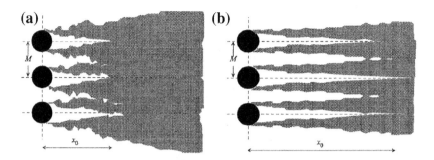

FIGURE 6.23 Schematic representation of: (a) incompressible and (b) compressible generation of grid turbulence. (Note that in this figure M is used to designate grid spacing.) From Briassulis et al. (2001) with permission.

and large eddy simulations of such complex configurations at relevant Reynolds numbers is not yet possible, the evolution toward hybrid computational methods, where time accurate solutions of RANS-type equations are obtained, should make it feasible to compute such base flows.

6.3.2 Flat Plate Wakes

In contrast to the base flows, supersonic slender body wakes are not subjected to strong compressibility effects from expansions and shocks, except those linked to the presence of the leading- and trailing-edge bevels. Since such bevels only correspond to small angles on the order of a few degrees of half-angle, the pressure drops are negligible. The pioneering experimental study in this area was performed by Demetriades (1970, 1976). Since then follow-on studies using Mie scattering (Bonnet & Chaput, 1986), hot-wires (Bonnet, Delville, Sapin, Sullivan, & Yeru, 1991), and schlieren visualizations and unsteady pressure measurements (Gai, Hughes, & Perry, 2002) have been made.

Flat plate wakes develop in a fairly stable manner with organized large scale structures convected downstream. As shown in the schlieren image in Figure 6.24, these structures have a marked two-dimensional character (Bonnet, 1982; Gai et al., 2002). (Recall that this optical method integrates the density variations in all the wake span, so that such marked traces should come from structures with a pronounced two-dimensional character.) The velocity gradient rapidly decreases downstream of the trailing edge in such a way that the convective Mach number is low and similarity is rapidly reached in these wakes. The relative Mach numbers in flat plate supersonic wake flows are then relatively low—on the order of a maximum of 0.5 (Gai et al., 2002). For example, at a free stream Mach number of 1.6 (see Nakagawa & Dahm, 2006), the convective Mach number is 0.5 at $x/\theta = 60$ and as low as 0.2 at $x/\theta = 150$ (θ is the momentum thickness of the wake). Nevertheless, such flows have weak, yet observable, compressibility effects. In contrast to subsonic

FIGURE 6.24 Schlieren image of $M = 2$ flat plate wake (Bonnet, 1982).

wakes, the virtual origin appears to be located upstream of the trailing edge (Gai et al., 2002). This upstream shift of the virtual origin relative to subsonic wakes of the same nature, has also been found in experiments on a thick flat plate wake (Nakagawa & Dahm, 2006). For blunt body wakes (base flows) the compressibility effects on the virtual origin are more difficult to analyze due to the strong expansion–recompression effect inducing the presence of a "neck" (cf. Figure 6.21). Additional parameters including Reynolds number may also influence this upstream shift (Johansson, George, & Gourlay, 2003).

Other compressibility effects have been observed (Bonnet et al., 1991). For external Mach numbers ranging between 1.6 and 3.3, it has been shown that the mean velocity and turbulent profiles follow similarity properties (see Figure 6.25) as is the case for subsonic flows. However, the value of the ratio between the minimum (axis value) and the maximum of the fluctuations and the location of the turbulence activity are affected by the Mach number (see Figure 6.26). For supersonic wakes, there is a larger minimum to maximum turbulence intensity difference, with this difference evolving from 50% down to 10% (subsonic case) as the Mach number decreases. The maxima are located farther from the wake axis (the maximum occurs roughly 20% farther away from the wake axis in the supersonic case) when compared with equivalent subsonic wakes at typically the same nondimensional distance from the trailing edge.

6.4 BOUNDARY LAYERS

The previous sections dealing with free shear flows have illustrated a commonality of structure and dynamics between jets, wakes, and mixing-layers, but at the same time have shown that a fundamental understanding of the dynamics is as yet incomplete and, in addition, difficult to assess. Wall-bounded flows are of crucial importance for numerous practical applications, but the presence of subsonic and low Reynolds number regions of the flow increases the complexity of the flow physics when compared with free turbulent

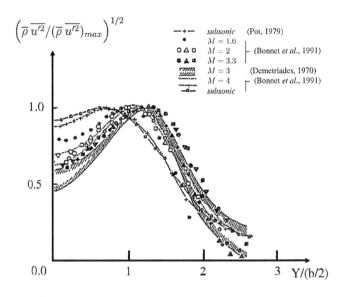

FIGURE 6.25 Transverse distribution of streamwise fluctuations. From Bonnet et al., (1991).

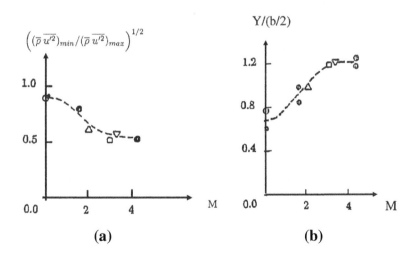

FIGURE 6.26 Turbulence characteristics of a supersonic flat plate wake: (a) minimum/maximum ratio of streamwise fluctuation intensity; (b) transverse location of maximum streamwise fluctuation intensity. From Bonnet et al. (1991).

shear flows. For such flows with mean Mach numbers up through the supersonic regime, $M_e \lesssim 5$, scaling laws (e.g. van Driest, 1951, 1955, 1956a, 1956b) and analogies (Morkovin, 1964) were investigated over a half century ago and have long been validated, by both physical and, now, numerical experiments. In fact, it has been over three decades since Bradshaw (1977) wrote "There is little point in taking turbulence measurements in constant-pressure boundary layers at $M_e < 5$ to check Morkovin's hypothesis, except possibly for intermittency measurements." Interpreted within the context of Morkovin's hypothesis, the statement accurately reflected the fact that assumptions of turbulent correlation behavior in constant-density flows also apply to the corresponding compressible correlations if properly scaled (see Section 3.5.2). Nevertheless, in the intervening period, wall-bounded flows have been further investigated and have been the subject of numerous reviews (Bradshaw, 1977; Spina, Smits, & Robinson, 1994), and collectively, thoroughly cover most of the research through their respective periods. In the last two decades, numerical experiments have moved to the forefront to both replicate and validate previous physical experiments as well as to investigate more deeply the flow structure. In contrast to the earlier recommendation (Bradshaw, 1977) dismissing the need for additional measurements, it is important in performing numerical simulations in constant-pressure boundary layers at $M_e < 5$, to include validation of the simulation data against Morkovin's hypothesis. With this background, numerical experimental results will be the primary focus in this section.

Compressible boundary layer flows have been the subject of numerous experimental studies that have validated the underlying theories associated with the thermal and velocity scale structure. These have been thoroughly vetted in Fernholz and Finley (1996) and a series of reports (Fernholz & Finley, 1977, 1980, 1981; Fernholz, Finley, Dussauge, & Smits, 1989). In the absence of shocks, and with a proper accounting of density and temperature variations, the mean and turbulent flow variables can be characterized in a manner similar to the incompressible flow. For this reason, the multitude of results for incompressible wall-bounded flows are applicable (the reader is referred to the classic texts Cebeci & Smith, 1974; Cebeci & Bradshaw, 1977, 1984). Numerical simulations of supersonic wall-bounded flows have now become common and have produced valuable databases that have allowed for a more in-depth probing of the flow physics than had been previously possible through physical experiments. Prediction methods, such as RANS, have also become more reliable due to increased resolution capabilities, algorithm development, and improved specification of both initial and boundary conditions.

The underlying basis for the development of a family of velocity and thermal profiles applicable to (two-dimensional) boundary layer flows is the preservation of the law of the wall in the equilibrium region of the turbulent boundary layer. The relevant velocity field is the van Driest density-weighted velocity (van Driest, 1951) which is extracted from a straightforward adaptation of the turbulent mixing-length model. When applicable, the influence of compressibility can be attributed mainly to mean density variations rather than fluctuating density or pressure fields. (Recall that this was the basis for Morkovin's hypothesis proposed ten years later and discussed in Section 3.5.2.)

A comparison of the conservation equations used by van Driest (1951), and the corresponding form in terms of Favre, or density-weighted, variables helps to quantify the relationship between the two formulations.

van Driest (1951) considered the turbulent flow in a zero pressure gradient turbulent boundary layer. As in Section 2.5 and for simplicity of notation, the flow is along a flat plate in the $(x_1, x_2) = (x, y)$ plane with x the streamwise direction, and y the direction of the mean shear (normal to plate located at $y = 0$), and with the respective velocities $(\bar{u}_1, \bar{u}_2)$ given by $(\bar{u}, \bar{v})$. If the following variable decomposition is used (cf. discussion in Section 4.2.1)

$$u = \bar{u} + u', \qquad v = \bar{v} + v', \qquad T = \bar{T} + T'$$
$$\rho u = \overline{\rho u} + (\rho u)', \quad \rho v = \overline{\rho v} + (\rho v)', \quad \sigma_{xy} = \bar{\sigma}_{xy} + \sigma'_{xy}, \quad (6.14)$$

then the mean mass, momentum and total energy equations, can be approximated as (van Driest, 1951)

$$\frac{\partial \overline{\rho u}}{\partial x} + \frac{\partial \overline{\rho v}}{\partial y} = 0 \tag{6.15a}$$

$$\overline{\rho u} \frac{\partial \bar{u}}{\partial x} + \overline{\rho v} \frac{\partial \bar{u}}{\partial y} = -\frac{\partial}{\partial y} \left[\overline{(\rho v)' u'} \right] + \frac{\partial \bar{\sigma}_{xy}}{\partial y} \tag{6.15b}$$

$$\overline{\rho u} \frac{\partial \bar{h}}{\partial x} + \overline{\rho v} \frac{\partial \bar{h}}{\partial y} = -\frac{\partial}{\partial y} \left[c_p \overline{(\rho v)' T'} \right] - \overline{(\rho v)' u'} \frac{\partial \bar{u}}{\partial y},$$
$$- \frac{\partial \bar{q}_y}{\partial y} + \bar{\sigma}_{xy} \frac{\partial \bar{u}}{\partial y}. \tag{6.15c}$$

Recall that in Chapter 3 the mean momentum and total energy equations, Eqs. (3.32) and (3.44), respectively, were derived in rather general forms. It is interesting to compare these equations with Eqs. (6.15b) and (6.15c) along with the reduced forms for the turbulent stress tensor, Eq. (3.41), and turbulent heat flux vector, Eq. (3.53). Using the decomposition in Eq. (6.14), it is easy to show that the van Driest formulation is directly applicable to density-weighted variables. For example, the variable $\overline{\rho u}_i$ is nothing more than the definition of the density weighted velocity $\bar{\rho} \tilde{u}_i$.

It is not uncommon to encounter some ambiguity in the literature related to the applicability of the van Driest relations to Reynolds or Favre variables. As the discussion suggests, the scaling and profile family developed are applicable to both Reynolds or Favre variables. (For the present purposes, and unless otherwise noted, the scaling relations will be in terms of the Reynolds variables. This is in keeping with the original derivation; however, it is important for the reader to recognize the interchangeability between the Favre and Reynolds variables in the discussion.) From these equations, it is now possible to construct the velocity and thermal fields associated with both the inner and outer layers of the compressible turbulent boundary layer.

6.4.1 Inner Layer: Mean Field Structure

As in the incompressible case, the inner layer of the turbulent boundary layer is composed of a viscous sublayer and an equilibrium layer, and distinguishing between the velocity and temperature distributions in the sublayer and equilibrium is useful in deducing composite velocity and thermal profiles in the inner layer.

6.4.1.1 Viscous Sublayer

In the sublayer, the advection effects on the left side of Eqs. (6.15b) and (6.15c) can be neglected, and the molecular effects dominate over the turbulent heat flux so that the velocity and temperature fields are governed by

$$\bar{\sigma}_{xy} = \bar{\mu}\frac{\partial \bar{u}}{\partial y} = \tau_w = \bar{\rho}_w u_\tau^2 \tag{6.16}$$

and

$$\bar{q}_y = \bar{k}_T\frac{\partial \overline{T}}{\partial y} = -q_w - \tau_w\bar{u}, \tag{6.17}$$

respectively, with the viscous stress assumed constant and equal to the stress at the wall τ_w ($u_\tau = \sqrt{\tau_w/\bar{\rho}_w}$ the friction velocity), and the wall heat flux $q_w = -(\bar{\mu}_w c_p Pr)(\partial\overline{T}/\partial y)|_w$. As customary, if wall units are used ($\overline{T}^+ = \overline{T}/\overline{T}_w, \bar{u}^+ = \bar{u}/u_\tau, y^+ = \bar{\rho}_w u_\tau y/\bar{\mu}_w$), Eqs. (6.16) and (6.17) can be written as

$$\bar{\sigma}_{xy}^+ = \bar{\mu}^+\frac{\partial \bar{u}^+}{\partial y^+} = 1 \tag{6.18}$$

and

$$\frac{\mu^+}{Pr}\frac{\partial \overline{T}^+}{\partial y^+} = -\left[B_q + (\gamma - 1)M_\tau^2\bar{u}^+\right] \tag{6.19}$$

with

$$M_\tau = \frac{u_\tau}{\sqrt{(\gamma - 1)c_p\overline{T}_w}} \tag{6.20a}$$

$$B_q = \frac{q_w}{\bar{\rho}_w c_p u_\tau \overline{T}_w}, \tag{6.20b}$$

where $Pr = \bar{\mu}c_p/\bar{k}_T$ is the molecular Prandtl number, M_τ the friction Mach number, and B_q the heat flux parameter. The analysis here closely follows that used by Huang and Coleman (1994). Equation (6.18) can be rewritten as

$$\frac{\partial \bar{u}^+}{\partial y^+} = \frac{1}{\mu^+} = \left(\frac{1}{\overline{T}^+}\right)^\varpi, \tag{6.21}$$

where a temperature power law is assumed ($\varpi \approx 0.7$) for the viscosity ratio, and which highlights the strong coupling between the velocity and thermal fields. However, if $\overline{\mu}$ varies slowly with distance from the wall, the sublayer velocity profile behaves in the usual fashion as a linear function of y^+. From Eq. (6.21), Eq. (6.19) can be easily integrated to obtain the temperature distribution in the sublayer as

$$\overline{T}^+ = 1 - Pr\, B_q \overline{u}^+ - Pr\, (\gamma - 1)\, M_\tau^2 \left(\frac{\overline{u}^{+2}}{2} \right). \qquad (6.22)$$

6.4.1.2 Equilibrium Layer

In the equilibrium part of the inner layer, a velocity law of the wall can be deduced (van Driest, 1951). This again requires the assumption that advection is small so the left sides of Eqs. (6.15b) and (6.15c) can be neglected, and now the turbulent correlations dominate over molecular effects. For the velocity field, the turbulent shear stress $-\overline{(\rho v)'u'}$ now dominates and it is assumed both constant and equal to the value at the wall, τ_w,

$$-\overline{(\rho v)'u'} = \tau_w, \qquad (6.23a)$$

and for the temperature field, the turbulent heat flux $\overline{(\rho v)'T'}$ dominates so that the enthalpy equation reduces to

$$c_p \frac{\partial}{\partial y} \left(\overline{(\rho v)'T'} \right) = \tau_w \frac{\partial \overline{u}}{\partial y}. \qquad (6.23b)$$

Both the turbulent shear stress and transverse heat flux can be modeled using a Boussinesq assumption so that eddy viscosity, $\overline{\mu}_t$, and diffusivity, $\overline{\kappa}_t$, relationships can be used,

$$-\overline{(\rho v)'u'} = \overline{\mu}_t \frac{\partial \overline{u}}{\partial y}, \qquad (6.24a)$$

and

$$\overline{(\rho v)'T'} = -\overline{\rho}\,\overline{\kappa}_t \frac{\partial \overline{T}}{\partial y} = -\left(\frac{\overline{\mu}_t}{Pr_t} \right) \frac{\partial \overline{T}}{\partial y}, \qquad (6.24b)$$

respectively, and where $\overline{\kappa}_t = \overline{k}_t/(\overline{\rho} c_p)$ with the turbulent Prandtl number $Pr_t = \overline{\mu}_t c_p/\overline{k}_t$. Using Eq. (6.24b), Eq. (6.23b) readily leads to an equation for the temperature gradient in the equilibrium layer given by (cf. Eq. (6.17))

$$c_p \left(\frac{\overline{\mu}_t}{Pr_t} \right) \frac{\partial \overline{T}}{\partial y} = -q_w - \tau_w \overline{u}. \qquad (6.25)$$

A mixing-layer hypothesis can be used as in the incompressible case (see Pope, 2000; Tennekes & Lumley, 1972, for a through discussion of the incompressible case) to represent both the turbulent eddy viscosity and diffusivities such that $\overline{\mu}_t = \overline{\rho} l^2 \partial \overline{u}/\partial y$ and $\overline{\kappa}_t = l_h \sqrt{\tau_w/\overline{\rho}}$, where l and l_h are the characteristic mixing-lengths associated with the velocity and thermal fields, respectively (Bradshaw, 1977). For each field, the mixing-length is assumed proportional to distance from the wall so that $l = \kappa y$ and $l_h = \kappa_h y$ with κ (κ usually takes its incompressible value 0.41) and κ_h the von Kármán constants for the velocity and thermal fields. Equations (6.23a) and (6.24a) can be combined so that for the velocity field

$$\left(\frac{\overline{\rho}}{\rho_w}\right)^{1/2} \frac{\partial \overline{u}}{\partial y} = \frac{u_\tau}{\kappa y} \tag{6.26}$$

and for the temperature field, Eq. (6.25) then assumes the form

$$c_p \frac{\partial \overline{T}}{\partial y} = -\frac{u_\tau}{\kappa_h y} \sqrt{\frac{\overline{\rho}_w}{\rho}} \left(\frac{q_w}{\tau_w} + \overline{u}\right), \tag{6.27}$$

where $\kappa_h = \kappa/Pr_t$. Since $Pr_t \approx 0.9, l_h > l$ implies that the thermal inner layer region is slightly larger than the velocity inner layer.

Equation (6.26) can be integrated to yield the van Driest transformed velocity

$$u_c = \int_0^{\overline{u}} \left(\frac{\overline{\rho}}{\rho_w}\right)^{1/2} du = \int_0^{\overline{u}} \left(\frac{\overline{T}_w}{\overline{T}}\right)^{1/2} du \tag{6.28}$$

or, rewritten in terms of wall units as

$$\overline{u}_c^+ = \int_0^{\overline{u}^+} \sqrt{\overline{\rho}^+} du = \int_0^{\overline{u}^+} \frac{du}{\sqrt{\overline{T}^+}} = \frac{1}{\kappa} \ln y^+ + C, \tag{6.29}$$

where the perfect gas law has been used and the pressure constant through the boundary layer, and the integration constant C can assume its incompressible value between 5.0 and 5.2. In general, however, both κ and C can be dependent on conditions at the wall so that implicit to these values may be a dependency on B_q and M_τ (Bradshaw, 1977) (see also So, Zhang, Gatski, & Speziale, 1994). The sensitivity of using a transformed van Driest velocity relative to the untransformed velocity is shown in Figure 6.27. As the figure shows, the untransformed velocity is significantly more sensitive to variations of wall heat flux and friction Mach number than the van Driest transformed velocity.

With Eq. (6.26), Eq. (6.27) can be integrated with respect to $\overline{u}$ to obtain the temperature distribution within the equilibrium layer,

$$\overline{T} = \overline{T}_w - Pr_t \left(\frac{q_w}{\tau_w}\right) \frac{\overline{u}}{c_p} - Pr_t \left(\frac{\overline{u}^2}{2c_p}\right), \tag{6.30a}$$

or in terms of wall units as

$$\overline{T}^+ = 1 - Pr_t B_q \overline{u}^+ - Pr_t (\gamma - 1) M_\tau^2 \left(\frac{\overline{u}^{+2}}{2}\right). \tag{6.30b}$$

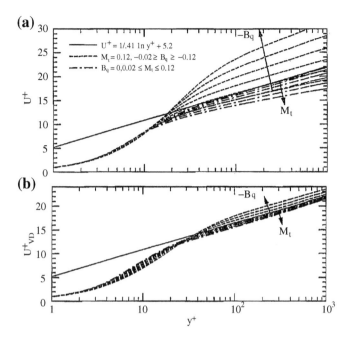

FIGURE 6.27 Effect of varying wall heat flux, B_q, and friction Mach number, M_τ; (a) untransformed and (b) transformed mean velocity. From Huang and Coleman (1994) with permission.

Note that van Driest (1951) obtained this same solution for the thermal field (enthalpy $\bar{h}$) by assuming a generalized Reynolds analogy, that is $c_p \bar{T} = f(\bar{u})$ and $Pr_t = 1$ throughout, but here Pr_t is only assumed constant to further generalize the analysis. It should be cautioned that the integration used in obtaining Eq. (6.30a) (and Eq. (6.30b)) assumed that the equations were valid to the wall, so that a viscous sublayer was not explicitly considered. This is not strictly true, but is the approach that is generally taken. Generalizations, such as accounting for pressure gradient effects, have been proposed in connection with the wall function approach that have been used in RANS methods (Viegas & Rubesin, 1983; Viegas, Rubesin, & Horstman, 1985). Since the mean pressure is assumed constant through the boundary layer, the perfect gas law yields an inverse relation between the mean density and temperature, that is, $\bar{\rho}/\bar{\rho}_w = \bar{T}_w/\bar{T} = 1/\bar{T}^+$. This is consistent with the Crocco–Busemann relation (e.g. White, 1991) that was discussed previously in Section 2.5.

Another feature of Eq. (6.30b) (and Eq. (6.30a)) can be gleaned from its wall-normal derivative. If there is no wall heat transfer, $B_q = 0$, and $\partial \bar{T}^+/\partial y^+$ vanishes at the wall with the temperature distribution showing a continuous decrease from its wall value with increasing distance from the wall. However,

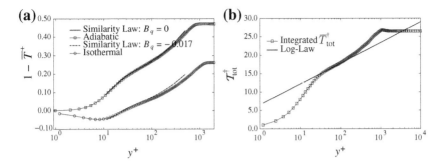

FIGURE 6.28 Mean thermal variable distribution across boundary layer: (a) mean temperature (solid and dashed lines are temperature similarity law distributions from Eq. (6.30b)); (b) integrated mean total temperature from Eq. (6.37) with parameter values $B_q = -0.017, Pr_t = 0.90$, and $T_w = 230\text{K}$ ($T_{aw} = 323\text{K}$). From Shahab et al. (2011) with permission.

if there is wall heat transfer, $B_q \neq 0, \partial \overline{T}^+/\partial y^+$ vanishes at a distance from the wall where $\overline{u}^+ = -B_q/[(\gamma - 1)M_\tau^2]$ ($\overline{u} = -q_w/\tau_w$). From this, the mean temperature will initially increase from its wall value $\partial \overline{T}^+/\partial y^+ > 0$ until the location of vanishing slope and then decrease with increasing distance from the wall (see, for example, Tamano & Morinishi, 2006 and Figure 6.28a).

Since both the wall heat flux and wall shear stress are assumed constant, the ratio q_w/τ_w appearing in Eq. (6.30a) can be related to conditions at the edge of the boundary layer

$$\frac{q_w}{\tau_w} = -\frac{1}{Pr_t}\left(\frac{c_p}{\overline{u}_e}\right)\left[\overline{T}_e - \overline{T}_w + Pr_t\left(\frac{\overline{u}_e^2}{2c_p}\right)\right]. \tag{6.31}$$

For an adiabatic wall condition $q_w = 0$, Eq. (6.31) can be rewritten as an equation for the adiabatic wall temperature, or recovery temperature, given by

$$\overline{T}_{aw} = \overline{T}_r = \overline{T}_e\left[1 + Pr_t\left(\frac{\gamma - 1}{2}\right)M_e^2\right] = \overline{T}_e + r\left(\frac{\overline{u}_e^2}{2c_p}\right), \tag{6.32}$$

which provides an estimate for the recovery factor in this two-dimensional turbulent flow as $r = Pr_t \approx 0.9$ which is consistent with the relation $r = Pr^{1/3}$ (for air $Pr = 0.72$) in the equilibrium layer (cf. White, 1991, pp. 504, 541). Equations (6.31) and (6.32) can be combined to yield the relation

$$\frac{q_w}{\tau_w} = \frac{(\overline{T}_w - \overline{T}_r)}{Pr_t}\left(\frac{c_p}{\overline{u}_e}\right), \tag{6.33}$$

which shows that if heat transfer to the fluid is to occur, the wall temperature must be higher than the recovery temperature (the adiabatic wall temperature).

If the expression for the temperature ratio extracted from Eq. (6.30a) is used, the van Driest transformed velocity, Eq. (6.28), can be written as (cf. Huang & Coleman, 1994)

$$\bar{u}_c = \sqrt{b} \left[\sin^{-1} \left(\frac{a_w + \bar{u}}{\sqrt{a_w^2 + b}} \right) - \sin^{-1} \left(\frac{a_w}{\sqrt{a_w^2 + b}} \right) \right], \tag{6.34}$$

with $a_w = q_w / \tau_w$ (see Eq. (6.33)) and $b = 2 c_p \overline{T}_w / Pr_t$.

Equation (6.30b) can be rewritten in terms of a total temperature variable, $\overline{T}_{tot}^+$, as (cf. Carvin, Debiève, & Smits, 1988; Debiève, Dupont, Smith, & Smits, 1997)

$$\overline{T}_{tot}^+ = \overline{T}^+ - 1 + Pr_t M_\tau^2 (\gamma - 1) M_\tau^2 \left(\frac{\bar{u}^{+2}}{2} \right) = -Pr_t B_q \bar{u}^+ \tag{6.35}$$

or, in a rescaled form using the friction temperature, $-B_q T_w$, as

$$\overline{T}_{tot}^\dagger = \frac{\overline{T}_{tot}^+}{-B_q T_w Pr_t} = \bar{u}_1^+. \tag{6.36}$$

Multiplying the differential form of Eq. (6.36) by $\sqrt{\overline{\rho}^+}$ and integrating yields

$$T_{tot}^\dagger = \int_0^{\overline{T}_{tot}^\dagger} \frac{d\overline{T}_{tot}^\dagger}{\sqrt{\overline{T}^+}} = \frac{1}{\kappa} \ln y^+ + C_{tot}. \tag{6.37}$$

The intercept coefficient C_{tot} may differ slightly from that of the van Driest transformed velocity given in Eq. (6.29) since the sublayer region is not accounted for in the integration. When the isothermal temperature distribution in Eq. (6.30b) (see Figure 6.28a) is recast in terms of $\overline{T}_{tot}^\dagger$ using Eq. (6.35), the integrated variable $T_{tot}^\dagger$ will then follow the log-law distribution. In Figure 6.28b the integrated temperature variable $T_{tot}^\dagger$ is shown along with a log-law distribution with slope κ^{-1}. The scalings used here are common, but there are alternate scalings for the van Driest transformation and (total) temperature similarity profile. For example, a new spatial variable dependent on the variation of (dynamic) viscosity through the layer can be used (Brun, Boiarciuc, Haberkorn, & Comte, 2008). It has the desirable effect of better collapsing the transformed mean velocity and temperature profiles over a wider range of parameters due to the improved accounting of wall-normal temperature variations.

6.4.1.3 Law of the Wall

With the structure of the mean velocity and temperature fields within the sublayer and equilibrium layer established, it is possible to consider a representation of these fields across the inner layer. Simply applying the van Driest transformation in Eq. (6.28) to the equation describing the velocity field in the viscous sublayer, Eq. (6.21), is inadequate since the resultant velocity field

$$\frac{\partial \overline{u}_{sc}^+}{\partial y^+} = \left(\frac{1}{\overline{T}^+}\right)^{1/2} \left(\frac{1}{\overline{T}_s^+}\right)^{\varpi},$$ (6.38)

where the subscript s has been inserted into the temperature variable associated with the sublayer. As such, it is still dependent on both the sublayer and equilibrium temperatures, and is unable to provide an adequate solution for the inner layer velocity field. It is necessary to make a more general assumption about the merging of the two layers.

Following van Driest (1956a) (see also Huang & Coleman, 1994), the stress in the inner layer is assumed constant, but is now the sum of the viscous stress and turbulent shear stress. This means that rather than Eq. (6.16) and Eq. (6.23a) each being equated to τ_w, it is now the sum $\overline{\sigma}_{xy} - \overline{\rho u'v'}$ equal to the stress at the wall. The mixing-length assumption for the eddy viscosity in Eq. (6.24a) is now used so that the normalized wall shear stress τ_w^+ can be written as

$$\tau_w^+ = 1 = \overline{\mu}^+ \frac{\partial \overline{u}^+}{\partial y^+} + \overline{\rho}^+ l^{+2} \left(\frac{\partial \overline{u}^+}{\partial y^+}\right)^2.$$ (6.39)

The solution to this quadratic equation gives the composite mean velocity gradient distribution for the inner layer as

$$\overline{\mu}^+ \frac{\partial \overline{u}^+}{\partial y^+} = \frac{2}{1 + \sqrt{1 + 4(\overline{\rho}^+ l^{+2}/\overline{\mu}^{+2})}}.$$ (6.40)

Equation (6.40) satisfies the respective sublayer ($l^+ \to 0$) and equilibrium layer ($l^+ \to \kappa y^+$, log-layer) limits, and the insertion of a suitable (van Driest) damping function into the length scale l^+ specification yields a continuous velocity gradient distribution across the inner layer. Although it is not possible to obtain an analytic solution to Eq. (6.40) for the velocity field, it can be integrated numerically to obtain a composite profile in the inner layer.

For incompressible boundary layer flows, various composite profiles have been proposed (Cebeci & Bradshaw, 1977, 1984). For smooth surfaces, a composite inner layer solution proposed by Reichardt (1951),

$$\overline{u}_c^+|_{\text{inner}} = C_1 \left(1 - e^{-y^+/11} - \frac{y^+}{11} e^{-0.33y^+}\right) + \frac{1}{\kappa} \ln\left(1 + \kappa y^+\right)$$ (6.41)

($C_1 = C - \kappa^{-1} \ln \kappa \approx 7.5$), has been used for validation of the van Driest transformed velocity from DNS results of isothermal (adiabatic wall) flow (Guarini, Moser, Shariff, & Wray, 2000; Pirozzoli, Grasso, & Gatski, 2004).

Similarly, Eqs. (6.17) and (6.25) show the temperature gradient in the sublayer and equilibrium layers individually equated to a right hand side composed of the wall heat flux q_w and wall shear stress τ_w that are assumed constant and valid through the inner layer. A composite distribution for the mean temperature gradient can now be constructed by assuming the heat flux distribution across the inner layer is the sum of the molecular (sublayer), Eq. (6.17), and turbulent (equilibrium layer), Eq. (6.25), contributions which are then equated to the wall heat flux and wall shear stress terms. This composite solution for the temperature gradient (in wall units) is then given by

$$\frac{\partial \overline{T}^+}{\partial y^+} = -\left[B_q + (\gamma - 1)M_\tau^2 \overline{u}^+\right]\left[\frac{Pr\,Pr_t}{\overline{\mu}_t^+ Pr + \overline{\mu}^+ Pr_t}\right]. \tag{6.42}$$

Although based on simple mixing-layer assumptions, Huang and Coleman (1994) were able to validate both the velocity and temperature distributions obtained from the composite models in Eqs. (6.40) and (6.42) against direct simulation channel flow data. They found that the models well replicated the DNS data from an isothermal (cold-wall) channel flow simulation (Coleman, Kim, & Moser, 1995) through most of the log-layer for the lower heat flux parameter case.

In contrast to the approach used in obtaining Eq. (6.42) that merged the sublayer and equilibrium layer solutions, van Driest (1954, 1955) (see also van Driest, 1952) defined a mixed Prandtl number Pr_m,

$$Pr_m = \left(\mu + \mu_t\right)\left[\frac{\mu}{Pr} + \frac{\mu_t}{Pr_t}\right]^{-1}, \tag{6.43}$$

that accounted for the migration from molecular to turbulent Prandtl number with increasing distance from the wall. The Crocco transformation was applied to the boundary layer equations, Eq. (6.15a–c), so that under the assumption of no streamwise variation in the flow the mean energy equation, Eq. (6.15c), could be written as (cf. Carvin et al., 1988)

$$\tau_t^+ \left[\frac{\partial}{\partial \overline{u}^+}\left(\frac{1}{Pr_m}\frac{\partial \overline{T}^+}{\partial \overline{u}^+}\right) + (\gamma - 1)M_\tau^2\right] + \frac{1 - Pr_m}{Pr_m}\frac{\partial \overline{T}^+}{\partial \overline{u}^+}\frac{\partial \tau_t^+}{\partial \overline{u}^+} = 0, \tag{6.44}$$

where τ_t^+ represents the sum of the viscous and turbulent stress. For the case when $Pr_m = 1 (= Pr = Pr_t)$, the solution to Eq. (6.44) is simply Eq. (6.22) with $Pr = 1$. If $Pr_m \neq 1$, or more generally, not constant (as in real flows) then the integration of Eq. (6.22) becomes more problematic. Finally, recalling the definition of the turbulent Prandtl number in Eq. (3.103) with the generalization of $Pr_m = Pr_t$, Eq. (6.44) can be integrated to obtain (see for example Smits & Dussauge, 2006) an integral relation for $\partial \overline{T}^+/\partial \overline{u}^+$ and yielding an SRA-type analogy between the turbulent shear stress $\overline{u'v'}$ and wall-normal heat flux $\overline{v'T'}$.

6.4.2 Outer Layer: Law of the Wake

Although the focus of mean flow analysis has been justifiably focused on the inner layer, replication of the outer layer dynamics from predictive methods and simulations, as well as experiments, is also useful. For example, predictive methods have inherent constraints that should be met for accurate flow predictions. Two such constraints for incompressible flows (Catris & Aupoix, 2000b) are the prediction of a correct logarithmic region slope and the smooth variation of flow variables and proper wake function behavior at the edge of the turbulent region (outer layer). While these constraints were developed for incompressible flows, their adaptation to the compressible case is clearly possible.

In the outer layer, effects of freestream boundaries on either experiments or numerical computations or on inflow conditions in numerical computations can be quantified when compared to a well-established functional form. As with the velocity law of the wall, an incompressible functional form can be used (at least for adiabatic conditions) to assess the compressible transformed van Driest velocity profile. Coles (1956) originally introduced a function $w(y/\delta) = 1 - \cos \pi y/\delta$ for incompressible flow that described the deviation of the outer layer profile from the law of the wall and is in general a function of streamwise distance. This law of the wake function was an empirical fit and did not yield zero slope for the velocity profile at $y/\delta = 1$. An alternate form proposed by Finley et al. (see Cebeci & Bradshaw, 1977),

$$\bar{u}_c^+ |_{\text{outer}} = \frac{1}{\kappa} \left[\left(\frac{y}{\delta_c} \right)^2 - \left(\frac{y}{\delta_c} \right)^3 + 6\Pi \left(\frac{y}{\delta_c} \right)^2 - 4\Pi \left(\frac{y}{\delta_c} \right)^3 \right], \quad (6.45)$$

(δ_c, boundary layer thickness based on van Driest velocity) remedies this deficiency and is a preferred form. Guarini et al. (2000) determined κ and C from the simulation data to be 0.40 and 4.70, respectively, with a corresponding value of $\Pi = 0.25$ for the wake parameter. As Figure 6.29 shows, the agreement between the composite velocity profile and their DNS data is very good. Similar agreement was obtained by Pirozzoli et al. (2004) who used the functional form in Eq. (6.45), but with values for κ and C chosen as 0.41 and 4.90, respectively. However, their value of the wake parameter was determined as $\Pi = 0.45$. Ringuette, Wu, and Martin (2007) performed a boundary layer simulation but did not compare to the functional form in Eq. (6.45), although the results shown for the wake parameter below suggest the level of agreement for the compressible velocity profile would be similar.

Spatial large eddy simulations of supersonic boundary layer flows have also been made at similar freestream Mach numbers as the direct simulations just discussed. Examples of these are the simulations of Urbin and Knight (2001) and Stolz and Adams (2003) who both adapted a rescaling and recycling method (originally proposed by Lund, Wu, and Squires, (1998)) for the imposition of inflow conditions. This allowed for a short spatial transient development of the

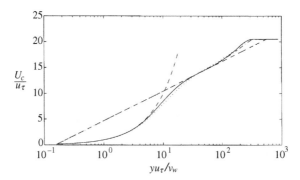

FIGURE 6.29 van Driest transformed velocity across boundary layer: ———, DNS; — — —, linear sublayer; — — —, log-law; ·····, Reichardt's profile with Finley's wake function; — · —, Coles' profile (Coles, 1956). From Guarini et al. (2000) with permission.

flow within the domain. Urbin and Knight (2001) compared results for both a Smagorinsky model and monotonically integrated LES approach for a Mach 3 boundary layer but at a relatively low Reynolds number $Re = 2 \times 10^4$ (based on freestream conditions), and Stolz and Adams (2003) used an approximate deconvolution method (ADM) at Mach 2.5 but at higher Reynolds numbers. Stolz (2005) also used the same inflow condition generation method, but for the subgrid scale modeling employed a high-pass filtered eddy viscosity model. Although both the direct and large eddy simulation studies were relatively close in freestream Mach number, the corresponding momentum thickness Reynolds number (based on freestream and wall viscosity) had larger variations. For these compressible simulation studies the wake parameter variations are shown in Figure 6.30. Overall, the simulation data shown are in good agreement with Coles' mean profile. Such sensitivity has also been documented for incompressible experimental and simulation data (Fernholz & Finley, 1996). It is rather common for numerical simulations of wall-bounded and free shear flows to be performed within a temporal frame rather than within a spatial frame, with the distinction being between "sustained" and "fully developed" turbulent flow (Spalart, 1988). It is not clear whether a quantity such as the wake parameter Π, which is strongly dependent on streamwise variation, and consequently the law of the wake can be adversely affected.

The fact that the van Driest transformed mean velocity field can be related to its incompressible counterparts allows for a thorough validation of numerical simulation results (e.g. direct or large eddy). Since relations between the mean thermal field and velocity field can be established this also provides another level of validation comparisons. Averaged methods, such as RANS, can also benefit from such validations, and can use such relations in calibration of closure models or at least an assessment of model consistency.

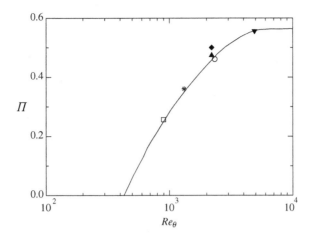

FIGURE 6.30 Wake parameter Π as a function of momentum thickness Reynolds number from numerical simulations: $\circ$, Pirozzoli et al. (2004); $\square$, Guarini et al. (2000); $\star$, $M = 2.92$, Ringuette et al. (2007); $\blacktriangle$, $M = 2.5 (Re_\theta = 2202)$ and $\blacktriangledown$, $M = 2.5$ ($Re_\theta = 4902$) Stolz and Adams (2003); $\blacklozenge$, $M = 2.5$ Stolz (2005); ———, Coles' mean curve (Coles, 1962).

In this latter case, a useful example is the consistency of the modeling of the diffusion term in the turbulent scale equation (e.g. ε, ω, etc.) needed in both two-equation and Reynolds stress models (these were discussed in Section 5.3.2 and a modified form for a length scale variable was given in Eq. (5.61)). Figure 6.31 shows the effect of this correction on some common two-equation models. The $K - \varepsilon$ model appears to benefit the most with this correction, although the models shown display improved approximation in the log-law region. Such a comparison quantifies the relative effect of the model modification and highlights the potential for poor predictions if no such density gradient modification is introduced.

6.4.3 Integral Parameters

Integral parameters are closely linked with the analysis of boundary layer flows. In this section, parameters related to the integral relations for displacement, momentum and (total) energy thicknesses are discussed. From the momentum and displacement thicknesses, the skin-friction and shape factor parameters can be extracted which are related to the mean density and velocity fields. From the enthalpy integral, the mean temperature field is introduced as well as the mean kinetic energy and turbulent kinetic energy fields. In low speed (incompressible) flows (perfect gas) this reduces to an energy integral and is related to a Stanton number (White, 1991), but in high speed compressible flows the kinetic energy contributions are retained. In incompressible flows, since general trends for these quantities have been well documented and quantified,

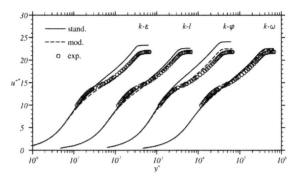

FIGURE 6.31 Effect of diffusive flux modification on two-equation prediction of van Driest mean velocity profile: ———, standard model; — — —, modified model (for $k - \ell$ model see Smith, 1995). From Catris et al. (2000a) with permission.

physical measurements, numerical simulations and RANS computations have utilized this information for validation and calibration purposes. Although not as extensively used, these same integral parameters can be similarly utilized in the compressible regime.

6.4.3.1 Skin-Friction

For zero pressure-gradient flows, the skin-friction along the wall is directly obtained from the momentum thickness, which for a compressible flow is given by

$$c_f = \frac{2\tau_w}{\overline{\rho}_e \overline{u}_e^2} = 2\frac{d}{dx} \int_0^{y_e} \frac{\overline{\rho}\,\overline{u}}{\overline{\rho}_e \overline{u}_e}\left(1 - \frac{\overline{u}}{\overline{u}_e}\right)dy, \qquad (6.46)$$

and where it is understood here that the limit of integration only extends to the boundary layer edge. The required density and velocity distributions are obtained from Eq. (6.30a) (recall the inverse relation between density and temperature) and Eq. (6.34). As might be expected, the integrand is rather complex and van Driest expanded it in a series using integration by parts (see p. 152 of van Driest (1951)) to obtain for van Driest I

$$\frac{1}{\sqrt{F_c c_f}} = 4.15 \log\left[Re_x c_f \left(\frac{\overline{\mu}_e}{\overline{\mu}_w}\right)\right] + 1.70 - \left(\frac{1}{2} + \varpi\right)\log\left(\frac{\overline{T}_w}{\overline{T}_e}\right), \quad (6.47)$$

or (see p. 30 of van Driest (1956b)), for van Driest II

$$\frac{1}{\sqrt{F_c c_f}} = 4.15 \log\left[Re_x c_f \left(\frac{\overline{\mu}_e}{\overline{\mu}_w}\right)\right] + 1.70 - \varpi \log\left(\frac{\overline{T}_w}{\overline{T}_e}\right), \qquad (6.48)$$

where the Reynolds number, $Re_x = \overline{\rho}_e \overline{u}_e x / \overline{\mu}_e$, is based on boundary layer edge values, and

$$F_c = \frac{\overline{T}_{aw}/\overline{T}_e - 1}{(\sin^{-1} A + \sin^{-1} B)^2} \tag{6.49}$$

with

$$A = \frac{\overline{T}_{aw}/\overline{T}_e + \overline{T}_w/\overline{T}_e - 2}{[(\overline{T}_{aw}/\overline{T}_e + \overline{T}_w/\overline{T}_e)^2 - 4(\overline{T}_w/\overline{T}_e)]^{1/2}}$$

$$B = \frac{\overline{T}_{aw}/\overline{T}_e - \overline{T}_w/\overline{T}_e}{[(\overline{T}_{aw}/\overline{T}_e + \overline{T}_w/\overline{T}_e)^2 - 4(\overline{T}_w/\overline{T}_e)]^{1/2}}, \tag{6.50}$$

and $\overline{T}_{aw}$ is defined in Eq. (6.32). In addition to the original papers by van Driest, the reader is referred to the texts by Cebeci and Bradshaw (1984) and White (1991) for additional discussion. As noted above, the Van Driest I derivation (van Driest, 1951) is based on Prandtl's mixing-length formula; whereas, Van Driest II is based on von Kármán's similarity law where $l = \kappa |(\partial \overline{u}/\partial y)/(\partial^2 \overline{u}/\partial y^2)|$, and provides the best comparison with experimental data. When compared with the Kármán-Schoenherr incompressible relation,

$$\frac{1}{\sqrt{c_{finc}}} = 4.15 \log (Re_x c_{finc}) + 1.70, \tag{6.51}$$

Equation (6.48) can be rewritten as

$$c_f = \frac{1}{F_c} c_{finc} \left(Re_{\theta inc} \right), \tag{6.52}$$

where $Re_{\theta inc} = Re_\theta (\overline{\mu}_e / \overline{\mu}_w) = Re_\theta F_\theta$ is the incompressible momentum thickness Reynolds number. Although the Van Driest II skin-friction relation is probably the most often employed, other formulas have been proposed, and many of these have been extensively assessed over three decades ago (Bradshaw, 1977; Spalding & Chi, 1964). Apparently never investigated, Bradshaw (1977) proposed a Van Driest III model that addressed some of the uncertainties introduced into either Van Driest I or II. For example, it was pointed out following Eq. (6.30a) that the temperature relation was obtained assuming that integration was valid to the wall—which is not strictly the case, and thus required a constant of integration other than $\overline{T}_w$. This proposed modification, coupled with a better assessment of the intercept coefficient C in the log-law and the wake parameter, could lead to better overall data collapse. More recently, a skin friction relation for internal flows with adiabatic and non-adiabatic walls and smooth and rough walls has been proposed (De Chant, 1998). Unfortunately, the skin friction relationship is implicit and an iterative method needs to be employed for a solution.

In general, the compressibility transformations have the functional dependencies $F_c = F_c(M_\infty, \overline{T}_w)$ and $F_\theta = F_\theta(\overline{T}_w)$ so that a plot of $F_c^{-1} - F_\theta$

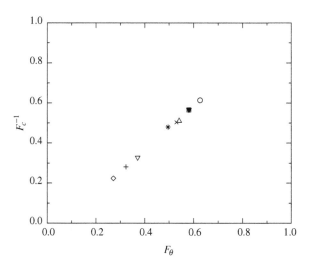

FIGURE 6.32 Influence of freestream Mach number on van Driest compressibility functions. ○, $M = 2.25$, Pirozzoli et al. (2004); □, $M = 2.5$, Guarini et al. (2000); ▼, $M = 2.5$, Stolz and Adams (2003); *, $M = 2.92$ and ×, $M = 2.98$, Ringuette et al. (2007); △, $M = 3$, ▽, $M = 4.5$, ◇, $M = 6$, Maeder et al. (2001); +, $M = 4.97$, Ringuette, Martin, Smits, and Wu (2006).

can be used to isolate the freestream Mach number effect on the transformation. With the relatively recently augmented capability of first large eddy and now direct numerical simulations of compressible boundary layer flows, the skin-friction formula can also be used to assess the relative contributions of the simulations. Figure 6.32 shows this variation through a collection of recent simulation studies covering a freestream Mach number range of 2.25 − 6. As the incompressible values of $F_\theta = F_c = 1$ are approached, one would expect the effects of compressibility on the mean flow (skin-friction) to be correspondingly reduced. Since the studies shown had wall temperatures either at or within 20% of the adiabatic wall temperature, Figure 6.32 shows the link between increased freestream Mach number and increased influence of the mean flow compressibility factors. If the freestream Mach number is fixed with a subsequent lowering of the wall temperature (e.g. a reduction of wall temperature to adiabatic wall temperature ratio), a lower viscosity in the near-wall region would result. A consequence of this, from a numerical standpoint, would be the need for improved numerical resolution requirements.

There have been innumerable studies of incompressible zero pressure gradient boundary layer flows over a wide range of Reynolds numbers (see Figure 6.33a). As numerical simulations become more prevalent and the numerical database matrix is filled, it is useful to determine the current range of investigation. Of course, numerous RANS calculations exist as well, but issues associated with choice of closure enter and such studies are intended

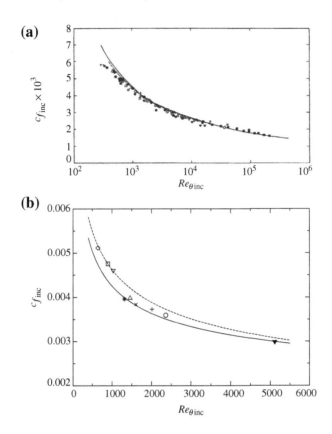

FIGURE 6.33 Skin-friction variation with momentum thickness Reynolds number. Incompressible data (a): ———, Eq. (9) of Fernholz and Finley (1996); - - - -, from Figure 4 of Coles (1964). From Fernholz and Finley (1996) with permission. Compressible data (b): ○, $M = 2.25$ Pirozzoli et al. (2004); □, $M = 2.5$ Guarini et al. (2000); ▼, $M = 2.5$ Stolz and Adams (2003); △, $M = 3$ and ▽, $M = 4.5$ and ◇, $M = 6$ Maeder et al. (2001); ★, $M = 2.92$ and ×, $M = 3.98$ and +, $M = 4.97$, Ringuette et al. (2006, 2007); ———, Kármán-Schoenherr (Eq. 6.51); - - - -, Blasius ($c_{f\text{inc}} = 0.026/Re_{\theta\text{inc}}^{1/4}$).

for "prediction" rather than being viewed as a database source and useful for validation. Figure 6.33b shows a comparison of some recent DNS and LES studies of supersonic boundary layer flows transformed using Eq. (6.52). With the exception of Pirozzoli et al., 2004, the DNS studies were temporal simulations so only one value of skin friction applies, and in the spatial developing case (Pirozzoli et al., 2004) only the last point in the simulation domain is used. For the LES of Stolz and Adams (2003) a short spatial transient region was obtained from their rescaling and recycling method. In this case, the location chosen for Figure 6.33b was midway in the region. The advantage of having a viable LES methodology is apparent from the figure since it is possible to achieve Reynolds number values more consistent with the abundance of incompressible data. As Figure 6.32 showed, increasing the freestream Mach number resulted in a decrease of the compressibility function F_θ and an increase in F_c^{-1}. This decrease in F_θ with increasing M_e results in increasingly lower Re_θ values. This suggests that in order to achieve higher Re_θ values at fixed M_e, the ratio $\overline{T}_w/\overline{T}_{aw}$ must be reduced. As noted previously, such a temperature ratio reduction then requires increased resolution for accurate wall proximity values.

6.4.3.2 Shape Factor
Another parameter useful in the assessment and comparison of compressible boundary layer simulations (and experiments) is the shape factor H defined as the ratio of the displacement thickness to the momentum thickness and written here for completeness as

$$
H = \frac{\delta^*}{\theta} = \left[\int_0^{y_e} \left(1 - \frac{\overline{\rho u}}{\overline{\rho}_e \overline{u}_e} \right) dy \right] \left[\int_0^{y_e} \frac{\overline{\rho u}}{\overline{\rho}_e \overline{u}_e} \left(1 - \frac{\overline{u}}{\overline{u}_e} \right) dy \right]^{-1}. \quad (6.53)
$$

Figure 6.34 shows the compressible shape factor H reported from the same studies shown in Figure 6.33. Unlike the transformed skin-friction distributions, the shape factors computed using the untransformed mean velocities show a large scatter across a rather broad Mach number range. As Stolz and Adams (2003) (their Figure 5b) and Stolz (2005) (his Figure 3) have shown this type of scatter is exhibited by a variety of experimental data (Coles, 1954;

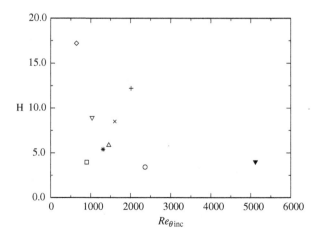

FIGURE 6.34 Shape factor variation with momentum thickness Reynolds number: $\circ$, $M_\infty = 2.25$, Pirozzoli et al. (2004); $\square$, $M_\infty = 2.5$, Guarini et al. (2000); $\blacktriangledown$, $M_\infty = 2.5$, Stolz and Adams (2003); $\triangle$, $M_\infty = 3$ and ∇, $M_\infty = 4.5$ and $\diamond$, $M_\infty = 6$, Maeder et al. (2001); $\star$, $M_\infty = 2.92$ and $\times$, $M_\infty = 3.98$ and $+$, $M_\infty = 4.97$, Ringuette et al. (2006, 2007).

Fernholz & Finley, 1977, 1981). While numerous reasons exist for measurement uncertainties, numerical simulations (particularly direct simulations) should minimize such scatter if associated with "measurement error" unless the results have an implicit dependence on other parameters that need to be taken into account in order to have a meaningful comparison—such as in the case of the transformed skin-friction. Other than the obvious general trend of a reduction of shape factor with increasing Reynolds number and the potential asymptotic limit at high Reynolds number, the data shown in Figure 6.34 is not readily amenable to quantitative assessment. However, Michel (see Cousteix, 1989, p. 500) has shown that when the shape factor is plotted against freestream Mach number there is indeed a correlated trend. The same cases shown in Figure 6.34 are now replotted in Figure 6.35 against freestream Mach number and compared with the correlation formula proposed by Michel,

$$H_m = H_{inc} + 0.4 M_e^2 + 1.222 \frac{\overline{T}_w - \overline{T}_{aw}}{\overline{T}_e}, \qquad (6.54)$$

for the case of adiabatic wall conditions with $H_{inc} = 1.4$. The correlation formula in Eq. (6.54) clearly provides an excellent representation for the simulation data over a broad range of Reynolds numbers (cf. Figure 6.34). In addition, it also provides a mean flow validation guide for subsequent numerical computations using direct, filtered, or averaged methods.

Once again, the underlying dynamic complexity of an apparently "simple" compressible flow is shown through an evaluation of a mean flow integral parameter. Clearly, in more complex cases where equilibrium layers cease to exist such as in flows with pressure gradients or significant wall heat transfer

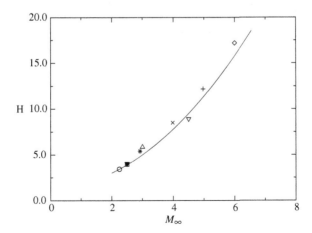

FIGURE 6.35 Shape factor variation with freestream Mach number: ○, Pirozzoli et al. (2004); □, Guarini et al. (2000); ▼, Stolz and Adams (2003); △, ▽, ◊, Maeder et al. (2001); ⋆, ×, +, Ringuette et al. (2006); Ringuette et al. (2007); ——, Cousteix (1989).

effects, the uncertainty in both measured and simulation data will increase and validation of results will become more difficult.

6.4.3.3 Total Energy Thickness

The total temperature variation across the boundary layer comprises both the mean field velocity and temperature as well as the turbulent kinetic energy, and is given by Eq. (3.47b). Analogous to the defining relations for the displacement and momentum thicknesses, the total energy thickness can be defined as,

$$\Delta = \int_0^{y_e} \frac{\overline{\rho u}}{\overline{\rho_e} \overline{u}_e} \left(\frac{\widetilde{T}_t}{\widetilde{T}_{te}} - 1 \right) dy. \tag{6.55}$$

A streamwise evolution equation can be correspondingly derived for Δ (see for example Cousteix (1989)) that can account for both freestream accelerations and wall heat flux q_w effects. In the absence of these accelerations and non-adiabatic conditions, Δ is constant. Recall in the discussion in Section 3.5.1 that the original SRA forms assumed negligible total temperature fluctuations as well as a constant mean total temperature across the boundary layer and equal to the wall temperature. While this low-order approximation does not hold in practice, the vanishing of Δ should approximately hold. Figure 6.36 shows the variation of the total energy across the boundary layer for both adiabatic and cold wall cases. For the adiabatic case, the overall trend is for an increase in total temperature with distance from the wall; although, this case is also characterized by a slight overshoot in the distribution in the outer layer of the boundary layer. Since the total temperature ratio $\widetilde{T}_t/\widetilde{T}_{te} < 1$ in the inner layer, there must be a region where it exceeds unity in order for total energy thickness to vanish. In contrast,

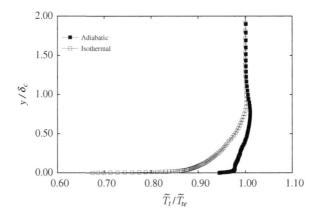

FIGURE 6.36 Mean total temperature distribution across boundary layer (δ_c is the van Driest boundary layer thickness for the adiabatic and isothermal cases respectively). From Shahab et al. (2011) with permission.

for the isothermal case, the variation of heat flux along the wall in the streamwise direction removes the restriction that the total energy thickness be constant and leads to a monotonic growth of total energy with distance from the wall. This behavior has also been observed experimentally (Smits & Dussauge, 2006).

6.4.4 Turbulent Field

Unlike the free shear flows discussed earlier in this chapter, the influence of compressibility in boundary layer flows is less pronounced on the statistical correlations involving the velocity and thermal fluctuations. This can be partly quantified through parameters such as the gradient Mach number M_g which is $\approx 0.5 M_t$ for boundary layers and $\approx 2.5 M_t$ for mixing-layers. Under adiabatic conditions and without shocks, Morkovin's hypothesis has been well established (see Section 3.5.2) and as such the vast data resource associated with incompressible wall-bounded flows (e.g. Fernholz & Finley, 1996) can be drawn upon for either model validation or analysis (see Saric et al., 1996, for a summary). In addition, the various forms of the strong Reynolds analogy discussed in Section 3.5 are useful in developing self-consistent closure models for averaged methods such as RANS and for assessment of numerical simulation database quality.

In the discussion focused on the development of the boundary layer structure using the mean field variables the intent was to replicate the historical development originating over a half-century ago. These derivations preceded the usage of density-weighted variables and the analysis was confined to the usual Reynolds variables. However, as the relationships

$$\overline{(\rho v)'u'} = \overline{\rho}\overline{u'v'} + \overline{v}\overline{\rho'u'} + \overline{\rho'u'v'}$$
$$= \overline{\rho}\widetilde{u''v''} + \widetilde{v}\overline{\rho'u'} \tag{6.56}$$

and

$$\overline{(\rho v)'T'} = \overline{\rho}\widetilde{v''T''} + \overline{v}\overline{\rho'T'} + \overline{\rho'v'T'}$$
$$= \overline{\rho}\widetilde{v''T''} + \widetilde{v}\overline{\rho'T'} \tag{6.57}$$

suggest, the discussion of the turbulent correlations may be best served with the presentation of the density-weighted, Favre, variable correlations; although, the assumption of neglecting the correlations involving the fluctuating density is required to obtain equality with the density-weighted moments. Correspondingly, the (general) relationship between the Favre and Reynolds averaged correlations is given by Eq. (5.35) for the fluctuating velocity field (see Eqs. (6.62) and (6.63) for analogous forms for the temperature), and is related through the appearance of higher-order correlations involving the density and velocity fluctuating quantities. Since LES simulations and RANS computations are expressed in terms of Favre variables with closure required of the associated stresses, it is the Favre or density-weighted correlations that are the more relevant quantity. In the absence of both shocks and strong wall heat transfer effects, the difference between the Reynolds mean and density-weighted correlations is nevertheless small.

6.4.4.1 Turbulent Velocity Field

As outlined in Chapter 4, measurements in the supersonic regime are challenging even with the latest advances in measurement technology. Reynolds number effects, for example, dictate the relative importance of the near-wall region, but here measurements are difficult. In particular, the stress components normal to the wall and along the span pose additional difficulties, so similarity scalings are even more challenging to identify for these components. Nevertheless, there has been an accumulation of information over a few decades so that some general consensus has been reached on some correlation quantities, and with the continued advances in computer power more DNS are becoming available; however, cost and time factors currently enter in to preclude extensive parametric simulation studies with such methods. Unfortunately, it is not as yet clear whether compressible LES, where subgrid or subfilter scales need to be accounted for, are sufficiently robust to be used for detailed analysis of flow physics. (Note that averaged RANS methods, with the turbulence dynamics completely modeled, are primarily tools for prediction, and analysis of dynamics with such methods is severely restricted.) While an accumulation of large independent simulation databases may be desirable, a negative consequence of this is the identification and quantification of differences between results from different studies, especially for turbulent correlations. Although general trends can be and have been deduced in some cases, it is difficult to precisely quantify the behavior of all statistical correlations across the boundary layer especially over a wide range of wall thermal conditions.

For the turbulent stresses τ_{ij} ($= \widetilde{u_i'' u_j''}$), there is no intrinsic scaling so the proper choice of normalization is dictated by a collapse of the correlation data. Fortunately, there is a similarity with the incompressible regime using Morkovin's scaling and without that scaling a clear decrease in the turbulence fluctuation level occurs with increasing Mach number. (Recall that reduced turbulent fluctuation levels were the origin of the reduced spreading rates in the supersonic mixing-layers.) As the Mach number increases into the hypersonic regime other factors arise such as the introduction of cold wall conditions that can have a stabilizing effect on the fluctuation levels.

For the mean variables, it sufficed to examine the inner layer behavior using the wall variable, y^+, normalization, and for the adiabatic case this wall variable scaling of the turbulent correlations also applies; however, for non-adiabatic isothermal cases, a semi-local scaling has been shown for both channel (e.g. Huang, Coleman, & Bradshaw, 1995; Morinishi, Tamano, & Nakabayashi, 2004) and boundary layer (e.g. Shahab, Lehnasch, Gatski, & Comte, 2011) to collapse the correlation data better than the wall scaling. This semi-local scaling

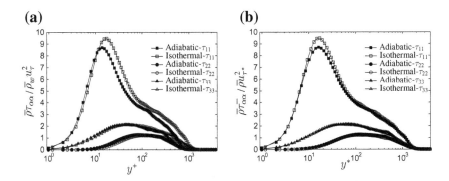

FIGURE 6.37 Distribution of turbulent normal stresses ($\tau_{\alpha\alpha}$, no sum): (a) wall unit scaling; (b) semi-local scaling. From Shahab et al. (2011) with permission.

involves a friction velocity given by $u_{\tau*} = \sqrt{\tau_w/\bar{\rho}}$ and local viscosity $\bar{\nu} = \bar{\mu}/\bar{\rho}$. For a boundary layer flow, Figure 6.37 shows the effect of change of scaling on the turbulent normal stresses by shifting the location of the peak values of the isothermal distributions toward the adiabatic results (although not shown, the shear stress component has both a collapse in wall-normal variation and amplitude across the boundary layer). As in the channel flow (Morinishi et al., 2004), the maximum value of the streamwise component in the adiabatic case lies below that of the isothermal case even with the semi-local scaling. The wall-normal and spanwise components show only a weak dependency on wall condition since small variations in both peak amplitude and peak location are found. The shift in peak amplitude location for the turbulent stresses is consistent with the van Driest transformed mean velocity result where the shift in intercept indicated a thickening of the inner layer region and which would correspond to the peak of the normal stress occurring at large value of y^+.

The above suggests that it is possible to provide for the turbulent velocity correlations a suitable rescaling of the compressible velocity second-moments through both Morkovin's hypothesis and the introduction of a semi-local scaling in non-adiabatic cases in order to establish a correspondence with incompressible scaling laws. Based on a series of studies by Perry and co-workers, a set of similarity expressions for the normal turbulent stress components were derived (Perry & Li, 1990) from scaling arguments applied to the turbulent energy spectra. The proposed expressions for incompressible flow were given in functional form by

$$\overline{u'^{+2}} = B_1 - A_1 \log\left(\frac{y}{\delta}\right) - V(y^+)$$

$$\overline{v'^{+2}} = B_2 - V(y^+)$$

$$\overline{w'^{+2}} = B_3 - A_3 \log\left(\frac{y}{\delta}\right) - V(y^+),$$

(6.58)

where the B_i are constants characterizing the large scales (see Perry, Henbest, & Chong, 1986), the A_i are assumed universal constants, and V represents a deviation from a logarithmic profile due to viscous effects in the wall region, given by

$$V(y^+) = \frac{5.58}{\sqrt{y^+}} - \frac{22.4}{y^+} + \frac{22.0}{(y^+)^{5/4}} - \frac{5.62}{y^{+2}} + \frac{1.27}{(y^+)^{11/4}}. \qquad (6.59)$$

An analysis using experimental data (Perry & Li, 1990) yielded $B_1 = 2.39$, $A_1 = 1.03$, $B_2 = 1.6$, $B_3 = 1.20$, and $A_3 = 0.475$. The similarity profile for the streamwise normal stress component has subsequently been improved from the original Perry and Li (1990) version, resulting in the form (Marusic, Uddin, & Perry, 1997; Marusic & Kunkel, 2003)

$$\overline{u'^{+2}} = B_1 - A_1 \log \left(\frac{y}{\delta}\right) - V_g \left(y^+, \frac{y}{\delta}\right) - W_g \left(y^+, \frac{y}{\delta}\right), \qquad (6.60)$$

where B_1 and A_1 retain their original values. The function V_g represents a viscous deviation that is composed of an isotropic Kolmogorov cut-off and an anisotropic part representing the action of dissipative motions, and the function W_g represents a wake deviation that is analogous of a form applied to a mean flow wake function by Lewkowicz (1982). Both V_g and W_g are somewhat complex and the reader is encouraged to review the article by Marusic et al. (1997) for the necessary background.

Using Morkovin's hypothesis, supersonic DNS data can be used to obtain corresponding profiles for these normal stresses. For an adiabatic supersonic boundary layer flow (Pirozzoli et al., 2004), a curve fit of the DNS data in the range $50 \leqslant y^+ \leqslant 300$, yielded the relations

$$\frac{\bar{\rho}}{\bar{\rho}_w} \frac{\overline{u'^2}}{u_\tau^2} \approx 1.350 - 1.145 \log \left(\frac{y}{\delta}\right) - V(y^+)$$

$$\frac{\bar{\rho}}{\bar{\rho}_w} \frac{\overline{v'^2}}{u_\tau^2} \approx 1.511 - V(y^+) \qquad (6.61)$$

$$\frac{\bar{\rho}}{\bar{\rho}_w} \frac{\overline{w'^2}}{u_\tau^2} \approx 1.189 - 0.518 \log \left(\frac{y}{\delta}\right) - V(y^+).$$

Clearly, differences exist between Eqs. (6.58) and (6.61), although the underlying functional form for comparison appears to be supported. It does appear that these incompressible similarity profiles can be useful in assessing the turbulent stress field. However, more such comparisons are necessary on both the incompressible and compressible sides to further validate the incompressible calibration and to assess whether the density scaled compressible turbulent stresses can be collapsed into the incompressible forms. Note that such comparisons have been limited to adiabatic conditions. Non-adiabatic isothermal wall conditions, for example, will further complicate the analysis and appear not to have been undertaken.

6.4.4.2 Turbulent Thermodynamic Fields

In contrast to the turbulent velocity field where an analysis of the turbulent correlations is often focused on the relationship with the corresponding incompressible correlations, the thermodynamic fields are often analyzed through (strong) Reynolds analogies with the velocity field. As pointed out previously, these analogies were based on Reynolds variables and initially assessed experimentally where the fluctuating correlations were also based on Reynolds variables. For the pressure and density fields, no ambiguities arise; however, for the temperature field represented by a density-weighted decomposition some differences may arise between correlations involving Reynolds variables and density-weighted variables. Nevertheless, the large amount of data available from numerical simulations has led to assessment of the thermal field using density-weighted variables u_i'' and T''. Analogous to the second-moment velocities given in Eq. (5.35), the heat flux and temperature variance can be related to the Reynolds correlations by

$$\widetilde{u_i''T''} = \overline{u_i'T'} + \frac{\overline{\rho'u_i'T'}}{\overline{\rho}} - \frac{\left(\overline{\rho'u_i'}\right)\left(\overline{\rho'T'}\right)}{\overline{\rho}^2} \tag{6.62}$$

$$\widetilde{T''^2} = \overline{T'^2} + \frac{\overline{\rho'T'^2}}{\overline{\rho}} - \frac{\left(\overline{\rho'T'}\right)^2}{\overline{\rho}^2}. \tag{6.63}$$

Once again the balance between the density-weighted and Reynolds variables is dictated by the higher-order correlations involving the density fluctuations. In the absence of shocks, there does not appear to be a sufficient difference between correlations formed from Reynolds variables and those formed from Favre variables to alter any qualitative assessments concerning the various strong Reynolds analogies, even in cases involving non-adiabatic isothermal wall conditions. Quantitatively, however, some caution should be exercised especially in the non-adiabatic cases since here the thermal layers can have a non-negligible influence. In the remainder of this section, either Reynolds or Favre variables will be used depending on the data availability in the examples used, but then necessarily compared to the analyses in Section 3.5.1 involving Reynolds variables. In addition, the focus will be on the thermal field since the pressure fluctuations are often small and the density and thermal fields are inversely related. In the presence of shocks such relationships may no longer hold and such quantitative validation of simulation or experimental data is more difficult.

Both the temperature variance and total temperature variance play different, but relevant, roles in the assessment of the turbulent thermal field. The temperature variance enters, through the temperature variance scale equation, into RANS-type closure models that attempt to account for variable turbulent Prandtl number effects. For the total temperature fluctuations, one of the underlying assumptions in the initial formulation of the strong Reynolds analogies was that the total temperature fluctuations were negligible (see Section 3.5). For a high speed boundary layer flow with isothermal (adiabatic) wall conditions, the *rms* temperature and total temperature variations are shown in Figure 6.38 for density-weighted variables. Across the wide speed range of the simulations examined, both thermal variables relative to the mean increase with Mach number. In addition, the figure clearly shows that the total temperature fluctuations and the temperature fluctuations are nearly the same magnitude which appears to contradict one of the underlying assumptions of the original SRA formulation. If a similar relationship for the density-weighted temperature and total temperature variance shown in Eq. (3.98) for the Reynolds variables holds, then Figure 6.38 shows that in regions where $\overline{T''^2} > \overline{T_t''^2}$ the correlation $\overline{T''T_t''} < 0$. In addition, while the vanishing of the total temperature fluctuations is a sufficient condition for obtaining a relation such as Eq. (3.95) between the temperature and velocity fluctuations it is not a necessary condition. As Eq. (3.98) showed, only the less restrictive condition, $\overline{T_t'^2} \ll \overline{T'^2} + 2\overline{T'T_t'}$ is required. A quantitative assessment of the influence of the correlation between the temperature and total temperature fluctuations can be extracted from Eqs. (3.98) and (3.105) and given by

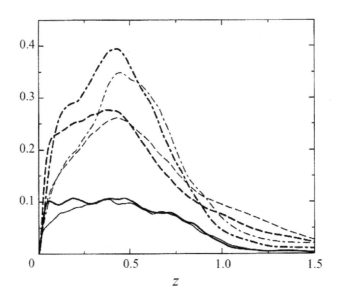

FIGURE 6.38 Distributions of temperature and total temperature *rms* values across boundary layer at different freestream Mach numbers: ——, $M_\infty = 3$; — — —, $M_\infty = 4.5$; — · —, $M_\infty = 6$. Thick lines represent $\sqrt{\overline{T''^2}}/\tilde{T}$ and thin lines represent $\sqrt{\overline{T_t''^2}}/\tilde{T}$. From Maeder et al. (2001) with permission.

$$\frac{2\overline{T'T'_t}}{\overline{T'^2}} \gg 1 + 2R_{u'T'}, \tag{6.64}$$

which confirms the fact that $\overline{T'T'_t} < 0$. The simulation results in Figure 6.38 affirm the fact that the correlation from the total temperature variance is less than the one from the temperature variance across a large portion of the boundary layer, and shows that this difference increases with increasing Mach number. A similar behavior has also been observed for the adiabatic case at $M_\infty = 2.25$ with a zero wall heat flux condition (Shahab et al., 2011); however, for the same Mach number but with isothermal cold-wall (relative to adiabatic) conditions the trends are reversed due to the suppression of the temperature fluctuations and the condition $\overline{T''^2_t} > \overline{T''^2}$ holds across most of the boundary layer.

There has been a significant focus on the quantitative evaluation of the various (extended) forms of the SRA initially from experiments (e.g. Gaviglio, 1987; Rubesin, 1990) and now with numerical simulations. The first extensive evaluation, using direct numerical simulations, of the SRA as well as other relationships between statistical correlations were by Coleman et al. (1995) and Huang et al. (1995). The analysis of the DNS cold-walled channel flow data provided significant insights into the turbulence dynamics that would impact the development of closure models. These studies also led to an improved and generalized SRA relationship between the temperature and velocity fluctuation levels given by Eq. (3.113) with the proportionality coefficient now including the turbulent Prandtl number Pr_t and rewritten in terms of the *rms* values as

$$\frac{\left(\sqrt{\overline{T'^2}}/\overline{T}\right)}{(\gamma - 1)M_\infty^2 \left(\sqrt{\overline{u'^2}}/\overline{u}\right)} = \left[Pr_t\left(1 - \frac{\partial\overline{T}_t}{\partial\overline{T}}\right)\right]^{-1}. \tag{6.65}$$

Equation (6.65) has been validated for both cold-wall channel flow simulations (Huang et al., 1995), and boundary layer flow with adiabatic wall conditions (Maeder, Adams, & Kleiser, 2001) (see Figure 6.39). These results show that over a large portion of the boundary layer Eq. (6.65) holds including the high Mach number case ($M_\infty = 6$). It should be cautioned, however, that even results from simulations may not, as yet, fully validate such relationships. Both Guarini et al. (2000) and Martin (2007) have compared their simulation results to the original SRA relation (see Eq. (3.99)) and found reasonable agreement across the boundary layer in contrast to the results shown in Figure 6.39. In all these comparisons density-weighted mean and fluctuating quantities were used although the original proposal by Huang (Huang et al., 1995) was predicated on Reynolds variables. In Huang et al. (1995) there was some quantification of the difference between density-weighted and Reynolds averaged (between 2% and 5%) turbulent shear stress and heat flux. This has been discussed previously in connection with comparison with experimental results where such percentage differences are not significant when compared to other uncertainties.

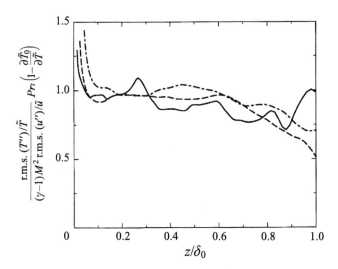

FIGURE 6.39 DNS data evaluation of extended form of SRA proposed by Huang et al. (1995) at different freestream Mach numbers: ———, $M_\infty = 3$; - - -, $M_\infty = 4.5$; — · —, $M_\infty = 6$. From Maeder et al. (2001) with permission.

Under adiabatic conditions, such differences between density-weighted and Reynolds-averaged results may occur as the freestream Mach number is increased to hypersonic speeds. In Figure 3.4, there were observable differences between the two averaging procedures. Once again, it is worth emphasizing that even in numerical simulations quantitative assessments of these various temperature and velocity analogies can be problematic.

A difficulty in assessing the validity of a relationship such as Eq. (6.65) against a constant value (here unity) is the inherent deviations of higher-order correlations that arise across the boundary layer from numerical simulations and certainly from experiments. This is especially true when peripheral factors, such as choice of fluctuating variables and numerical differentiation enter into consideration. Rather than compare against a constant value, it may be easier to provide an assessment when the comparison is against a quantity that also has a non-constant behavior across the boundary layer as well. For example, Eq. (6.65) can be used to obtain the turbulent Prandtl number Pr_t, and then this value can be compared with the usual Pr_t definition of $(\widetilde{u''v''}\partial\widetilde{T}/\partial y)/(\widetilde{v''T''}/\partial\widetilde{u}/\partial y)$. As an alternative to the turbulent Prandtl number Pr_t, the mixed Prandtl number Pr_m given in Eq. (6.43) can be used in such validations and assessments. As Figure 6.40a shows, this composite parameter behaves as the molecular Prandtl number Pr near the wall and as the turbulent Pr_t away from the wall. Near the wall $Pr_m \approx 0.72$ and away from the wall in the log-layer region, a value of $Pr_m \approx 0.9$ is obtained and with a further decay in the outer layer as the edge of the boundary layer is approached.

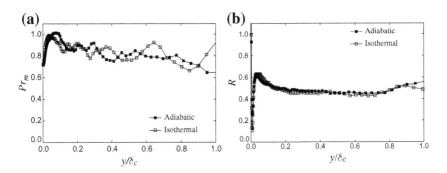

FIGURE 6.40 Variation of (a) mixed Prandtl number Pr_m and (b) time scale ratio parameter R across a supersonic boundary layer under adiabatic and isothermal (cold) wall conditions. From Shahab et al. (2011) with permission.

As mentioned in Section 5.3.4, the turbulent time scale ratio between the thermal and velocity fields is often introduced (Abe, Kondoh, & Nagano, 1996; Nagano & Shimada, 1996; Nagano, 2002), and provides a further quantification of the coupling between these fields. This ratio is defined here as

$$R = \frac{K_T}{\varepsilon_T} \frac{\varepsilon}{K}. \qquad (6.66)$$

In homogeneous flows, this timescale ratio is found to be constant (Jin, So, & Gatski, 2003), and implies that in an equilibrium layer of a boundary layer flow this timescale parameter should also be constant. As Figure 6.40b shows for a supersonic boundary layer flow, this is indeed the case where the value in the log-layer region is relatively constant at $\simeq 0.49$ (Abe, Kondoh, & Nagano, 1995; Béguier, Dekeyser, & Launder, 1978; Nagano, 2002).

Chapter 7

Shock and Turbulence Interactions

The existence of a shock in a turbulent flow can significantly affect both the mean field and turbulence dynamics in the vicinity of the shock as well as have a potentially large downstream dynamic range. Over a half-century of experimental, numerical, and theoretical studies have attempted to quantify the dynamic characteristics of such flows. Although it is not possible to identify all such studies, there have been some identifiable trends in these studies that have led to significant advances in the understanding of compressible turbulence dynamics. Some of these studies have also impacted the development of closure models for Reynolds-averaged type formulations as well. As might be expected, the studies involving homogeneous turbulence and shocks have inherently focused on the fundamental aspects of compressible turbulence dynamics; whereas, the studies involving inhomogeneous turbulent flows and shocks have additionally focused on flow prediction and subsequent control. In the latter case, the shock is evanescent and other effects associated with spatial variations can enter.

7.1 HOMOGENEOUS TURBULENCE INTERACTIONS

The interaction of homogeneous turbulence with a shock structure has been the subject of study for almost half a century starting with the analytical studies of Ribner (1954a) and Moore (1954). It is in many ways a fundamental building block flow in understanding the dynamics of inhomogeneous turbulence and (oblique) shock interaction. The main advantage of investigating such a flow is to isolate the net effect of the strong gradients of mean quantities imposed by the shock wave. In addition, it is amenable to linear analysis which, when coupled with the modal description of compressible turbulence proposed by Kovasznay (1953) (see Section 3.4), provides estimations for turbulence amplification across the shock. Such estimates, both qualitative and quantitative, are invaluable guides in assessing the fidelity of numerical simulations, and subsequently on closure model development and validation. While the number of geometrical parameters in the problem is reduced, experimental studies of such flows still require care in both setup and data collection.

7.1.1 Application of Linear Theory

The theoretical analysis of shock–turbulence interactions is based on the linearization of the turbulence dynamics. This involves either a rapid distortion analysis, or a linear interaction analysis (LIA). The rapid distortion theory (RDT) accounts for a (homogeneous) mean flow compression and the LIA includes both vorticity and turbulent kinetic energy generation due to shock-front distortion and unsteady movement of the shock, as well as the mean flow compression. Thus, the LIA is a closer representative to the flow field physics. Nevertheless, RDT has been a useful tool in studying a variety of shock–turbulence interaction problems.

Some of these RDT studies that focused on compressed compressible (as well as incompressible) flows were discussed in Chapter 5. While applicable to the general shock–turbulence interaction problem, the primary motivation for those studies was to investigate the effects of dilatation and to better understand the eddy shocklet problem. The more relevant analysis is to focus on the response of the anisotropic turbulent flows to such rapid compressions. Such studies were carried out on both shear and axisymmetric flows (Mahesh, Lele, & Moin, 1994) where it was found that the response to compression was different than the isotropic flows studied previously. These results, though inherently constrained by the RDT analysis to homogeneous and solenoidal turbulence, motivated further investigation of the shock–turbulence interaction problem using the linear interaction analysis where thermodynamic effects could be accounted for.

The early studies using a linear interaction analysis were applied to a plane disturbance interacting with a shock wave. Ribner (1954a, 1954b) investigated the interaction of a normal shock wave and a weak vorticity wave, Moore (1954) investigated the interaction of a normal shock wave and a sound wave as well as a vorticity wave, and Chang (1957) investigated the interaction problem of a normal shock wave and an entropy wave. The theory, often called Ribner's theory, is mostly based on the mode decomposition developed by Kovasznay (1953) and applied to the shock problem by focusing mainly on the acoustic field generated by the interaction with the shock wave. The focus here will be on the LIA since it is ideally suited for analyzing the interaction dynamics and has been the primary theory for comparison with the numerical simulation results (see Section 7.1.2). It provides amplification information for the turbulent kinetic energy and fluctuating vorticity as well as for the thermodynamic fluctuations through the pressure, temperature and density variances. Additionally, the analysis provides information on turbulent length scale reduction and anisotropy behind the shock. All of the (direct) numerical simulation studies include in the upstream turbulence field a vortical mode (e.g. Hannappel & Friedrich, 1995; Jamme, Cazalbou, Torres, & Chassaing, 2002; Lee, Lele, & Moin, 1993, 1997; Mahesh, Lele, & Moin, 1997), and other simulations (Hannappel & Friedrich, 1995; Jamme et al., 2002; Mahesh, Lee, Lele, & Moin, 1995) have also included an acoustic mode, an entropy mode, and both an entropy and acoustic mode, respectively. Details of the individual cases just cited can be found in the various references.

The theory results from a linearization of the Navier–Stokes equations (Euler equations) for moderately supersonic flows. For the shock wave-turbulence interaction, the physics of the problem can be modeled by using the modal decomposition for both the incoming and the transmitted turbulence across the shock wave. In the general case, the incoming turbulent flow is three-dimensional and consists of many waves with a continuous spectrum representative of a high Reynolds number turbulence; however, for illustrative purposes the linear interaction theory will be applied to some simplified cases.

The analysis of a two-dimensional incoming wave structure, composed of vorticity and entropy modes interacting with a normal shock, is the model problem of interest. It can be used as a validation of numerical simulations and as a guide for estimating the effects of shock interactions in inhomogeneous flows. Morkovin (1962) highlighted the vorticity–entropy modal composition of the fluctuation field for supersonic boundary layer flows. A good illustration of the interaction of such an upstream disturbance field with a shock is shown in Figure 7.1. The undistorted shock is aligned along the y-axis and the streamwise direction is along the x-axis. In the figure an incoming entropy wave field (wavenumber $\varkappa$) generates behind the shock a modal distribution of waves that propagate differently depending on the angle of incidence and Mach number. In Figure 7.1a, behind the shock the incoming entropy wave is refracted (wavenumber $\varkappa^s$), a shear wave (vorticity mode) generated (with the same wavenumber), and a pressure wave propagates (wavenumber $\varkappa^p$) away as an acoustic mode. In Figure 7.1b, behind the shock the incoming entropy wave is refracted with a phase shift, a shear wave (vorticity mode) generated and an evanescent pressure wave decays exponentially. In the latter case, there is also a phase shift of the waves behind the shock.

The incoming (region 1) turbulence field ahead of the shock can be, in general, described by the modal contributions

$$
\text{Vorticity Mode}: \begin{cases} \dfrac{u_1'}{\overline{u}_1} = \dfrac{\varkappa_y}{\varkappa} A_v e^{i\left[\varkappa_x (x-\overline{u}_1 t)+\varkappa_y y\right]} \\[3mm] \dfrac{v_1'}{\overline{u}_1} = -\dfrac{\varkappa_x}{\varkappa} A_v e^{i\left[\varkappa_x (x-\overline{u}_1 t)+\varkappa_y y\right]} \end{cases} \tag{7.1a}
$$

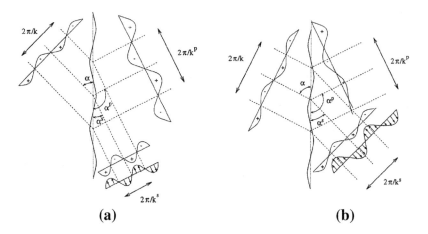

FIGURE 7.1 Sketch of an entropy mode interaction with a shock: (a) pressure wave propagation; (b) no pressure wave propagation. From Fabre et al. (2001) with permission.

$$
\text{Acoustic Mode} : \left\{ \frac{p_1'}{\overline{p}_1} = A_p e^{\iota \varkappa \overline{c}_1 t} e^{\iota [\varkappa_x (x - \overline{u}_1 t) + \varkappa_y y]} \right. \tag{7.1b}
$$

$$
\text{Entropy Mode} : \left\{ \begin{array}{l} \dfrac{\rho_1'}{\overline{\rho}_1} = A_s e^{\iota [\varkappa_x (x - \overline{u}_1 t) + \varkappa_y y]} \\[2mm] \dfrac{s_1'}{c_p} = \dfrac{p_1'}{\gamma \overline{p}_1} - \dfrac{\rho_1'}{\overline{\rho}_1}, \end{array} \right. \tag{7.1c}
$$

where $\varkappa_x$ and $\varkappa_y$ are the wavenumber components in the respective directions of the incoming wave with $\varkappa^2 = \varkappa_i \varkappa_i$, A_v incoming vorticity mode amplitude, A_s incoming entropy mode amplitude, and A_p incoming acoustic mode amplitude (the subscript 1 refers to upstream conditions relative to the shock). Within this framework, the amplitudes A_i can be complex. The modified frequency in the incident acoustic field, $\varkappa \overline{u}_1 (\cos \alpha + 1/M_1)$, is necessitated by the fact that pressure waves can propagate in the same and in opposite directions relative to the mean flow (Fabre, Jacquin, & Sesterhenn, 2001; Moore, 1954). In addition, the shock is allowed to distort in response to the passage of the wave field and this is included in the analysis through the function $x_d' = \xi(y,t)$ given by

$$
x_d' = A_d e^{\iota [\varkappa_y y - \varkappa_x \overline{u}_1 t]}. \tag{7.2}
$$

Behind the shock (region 2), the flow field is assumed to be governed by the linearized Euler equations that can be written as

$$
\frac{\partial \rho_2'}{\partial t} + \overline{u}_2 \frac{\partial \rho_2'}{\partial x} = -\overline{\rho}_2 \left(\frac{\partial u_2'}{\partial x} + \frac{\partial v_2'}{\partial y} \right) \tag{7.3a}
$$

$$
\frac{\partial u_2'}{\partial t} + \overline{u}_2 \frac{\partial u_2'}{\partial x} = -\frac{1}{\overline{\rho}_2} \frac{\partial p_2'}{\partial x} \tag{7.3b}
$$

$$\frac{\partial v_2'}{\partial t} + \bar{u}_2 \frac{\partial v_2'}{\partial x} = -\frac{1}{\bar{\rho}_2} \frac{\partial p_2'}{\partial y} \tag{7.3c}$$

$$\frac{\partial s_2'}{\partial t} + \bar{u}_2 \frac{\partial s_2'}{\partial x} = 0 \tag{7.3d}$$

$$\frac{p_2'}{\bar{p}_2} = \frac{\rho_2'}{\bar{\rho}_2} + \frac{T_2'}{\bar{T}_2} \quad \text{and} \quad \frac{s_2'}{c_p} = \frac{p_2'}{\gamma \bar{p}_2} - \frac{\rho_2'}{\bar{\rho}_2}. \tag{7.3e}$$

The mean field variables behind the shock are determined from the Rankine–Hugoniot jump conditions and can be expressed in terms of the upstream Mach number M_1 as (cf. Fabre et al., 2001)

$$\frac{\bar{\rho}_2}{\bar{\rho}_1} = \frac{\bar{u}_1}{\bar{u}_2} = \frac{(\gamma + 1)M_1^2}{2 + (\gamma - 1)M_1^2} \tag{7.4a}$$

$$M_2 = \sqrt{\frac{2 + (\gamma - 1)M_1^2}{2\gamma M_1^2 - (\gamma - 1)}} \tag{7.4b}$$

$$\frac{\bar{p}_2}{\bar{p}_1} = 1 + \frac{2\gamma}{\gamma - 1}\left(M_1^2 - 1\right) \tag{7.4c}$$

$$\frac{\bar{T}_2}{\bar{T}_1} = \frac{\bar{c}_2^2}{\bar{c}_1^2} = \frac{\left[2\gamma M_1^2 - (\gamma - 1)\right]\left[(\gamma - 1)M_1^2 + 2\right]}{(\gamma + 1)^2 M_1^2}. \tag{7.4d}$$

Using Eq. (7.4) to determine the mean field behind the shock, Eq. (7.3) can be used to solve for the (linearized) disturbance field behind the shock. The boundary conditions are then determined from a linearization of the Rankine–Hugoniot equations (in a frame of reference fixed to the speed of the shock) to complete the specification of the problem. The three-dimensional generalization of this theory can be obtained by integrating (over physical and Fourier space) the results from all the contributing waves included in the incoming "turbulent" field, and which can include initially acoustic fields. For additional details on implementation particularly focused on the important case of an incoming wave field composed of vorticity–entropy modes, the reader is referred to Mahesh, Moin and Lele (1996), Mahesh et al. (1997), and Jamme et al. (2002) and for a more general analysis Fabre et al. (2001).

From the discussion up to this point, it is clear that the composition and orientation of the incoming turbulence field will have a significant influence on the disturbance field behind the shock. It is useful to correlate the amplification factors associated with some key parameters and the modal structure of the incoming turbulence field. Unless otherwise noted, only the case of propagating pressure waves will be considered and where the strong Reynolds analogy (SRA) holds and the velocity–temperature correlation is negative. Figure 7.2 shows the amplification of the incoming turbulence field through the shock at $M_1 = 1.5$. The quantities displayed are normalized by the values just ahead of

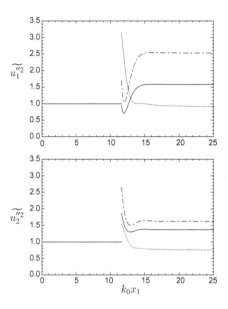

FIGURE 7.2 Amplification from LIA of different incoming modes of turbulence on streamwise and transverse (normalized) stress components $M_1 = 1.5$. (Note that in this figure the subscripts denote normal stress component and not region ahead or behind shock.) —— Pure vorticity case; — · — vorticity/entropy mode; · · · · · · pure acoustic mode. From Jamme et al. (2002) with permission.

the shock, and k_0 is the most energetic wavenumber of the spectrum associated with the initial random velocity field (Jamme et al., 2002). The figure shows that the introduction of entropy waves augments the amplification of both the streamwise and transverse normal stress components by factors of 2.54 and 1.59, respectively, relative to the incoming level. For the pure acoustic case, there is an attenuation for both components behind the shock, with a 43% reduction for the streamwise component and a 46% reduction for the transverse component, relative to the far field level of the pure vorticity mode.

Some care has to be exercised in assuming these trends over a broader Mach number range. This is clearly illustrated in Figure 7.3 which shows a somewhat different picture for an incoming acoustic wave. After an incoming Mach number of ≈ 1.8, a further increase in Mach number is associated with a corresponding increase in amplification rate for the kinetic energy and its streamwise and transverse components. This further highlights the potential of the LIA since it provides an accurate (qualitative) guide to the behavior behind the shock at various Mach numbers (as well as other parameters such as incidence angle) without expending unnecessary resources. This is extremely useful in delimiting parameter ranges for more complex inhomogeneous flows.

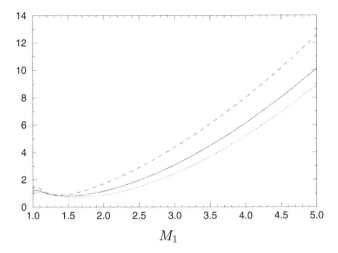

FIGURE 7.3 Far field amplification factor from LIA for pure acoustic incoming mode of: ——— turbulent kinetic energy; – – – streamwise normal stress component; and · · · · · · transverse normal stress component. From Jamme et al. (2002) with permission.

As mentioned previously, a critical angle exists that delineates between the propagation and evanescent behavior of the pressure wave behind the shock. The behavior of the mean square (transverse) vorticity amplification factor is shown in Figure 7.4. (Note that the angle ψ_1 is equal to the angle α in Figure 7.1.) Once again it is shown that the addition of the entropy mode augments the amplification relative to the pure vorticity mode case. As shown in Section 5.3.2, the enstrophy plays an important role in the turbulent energy dissipation rate dynamics, and an understanding of the influence of shocks will be beneficial in developing accurate transport equations for it.

Since the LIA theory also predicts the evolution of the turbulent thermodynamic variables across the shock, it is useful to return to some thermodynamic relations derived previously that can quantify the influence of the fluctuating pressure field on the fluctuating density. In Section 3.5, it was shown how the temperature and density fluctuations are related to the pressure and entropy fluctuations. A ratio Γ_B between polytropic exponents was constructed, given by

$$\Gamma_B = \left(\frac{2-\gamma}{\gamma}\right)\frac{n_{p\rho}^2}{n_{\rho T}}, \tag{7.5}$$

with

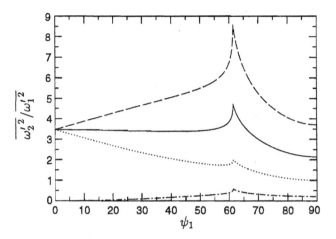

FIGURE 7.4 Amplification factors for mean square vorticity across the shock at different angles of incidence from linear analysis: —— pure vorticity mode; – – – (propagating) and · · · · · · (non-propagating) vorticity/entropy mode; —·— pure entropy mode. From Mahesh et al. (1996) with permission.

$$n_{p\rho} = \frac{\sqrt{\overline{p'^2}/\overline{p}}}{\sqrt{\overline{\rho'^2}/\overline{\rho}}} \tag{7.6a}$$

$$n_{\rho T} = 1 + \frac{\sqrt{\overline{T'^2}/\overline{T}}}{\sqrt{\overline{\rho'^2}/\overline{\rho}}}. \tag{7.6b}$$

For a pure isobaric turbulence, $p' = 0$, only the vorticity and entropy modes are active so that $\Gamma_B = 0 = n_{p\rho}$ and $n_{\rho T} = 2$, and for a pure acoustic field, $s' = 0, \Gamma_B = 2 - \gamma$, and $n_{p\rho} = \gamma = n_{\rho T}$. Since the boundary layer is composed primarily of vorticity/entropy modes, the isentropic limit $\Gamma_B = 2 - \gamma$ is an upper bound below which, $\Gamma_B \ll \Gamma_B^{(\mathrm{isen})}$, many of the scaling relations, such as Morkovin's hypothesis apply.

Figure 7.5 shows the evolution of the LIA computed polytropic exponents (downstream far field conditions) as a function of the incoming Mach number. The upstream turbulence was composed of pure vorticity waves in this computation. For small Mach number ($M_1 < 1.2$), the downstream turbulent field is isentropic, but as the Mach number increases the polytropic exponents reach asymptotic values of $n_{\rho T} = 1.9$ and $n_{p\rho} = 0.7$. The resulting field now exhibits non-negligible entropy fluctuations that were not present in the incoming field. In the case of upstream turbulence composed of vorticity and entropy modes (with $\overline{u_1' T_1'} < 0$) passing through a $M = 1.5$ shock, the

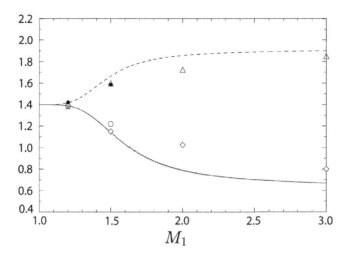

FIGURE 7.5 Downstream value of polytropic exponents as a function of incoming Mach number: —— $n_{p\rho}$ and – – – n_{pT}, LIA; ● $n_{p\rho}$ and ○ $n_{p\rho}$, Jamme et al. (2002); × $n_{p\rho}$, Lee et al. (1993); ◇ $n_{p\rho}$ and △ n_{pT}, Lee et al. (1997). From Jamme et al. (2002).

corresponding values for $n_{p\rho}$ and $n_{\rho T}$ are 0.4 and 1.96, respectively. This gives $\Gamma_B = 0.058 \Gamma_B^{(isen)}$, and shows that the fluctuating field after the shock is only about 6% pure acoustic mode even when the pressure fluctuations represent 40% of the density fluctuations. The turbulence then satisfies Morkovin hypothesis even though it is not isobaric. In this case p'/p is not negligible since

$$\frac{\sqrt{\overline{p'^2}}}{\overline{p}} = 0.4 \frac{\sqrt{\overline{\rho'^2}}}{\overline{\rho}}. \tag{7.7}$$

7.1.2 Numerical Simulations

As was discussed in Chapter 5, early simulations of compressible flows were strongly motivated by the discovery and subsequent investigation of eddy shocklets (Passot & Pouquet, 1987) and the assessment of dilatational effects on the turbulence dynamics. Subsequent direct simulations of compressible turbulence then focused on the fundamental case of isotropic, decaying turbulence (Blaisdell, Mansour, & Reynolds, 1993; Erlebacher, Hussaini, Kreiss, & Sarkar, 1990; Kida & Orszag, 1990b, 1992; Lee, Lele, & Moin, 1991; Passot & Pouquet, 1987, 1991; Sarkar, Erlebacher, Hussaini, & Kreiss, 1991). Concurrently, simulations were also performed on forced compressible turbulence (Dahlburg, Dahlburg, Gardner, & Picone, 1990; Kida & Orszag, 1990a; Staroselsky, Yakhot, Kida, & Orszag, 1990). Although not motivated by the existence of shocklets, such studies were relevant to an improved understanding of the interactive spectral dynamics in compressible flows. In light of the current trend toward hybrid Reynolds-averaged Navier–Stokes/large eddy simulation (RANS/LES) models, such simulation results can have a direct impact on both time-dependent RANS-type model and sub-filter scale model calibration and validation.

Even though two-dimensional simulations were performed (Passot & Pouquet, 1987) and focused on the small scale turbulent motions, shock structures (eddy shocklets) at these scales were identified that had a direct impact on the dynamics. These eddy shocklets were found to permeate the flow field and were strongly influenced by initial levels of density fluctuations and rms Mach number ($M_{rms} > 0.3$). Unfortunately, in a later three-dimensional simulation (Passot & Pouquet, 1991) such structures were not observed. The study of decaying isotropic turbulence continued with simulations performed in two-dimensions (Erlebacher et al., 1990) and three-dimensions (Sarkar et al., 1991, see also Zang, Dahlburg, & Dahlburg, 1992) as well as an asymptotic analysis of applicable governing equations. In both cases, the parameters of interest were the ratio of compressible kinetic energy to total kinetic energy and the ratio of compressible kinetic energy to compressible potential energy. In addition, the role of initial conditions was analyzed, and it was found that a necessary condition for shocks to form was that the ratio of the compressible kinetic energy to the compressible potential energy be less than 1 (Erlebacher et al., 1990). Shocks were also observed in two-dimensional mixing-layer simulations (Sandham & Reynolds, 1991), but not in the corresponding three-dimensional simulations.

Additional simulations were performed (Blaisdell et al., 1993; Lee et al., 1991) that continued to address several of the issues raised by these previous studies, including attempts to reconcile some of the differences found in the formation of shocklets (Lee et al., 1991). In both cases, three-dimensional numerical simulations were performed, and the flow dependence on dimensionality, initial turbulent Mach number and Reynolds number were analyzed. The existence of the eddy shocklets was confirmed and had the characteristics of typical shock waves. Importantly, it was found that by increasing the fluctuation Mach number and turbulent Reynolds number, regions of shocklet formation increased. However, three-dimensional turbulence is less susceptible to such formations and can require higher fluctuation Mach numbers for their occurrence. Nevertheless, initial conditions can play an important role in the overall dynamics of such flows (Blaisdell et al., 1993). Decaying compressible turbulence has also been simulated and analyzed (Kida & Orszag, 1990b, 1992). Although not directly addressing the issue of shocklets, the occurrence of shocks was observed and the enstrophy budgets and vorticity generation due to baroclinic torque in the vicinity of the fronts analyzed. This association with the baroclinic torque will arise again in the discussion in Section 7.2.1 on inhomogeneous free shear flows. With the conclusion of these studies, the issue of shocklet occurrence and the role of dilatation effects in turbulence dynamics, at least for this simple class of flows, appears to have been thoroughly addressed.

It is clear that in a little over a decade after 1981, the issue of shock formation (shocklets) in shear free turbulence was thoroughly investigated within the limitations of the computational resources available. Although these numerical resolution constraints prevented simulations at high values of Taylor microscale Reynolds numbers ($Re_\lambda < 10^2$), a wide range of initial conditions were utilized at turbulent (or *rms*) Mach numbers ≈ 0.5. Even though these turbulent Mach numbers were subsonic, it was pointed out in Chapter 1 (see discussion associated with Figure 1.8) that instantaneous fluctuating Mach number values can (and do) reach supersonic values. It should be pointed out that almost a decade later, new simulations were performed (Samtaney, Pullin, & Kosović, 2001) on decaying isotropic turbulence and shocklet statistics were investigated. With a decade of improvements in computational speed and memory, it was possible to perform simulations at Taylor microscale Reynolds numbers between 50 and 100; however, at the parameter values studied, the decay of kinetic energy was minimally influenced by the level of thermodynamic fluctuations or velocity dilatation in the initial conditions.

The simulation of shock/turbulence interactions has been ongoing for two decades. The initial simulations focused on two-dimensional turbulent fields (Hannappel & Friedrich, 1992; Rotman, 1991) interacting with a normal shock, but were shortly followed by three-dimensional simulations of interactions with weak shocks (Hannappel & Friedrich, 1995; Lee et al., 1993). The case of quasi-incompressible turbulence interacting with a weak normal shock (with an upstream Mach number $M_u \sim 1.1$; Lee et al., 1993), and the case of compressible turbulence interacting with a normal shock (with an upstream Mach number $M_u \sim 2.0$; Hannappel & Friedrich, 1995) were initially studied. Resolution requirements dictated the limitation to low Reynolds numbers (that is, microscale Reynolds number of $Re_\lambda \sim 18$ (Lee et al., 1993) and $Re_\lambda \sim 6$ (Hannappel & Friedrich, 1995)). Although the overall trends from the simulations (Hannappel & Friedrich, 1995; Lee et al., 1993) were similar, the two studies differed in the makeup of the upstream turbulence field. In the earlier study (Lee et al., 1993), only upstream vortical turbulence was considered, and in the later study (Hannappel & Friedrich, 1995) upstream turbulence fields composed solely of a vortical contribution as well as both a vortical and acoustic contribution were considered.

In such simulations with these upstream Mach numbers, it was possible to perform calculations within the shock. In these cases, the simulation data was assessed relative to the linear theory (discussed in the previous section), and it was found for the velocity fluctuation field that both turbulent kinetic energy and the transverse vorticity components increased consistent with the LIA. In addition, due to strong amplification of the streamwise Reynolds stress component, the upstream isotropic Reynolds stress components then showed an axisymmetric character that was also consistent with the linear analysis.

The strong mean flow compression in the direction normal to the shock led to vorticity generation (shock curvature effect) and turbulent kinetic energy generation (caused by unsteady shock movement). Similar amplification of the fluctuating thermal field also occurred. It was also shown (Lee et al., 1993) that the turbulent spectra across the shock exhibits a bigger amplification at the large wavenumbers compared to lower ones which means that the longitudinal integral length scales were decreased while the lateral ones were unaffected by the shock interaction. Compressible inflow conditions consisting of both vortical and acoustic contributions have also been investigated (Hannappel & Friedrich, 1995). It was found that qualitatively similar results to the purely vortical case were obtained but with reduced amplification levels and weaker length scale reductions.

A relevant finding for modeling purposes (Hannappel & Friedrich, 1995; Lee et al., 1993) was the conclusion that the pressure work term was the main contributor to the increase in turbulent kinetic energy. A decomposition of this term into a pressure–dilatation term $\overline{\rho}\Pi_{ij}^{dl}$ and a pressure transport term $\overline{p'(u_i'\delta_{jk} + u_j'\delta_{ik})}_{,k}$ showed that the latter was the main contributor to the pressure work downstream of the shock. The pressure–dilatation term acted to convert the mean internal energy into turbulent kinetic energy.

For this weak shock case, it is possible to get some insight into the effect of the turbulence on the shock. Simulations (Lee et al., 1993) have shown that this causes the *rms* value of the peak compression to vary significantly across the transverse plane. The instantaneous non-uniformities associated with the turbulence caused both the location and thickness of the shock to vary in the transverse direction. For weak upstream turbulent intensities ($M_t < M_u - 1$), well-defined fronts were present with a single mean compression peak. For strong upstream turbulent intensities ($M_t > M_u - 1$), shock-front distortions occurred which caused multiple compression peaks inside the shock and no well-defined fronts in the transverse direction.

There have also been some spatial large eddy simulation studies performed on these flows for both validating numerical methods and subgrid scale (SGS) models. For example, an LES of the weak turbulence shock interaction case (Lee et al., 1993) has been performed (Ducros et al., 1999). An over-estimation of the amplification rates for both the kinetic energy and vorticity fluctuation fields was found. Since the direct numerical simulation (DNS) had revealed such a complex array of dynamic interactions, it is not clear what were the underlying causes of these differences. Another assessment of a SGS method, at a similar upstream Mach number ($M_u \approx 1.29$) was performed (Dubois, Domaradzki, & Honein, 2002). It was found that the various correlation values obtained were generally less than the DNS values in the *a priori* tests, but were the same when actual large eddy simulations were performed. This highlights an inherent dilemma with using filtered (LES) or even averaged (RANS) methods in extracting underlying physics. The methods employ models (to a greater or lesser extent) and as such are primarily designed for replication of physical behavior.

Although the weak shock interaction study yields important information on the effect of the shock on the turbulence as well as the turbulence on the shock, stronger shock interaction studies are needed to more closely replicate practical engineering situations. For the stronger shock interactions, however, it is necessary to employ shock capturing approaches in the simulations. It should be recognized that the necessity for a shock capturing approach, for example, is not just predicated on resolution issues. As was discussed in Section 4.1.1 and shown in Eq. (4.9), the corresponding Kn_{δ_s} has a value (0.5 at a M_u of 2) where the applicability of continuum equations is questionable.

The earlier isotropic vortical turbulence interaction simulations (Lee et al., 1997) were later extended to interactions with a stronger shock. Once again, turbulent kinetic energy was amplified through the shock; although, now with a possible saturation level for upstream Mach numbers around 3, but with the stress components evolving from isotropic to axisymmetric as in the weak shock case. Behind the shock, the time scale associated with the return to isotropy was much larger than the decay rate time scale. Vorticity fluctuations were also amplified consistent again with linear theory and the weaker shock simulations, and most length scales were found to decrease across the shock wave. An exception to the length scale decrease was the dissipation length scale which was found to decrease for some shock strengths. An assessment of SGS models using this stronger shock/turbulence simulation data (as well as the weak shock/turbulence interaction data discussed previously) was also performed (Garnier, Sagaut, & Deville, 2002). In assessing the SGS models, it was found that the dynamic Smagorinsky model performed as well as the dynamic mixed model for less cost, and that it was important to adequately resolve the shock corrugations in the simulations. This last point is also applicable to direct simulations where higher resolution in the vicinity of the shocks is necessary in order to maintain the accuracy of the solution.

Although the interaction of an incoming vortical field leads to the generation and alteration of all three compressible modes downstream of the shock, the interaction of an incoming vortical and entropy (temperature) mode with a shock more closely approximates the dynamic situation that might occur in flows such as shock and boundary layer interactions. Direct simulations of this flow, at upstream Mach numbers of 1.29 and 1.8, have been performed (Mahesh et al., 1997). The presence of the upstream entropy fluctuations as well as the sign of the upstream velocity–temperature correlation had an impact on the turbulence behind the shock. Higher amplification of the turbulence kinetic energy and vorticity occurred when the upstream entropy modes were present and when the upstream velocity–temperature correlation was negative rather than positive. Two important contributions to the evolution of vorticity across the shock were identified. One was the bulk compression of incident vorticity and the other was the baroclinic production. The upstream correlation between vorticity and entropy fluctuations determined the net level of production between these two source contributions. An enhanced level of vorticity fluctuations implies an increase in the turbulent dissipation rate levels downstream of the shock as well. Recall that in the discussion of the solenoidal dissipation rate (enstrophy equation) in Section 5.3.2, the baroclinic term was also considered an important contributor to the enstrophy balance and in the absence of shear the bulk compression is the sole contributor to the mean production mechanism. Not unexpectedly, it was shown that Morkovin's hypothesis did not hold immediately behind the shock.

It is clear from these earlier simulations, that the composition of the upstream disturbance field can have a significant impact on the flow field behind the shock. In the previous studies, the upstream fields have been composed of pure vortical (Lee et al., 1993, 1997), vortical and acoustic (Hannappel & Friedrich, 1995), and vortical and entropic (Mahesh et al., 1997) modes. The remaining combination of vortical, entropic and acoustic modes have also been studied (Jamme et al., 2002). An assessment of the modal makeup of the incoming turbulence field (Jamme et al., 2002), found that the velocity and vorticity fluctuations were more amplified when entropic fluctuations were included relative to the pure solenoidal (vortical) fluctuations and the (longitudinal) velocity and temperature fluctuations were negatively correlated. Correspondingly, the amplification was less when the (longitudinal) velocity and temperature fluctuations were positively correlated. Similar behavior has been found (Mahesh et al., 1997) and attributed to the baroclinic term in the enstrophy equation. Another consequence is the increased reduction of the microscale when the velocity and temperature correlation is negative (relative to when it is positive). The addition of the acoustic mode causes less amplification relative to the pure vortical mode case, at least for low Mach numbers; whereas, the opposite occurs for $M_1 > 3$. Figure 7.6 shows a distribution of amplification factors (ratio of correlation values behind and ahead of the shock) for the streamwise and transverse components of the density-weighted stress tensor components, $\widetilde{u''^2}$ and $\widetilde{v''^2}$, respectively. As the Mach number increases up to ≈ 1.7, the longitudinal fluctuations are amplified, but for higher Mach numbers a slight decrease is observed. For the transverse component, a monotonic increase is observed over the entire Mach number range up to $M = 3$. It appears from the figure that the agreement is generally good between LIA theory and the simulations and experiments if differences in Reynolds numbers and underlying physics are taken into account. As will be discussed in the next section, the theoretical and computational results also agree quite well with the experimental results obtained (Barre, Alem, & Bonnet, 1996) for an incoming Mach number of 3.

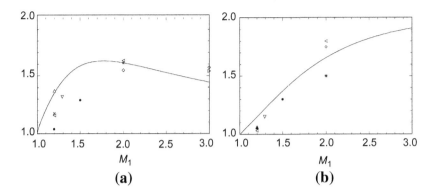

FIGURE 7.6 Amplification factor of the streamwise (a) and transverse (b) velocity components as a function of incoming Mach number: ——, LIA; •, Jamme et al. (2002); ×, Lee et al. (1993); ◇, Lee et al. (1997); ▽, Mahesh et al. (1997); *, Hannappel and Friedrich (1995); △, Ducros et al. (1999); ◁, Garnier et al. (2002); ▷, Barre et al. (1996). From Jamme et al. (2002) with permission.

7.1.3 Experiments

There have only been a few attempts experimentally at generating either compressible, homogeneous turbulent flows or compressible, homogeneous turbulence interacting with a shock. In attempting to generate such "simple flows," there are two possible experimental approaches. The first approach is to generate a homogeneous turbulent flow in a supersonic regime, as close as possible to isotropic conditions in a conventional wind tunnel. This supersonic turbulent flow can be "shocked" by a shock wave that can be either oblique or normal (as depicted in Figure 4.4), depending on the desired shock strength. However, in order to replicate a pure interaction, the shock has to be as stable as possible. The second approach utilizes shock tubes in which a grid is installed in such a way that when it is traversed by an incident shock during the first passage, a weakly compressible turbulent flow is generated, whose Mach number M_u is of the order of 0.6 or less. When the shock reflects from the end wall of the shock tube, it is characterized by a supersonic Mach number, M_s and subsequently interacts with the generated turbulent flow, resulting in a supersonic interaction. The variation of the damping properties of the end wall allows for the adjustment of the strength of the interaction. Among the different measurement methods available, hot-wire anemometry (HWA; see Section 4.2.1) is often preferred, despite its intrusive character, due to the required time resolution. For shock tube flows, a single hot-wire can be used (Troiler & Duffy, 1985); although, multiple (up to 12) hot wires have also been used in order to measure several velocities and instantaneous spatial gradients so that quantities such as the vorticity and the dissipation rate (Briassulis, Agui, & Andreopoulos, 2001) can be obtained. LDA has also been used, but as mentioned in Section 4.2.2, great care has to be taken due to the inertia of the seeding particles in the vicinity of the shock.

In addition, for small turbulent scales (that arise when the grid is sufficiently small), the measurement volume can induce a filtering effect, and low levels of turbulence behind grids can approach the minimum sensitivity requirements of LDA. In shock tubes, optical methods are also used to provide information on the density field (Keller & Merzkirch, 1990).

In the case of wind tunnel experiments, the main difficulty is to obtain well controlled initial turbulent flow with minimal extraneous perturbations (e.g. shock waves) and a sufficient level of fluctuations in order to be measurable. Typical values of the *rms* incoming longitudinal fluctuation velocity are 2% of the mean velocity (Jacquin, Blin & Geffroy, 1993) or, expressed in terms of longitudinal *rms* mass flux fluctuations, of order of 1% (Barre et al., 1996). For wind tunnel experiments, the measurement procedure is the same as for any aerodynamic configuration. For the shock tube experiments, the major difficulties are linked to boundary condition (wall) effects and to the fact that the experiments are one shot making the statistics more difficult to obtain. Another complicating issue is the inherent difficulty in identifying the origin of the incoming fluctuations in terms of the modes described in the previous sections. In most cases, the fields are assumed to be solely comprised of solenoidal modes; however, particularly for wind tunnel experiments, parasitic acoustic modes can be created either by aerodynamic interactions close to the solid boundaries generating the turbulence or emanating from the wind tunnel walls. Of course, such tunnel studies are also limited by the fact that it is only possible to achieve a quasi-homogeneous turbulent flow. Even though inherent difficulties exist in performing such turbulence-shock interaction studies in wind tunnels, there have been a few attempts that should be highlighted due to their different turbulence/shock interaction setups.

A (quasi-) homogeneous turbulence can be obtained by generating fluctuations upstream of the sonic throat in the settling chamber of a supersonic wind tunnel (Debiève & Lacharme, 1986); however, after the expansion in the nozzle, the turbulent field can become strongly anisotropic with a significant decrease in the turbulence level that can lead to experimental difficulties. A (quasi)homogeneous turbulent flow can also be generated using a simple grid as a sonic throat so that the flow downstream of the grid is supersonic (with a relatively low Mach number, $M = 1.7$; Blin, 1993; Jacquin, Blin & Geffroy, 1993); however, parasitic shock waves that significantly deteriorate the quality of the turbulent flow can be present. In this case, it is necessary to "shock" the flow by means of a second throat creating a pure normal shock wave/free turbulence interaction. Unfortunately, this shock is particularly difficult to stabilize and a boundary layer control may be necessary to obtain a stable configuration. Also, experimental difficulties, due to the low turbulence level (close to the noise level of a laser Doppler velocimeter (LDV), for example), can lead to significant difficulties in interpreting the results. Yet another type of experiment has been performed (Barre et al., 1996) with a multi-nozzle (625 individual nozzles) system to generate a homogeneous and isotropic turbulent field in a Mach 3 flow. The turbulent field then interacted with a normal shock wave generated at the center of the test section of a supersonic wind tunnel by means of a Mach effect that was described in Section 4.2.2 and plotted in Figure 4.5b. The main advantage of such a device is the ability to obtain a shock wave of maximum strength and free from parasitic excitation. This makes the interaction close to an ideal interaction, and in agreement with the underlying assumptions of the linear theories. Nevertheless, limitations are associated with this configuration as well: (i) a series of weak shock waves are embedded in the turbulent field at the nozzle exit of the multi-nozzle configuration; (ii) acoustic waves are generated at the trailing edge of the multi-nozzle, and even if these waves are damped when they reach the shock wave, it is increasingly difficult to interpret the results due to a large influence on the interaction with the initial turbulent field; (iii) as outlined in Section 4.2.2 the generation of the normal wave induces an acceleration of the flow immediately downstream of the grid due to the presence of mixing layers at the edges of the shock surface which influences the amplification of the turbulent field; (iv) the turbulence level is quite low immediately upstream of the shock wave, and this leads to experimental difficulties—even with the high sensitivity of an HWA, for example. Nevertheless, even with these constraints experimental results (Barre et al., 1996; Jacquin, Blin & Geffroy, 1993) have been analyzed in detail and given as a database in numerical form (Leuchter, 1998).

For experiments in a shock tube, studies have been performed to investigate either the (pure) interaction between a traveling normal shock wave reflected on the end wall of the tube and the subsequent interaction with the flow induced by the incident shock (Hartung & Duffy, 1986; Troiler & Duffy, 1985), or the interaction of wake generated turbulence by perforated plates or grids (Briassulis & Andreopoulos, 1996; Briassulis et al., 2001; Honkan & Andreopoulos, 1992; Keller & Merzkirch, 1990; Poggi, Thorembey, & Rodriguez, 1998). These experiments show a strong increase in density fluctuations through the shock, but there is less data describing the behavior of the velocity field (see Andreopoulos, Agui, & Briassulis, 2000, for a review). The characterization of the turbulence generated by several grids in a shock tube is a difficult task, but has been done in detail in a large-scale shock tube (Briassulis et al., 2001) where some compressibility effects on power law decay have been shown, and where the interaction of grid turbulence with the reflected shock wave has been studied (Agui, Briassulis, & Andreopoulos, 2005). It is important to recognize that the procedures used in operating shock tubes is quite different from the usual wind tunnels. It is necessary to draw a time/space chart (for the principles of operation, see Tropea, Yarin, & Foss, 2007c, chap. 16) to correctly identify the exact meaning of the parameters involved. In the case of the incoming grid turbulence for example, the value of the Mach number M_u

has a direct impact on the turbulence level and scales; whereas, the shock Mach number, M_s, is more representative of the usual Mach number of wind tunnel tests.

Generally, both the wind tunnel and shock tube experiments provide limited information when compared with either theoretical or numerical studies. However, in order to highlight some of the major effects the shock has on homogeneous turbulence, the remainder of this section will focus on two complementary experiments. This will also allow for some comparisons with the results given in the previous section, particularly for features such as turbulence amplification and evolution of turbulent scales through the shock. The first set (hereafter Case A) correspond to the wind tunnel experiments (Barre et al., 1996). The upstream Mach number is $M = 3$, with a fixed grid configuration of equivalent mesh size $\Delta = 6$ mm. The grid (mesh size) and Taylor microscale Reynolds numbers were $Re_\Delta = 6 \times 10^4$ and $Re_\lambda = 15.5$, respectively. The second set (hereafter Case B) correspond to shock-tube experiments (Agui et al., 2005). In this case, several grid sizes have been used (from $\Delta = 2.54$ mm to 19 mm), with incoming flow Mach numbers, M_u, ranging from 0.3 to 0.6. The resulting shock Mach numbers, M_s, then range from 1.0 to 1.2. This configuration induces a variety of grid and microscale Reynolds numbers, ranging between 5×10^4 and 40×10^4 for Re_Δ, and 260 and 1300 for Re_λ. These two sets of physical experiments ideally complement one another and serve to illustrate also the current state-of-the-art. Combined, both Cases A and B cover a wide range of parameter values and complementary flow variable information: Case A, two velocity components and spectra; Case B, multi-components and space statistics.

An obvious characterizing feature of a shock–turbulence interaction is the amplification of the turbulence measured by the ratio $G_{u^2} = \overline{u_i'^2}/\overline{u_0'^2}$ ($\overline{u_0'^2}$ is the initial level just before the shock). For the higher Mach number Case A, Figure 7.7 shows a typical amplification ratio for the streamwise component of the velocity through the normal shock, and where as usual, after a strong amplification, a relaxation is observed. Also on this figure is a comparison to Ribner's theory (LIA) which shows some disagreement in the near field. This is probably due to the fact that it was not possible, from the hot-wire anemometry, to isolate the contribution of the pressure fluctuations to the total measured fluctuation levels. The measured fluctuations were thus assumed to be solely due to velocity fluctuations which could easily cause the over-estimation of the experimental results shown in the near-field of Figure 7.7. In the "far" field (sufficiently downstream of the shock), however, the agreement with linear theory is quite good with an amplification ratio of $G_{u^2} \approx 1.47$. This is also in good agreement with the results from LIA and DNS shown in Figure 7.6a where a value of 1.45 was obtained at a Mach number of 3.

For the other normal stress components, results from Case B are given in Figure 7.8 for the amplification ratio G_{u^2} across the shock for several values

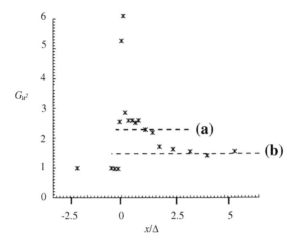

FIGURE 7.7 Amplification of the streamwise turbulent velocity component across a Mach 3 normal shock, Case A. ——— Ribner theory, (a) near field, (b) far field ($x = 0$ corresponds to shock location).

of shock wave strength M_u. Consistent with the higher Mach number Case A presented previously, the global trend is an amplification of the longitudinal component; although the evolution exhibits a large scatter. As Figure 7.8 shows, the amplification ratio for the longitudinal component can evolve from 3 to 1.2 between $M_s = 1.1$ and $M_s = 1.15$ and then rise to a value of 4 at $M_s = 1.19$ (Andreopoulos, Agui, & Briassulis, 2000). It is apparent that the scatter between experimental results is not only due to the experimental difficulties linked with the statistical convergence for the shock tube results, but is also due to the influence of size of the grid (and then on the microscale Reynolds number) that varies between the different results presented in the figure. Currently, this influence is not clearly understood. Lastly, the amplification ratio computed from DNS, LIA and RDT underpredict the experimental results. As has been shown for RDT (Jacquin, Blin, & Geffroy, 1993; Jacquin, Cambon, & Blin, 1993), the amplification rate is proportional to the density ratio across the shock squared, $(\rho_d/\rho_u)^2$. The lateral components remain unaffected or slightly decrease through the shock. In terms of vorticity components, this corresponds to an amplification in *rms* values by a factor of 25% for the transverse vorticity components with a corresponding slight decrease in the longitudinal component (Agui et al., 2005).

Another important feature of the interaction is the influence on the turbulent length scale. The variation across the shock also provides a quantitative estimate of the effect of the shock on the turbulence. Unfortunately, there is only a limited amount of experimental data available for the length scale evolution for such shock interactions. Values for the longitudinal integral scale amplification

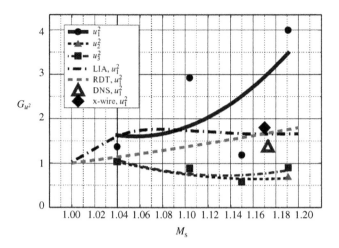

FIGURE 7.8 Amplification ratio of various velocity components across a shock from shock tube experiments, LIA and DNS (Lee et al., 1993), and RDT (Jacquin, Blin, & Geffroy, 1993). From Agui et al. (2005) with permission.

rate $G_{L_{11}}(\zeta_1)$ across the shock are plotted at different positions (x/Δ) (Δ is the mesh spacing of the grid that acts as the turbulence generator) in Figure 7.9. Here G is the ratio of the longitudinal integral scales L_{11} measured downstream and upstream of the shock. These scales are expressed in terms of the separation distance in the streamwise direction (ζ_1). Note that the integral scales are actually computed from the autocorrelation of the temporal signals (imposing time delays τ between velocity signals), along with invoking the Taylor hypothesis. The mean velocity is multiplied by the time delay to recover the length scale.

For Case A at Mach 3, the amplification rate (Figure 7.9a) shows a dramatic decrease of 85% of the longitudinal length scale across the shock, coupled with a slight increase further downstream. This latter increase is related to the flow acceleration already mentioned. For the lower Mach numbers in Case B, Figure 7.9b shows the amplification characteristics for different combinations of upstream and shock Mach numbers. This figure illustrates the complexity of the phenomenon and the sensitivity to initial conditions. A first level of difficulty is the interpretation of the shock tube experiments when compared with those performed in conventional wind tunnels. In Case B, the interaction can take place at any distance from the grid which means the shock can interact with different levels and scales of turbulence. The distance x/Δ corresponds to the distance from the grid where G is measured (the probe is fixed). In contrast, in wind tunnel experiments, the distance x/Δ is the distance from the shock interaction (i.e. the probe is moved downstream). Thus, the information gained from the two plots in Figure 7.9 is not the same even with the same

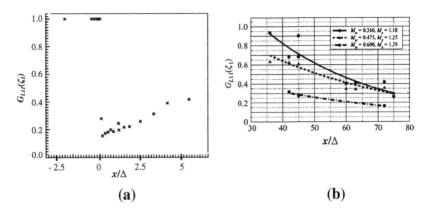

FIGURE 7.9   Ratio of the longitudinal integral length scale across the shock. (a) Case A: interaction with a Mach 3 normal shock (Barre et al., 1996). (b) Case B: evolution for different upstream integral scales (related to the value of upstream M_u and shock M_s Mach numbers). From Agui et al. (2005) with permission.

horizontal axis. For Case B, the distance x/Δ has to be interpreted in terms of the level of incoming fluctuations and corresponding integral scales that are affected by the shock. More precisely, the different curves plotted in Figure 7.9b correspond to different initial values of the longitudinal length scales. (The reader can refer to Figure 23 of Briassulis et al. (2001) to obtain the corresponding values of L_{11} in the unperturbed flow.) One main result is that, in weakly compressible grid turbulence, the incoming integral scales increase downstream of the grid but do not behave in a monotonic manner as a function of the flow Mach number. When a shock is interacting with these turbulence fluctuations, a strong reduction of the integral scale is observed for all cases. For all the configurations, the reduction is greater for high values of x/Δ, that is the reduction is greater when the scale is higher. This reduction is even higher for the larger shock strengths. For example, in Figure 7.9b, the results obtained at $M_s = 1.39$ have the highest reduction (between 70% and 85%) for the two extreme postions ($x/\Delta = 40$ and 70).

Summarizing, it should be recalled that the determination of integral length scales in Cases A and B are different, which can contribute to discrepancies in the amplification rates. However, when comparing the Mach 3 wind tunnel results of Case A and the Mach 1.39 shock tube results of Case B, the maximum attenuation is about 0.85 in both cases. Even with this limited experimental data, it can be assumed that beyond Mach 1.5, an asymptotic level is reached for the reduction in longitudinal integral scale across the shock. It should be mentioned that there has been no systematic study on the possible effect of the ratio between the upstream integral scale and the shock thickness. For the evolution of the lateral integral scale, it has been shown in Case A that this integral scale is practically unaffected by the shock interaction; whereas in Case B, the turbulent

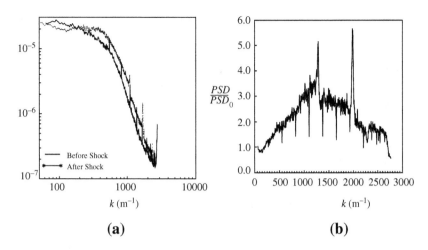

FIGURE 7.10 Hot-wire signal spectra for a Mach 3 normal shock: (a) upstream and downstream spectra; (b) ratio of upstream and downstream spectra. From Barre et al. (1998) with permission.

eddies appear to be compressed in both directions when the shock strength is large. This behavior is believed to be caused by the variation in the state of the incoming turbulence; although, this has yet to be thoroughly assessed.

From Case A, the behavior of the turbulent scales across a normal shock can be more precisely analyzed in terms of the fluctuating longitudinal mass flow rate spectrum. Figure 7.10a shows the upstream and downstream hot-wire spectra obtained across a Mach 3 normal shock (spectra are plotted in dimensional form). Their integral is the local velocity variance, so that this representation takes into account the turbulent kinetic energy amplification across the shock. It is clear that the downstream spectrum is shifted towards the high frequencies, which confirms the observed reduction in longitudinal integral scale across the shock (cf. Figure 7.9a).

The amplification ratio of each turbulent scale resolved by the experiment can be obtained by computing the ratio of the two spectra as shown in Figure 7.10b. It is clear that the most amplified wavenumbers are higher than the ones corresponding to the incoming longitudinal integral scale (which typically lies around $300-400$ m^{-1}) (see Figure 7.10a). Thus, the shock induces an energy transfer from fluctuations associated with the incoming integral scale towards smaller scale fluctuations. In the Case B shock tube experiments (as well as DNS), an estimate of the dissipation rate through the measurement of all the instantaneous velocity components was possible, and it was found that the dissipation rate increased through this interaction (e.g. Agui et al., 2005; Barre et al., 1996; Hannappel & Friedrich, 1995; Jamme et al., 2002; Lee et al., 1993). This result is consistent with the evolution of the spectra, although the dissipative range encompasses a larger spectral range than the one given in Figure 7.10.

Currently, there is still an insufficient amount of experimental, or numerical, data for such homogeneous turbulence/shock interaction cases. While experimental progress has been slow, it should not minimize the importance of understanding the dynamics in this simplified turbulence interaction problem. The understanding of such flows is an important element in attempting to develop models for much more complex shock/turbulence interactions such as those to be discussed next for inhomogeneous flows.

7.2 INHOMOGENEOUS TURBULENCE INTERACTIONS

As was alluded to in the previous section, the investigation of shock interactions with homogeneous turbulence provided important insight into the underlying dynamics of shock–turbulence interactions and also served as a type of building block case from which information about dynamic interactions could be applied to the more complex inhomogeneous flows. Both free shear flows and wall-bounded flows have, of course, been investigated in both physical and numerical experiments. Unlike the homogeneous flows where a rather thorough understanding of the dynamics could be achieved through a combination of tri-modal decomposition of the incoming fluctuation field, and then validated by a linear theory, inhomogeneous flows have a much broader set of flow field effects that have to be accounted for. In addition, limitations on both measurements and numerics due to demanding parameter ranges prevent all the various solution tools from being applied to all the flows.

7.2.1 Free Shear Flows

As outlined in Section 6.2.2, one of the most important consequences of compressibility in supersonic flows is the lack of efficient mixing. For several practical applications this can have a strong negative impact. In particular, this is the case for combustion in supersonic regimes where scramjet efficiency is limited when the Mach number increases, and several studies have been devoted to the use of shock waves to enhance mixing.

The previous section has illustrated how homogeneous turbulence can be amplified through different shock-wave interaction mechanisms. In shear flows, the shock wave has the same intrinsic turbulence effect described for homogeneous flows; however, the presence of shear induces an additional effect on the dynamic and thermodynamic variables, that is, to the so-called baroclinic effect. Specifically, the combined effects of pressure and density gradients influence the evolution of the vorticity.

In order to illustrate this effect, the vorticity transport equation is often used. Equation (2.11) for the instantaneous vorticity can be rewritten in the form

$$\frac{D\omega_i}{Dt} = S_{ik}\omega_k - \omega_i S_{kk} + \underbrace{e_{ijk}\left(\frac{1}{\rho^2}\frac{\partial\rho}{\partial x_j}\frac{\partial p}{\partial x_k}\right)}_{B_{\omega_i}} + e_{ijk}\frac{\partial}{\partial x_j}\left(\frac{1}{\rho}\frac{\partial\sigma_{kl}}{\partial x_l}\right), \quad (7.8)$$

where the velocity gradient has been replaced with the strain rate, and Eq. (2.22) has been used with a constant pressure gradient assumed, and no effects of body forces included. Since in most cases the pressure gradient induced by the shock occurs in the streamwise direction (particularly true for normal shock waves), the instantaneous spanwise (or azimuthal) component of the baroclinic production term, B_{ω_3}, is of interest and can be written as

$$B_{\omega_3} = \frac{1}{\rho^2}\left[\frac{\partial\rho}{\partial x_1}\frac{\partial p}{\partial x_2} - \frac{\partial\rho}{\partial x_2}\frac{\partial p}{\partial x_1}\right]$$
$$\approx -\frac{1}{\rho^2}\frac{\partial\rho}{\partial x_2}\frac{\partial p}{\partial x_1}. \quad (7.9)$$

Depending on the relative signs of the streamwise pressure gradient through the shock and the transverse density gradients, the contribution to the vorticity development can be either positive or negative, inducing either an amplification or a damping of the initial vorticity, with the associated impacts on mixing and turbulent fields.

7.2.1.1 Jet/Shock Wave Interactions

The interaction of a shock wave with a turbulent jet can be illustrated in the co-flowing jet example given in Figure 7.11 The figure shows the velocity distribution of a conventional jet in a co-flow, inducing a vorticity distribution that is positive in the upper part and negative in the lower part. Note that, as will be illustrated later, the density gradients can be changed according to the flow arrangement. As with the homogeneous turbulence/shock wave experimental studies, turbulent free shear flow/shock wave interactions can be analyzed either in supersonic wind tunnels with a steady shock, or in shock tubes with traveling shock waves.

In the wind tunnel studies, the entire flow should be supersonic so that shock waves can be generated. Unfortunately, only a few experiments are available due to the complexity of the flow configuration. Two such experiments that have been performed (Jacquin, Blin, & Geffroy, 1993; Jacquin & Geffroy, 1997) have a jet in the central part of a supersonic flow at the sonic nozzle location, and the flow was shocked through a second throat. If the total (settling chamber) temperatures of the two (supersonic) co-flows are the same, the static temperature is lower and the density is higher in the high speed region. This was the case in the first isothermal experiment (Jacquin, Blin, & Geffroy, 1993) (denoted by case J-WT1) which yielded the inverse of the density distribution shown in Figure 7.11a. This flow corresponds to a jet Mach number of 2 and a convective Mach number of 0.26.

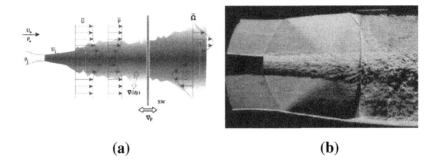

(a) **(b)**

FIGURE 7.11 Jet/shock wave interaction. (a) Schematic description of a jet hotter than the ambient ($\overline{\Omega}$, mean vorticity; SW, shock wave) (from Andreopoulos, Agui, & Briassulis, 2000, with permission); (b) schlieren visualization of the heated jet/shock wave interaction case J-WT2 (from Jacquin & Geffroy, 1997, with permission).

The sign of the density gradient can be changed by changing the gases or heating one of the two streams. This was the case in the second experiment (Jacquin & Geffroy, 1997) (hereafter referred as case J-WT2) in which the central jet was heated. A schlieren visualization of the flow is shown in Figure 7.11b. The Mach number was 1.6, and the convective Mach number was less than 0.5, corresponding to weak compressibility effects in the incoming shear layer. In this case, the density is lower in the hotter stream (the density ratio being 0.33) and the density distribution is as shown in Figure 7.11a. Nevertheless, in both cases, the steady shock induces a positive pressure gradient.

For shock tube experiments, a subsonic jet in co-flow (the co-flow is needed to balance the jet entrainment) has been used (for example Andreopoulos, Agui, Wang, & Hermening, 2000; Hermanson & Cetegen, 2000). The direction of the jet can be toward one or the other end of the shock tube. The shock can have either the same direction as the jet (hence the pressure drop is negative as seen by an observer fixed with the shock—and denoted as J-ST1), or be propagating upstream of the jet flow (the pressure gradient is then positive—and denoted as J-ST2). Since the velocity of the jet is low, the density gradient in this case is not imposed by the jet velocity but by the gas that is used. When the jet gas discharging into the air is heavier than the ambient (as it is for Krypton, Andreopoulos, Agui, Wang, & Hermening, 2000), the density gradient is positive. This case will be denoted as J-ST1-1 or J-ST2-1 depending on the direction of the shock wave. When the jet gas is lighter (as in the Hermanson & Cetegen, 2000, experiment where helium or CO_2 was used), the density gradient is negative, corresponding to the distribution shown in Figure 7.11. This case will be denoted as J-ST1-2 or J-ST2-2, again depending on the direction of the shock wave.

Any jet/shock interaction can correspond to either an amplification or damping of the vorticity due to the baroclinic effect. In the experiments

J-WT1, J-ST1-2 and J-ST2-1, the pressure gradient is of an opposite sign when compared with the density in the upper part of the flow. In this region, the initial vorticity is positive, so the baroclinic term will contribute to the production of vorticity. The same trend will be observed in the lower part of the jet. Thus, for the experiments J-WT1, J-ST1-2, and J-ST2-1, there will be an amplification of the vorticity in the shock interaction. The other density/pressure distribution experiments, J-WT2, J-ST1-1, and J-ST2-2, will correspond to a baroclinic damping effect provided the sign of B_{ω_3} is opposite to the sign of $\overline{\Omega}$. The baroclinic amplification effects have been clearly demonstrated in shock tube experiments for shock Mach numbers ranging from 1.23 to 1.45 interacting with a subsonic jet ($M = 0.9$) using Mie scattering visualizations (Andreopoulos, Agui, & Briassulis, 2000; Hermanson & Cetegen, 2000). In addition, after the shock passage, large scale energetic events were observed that increased the overall turbulence and mixing.

More quantitative results are available from the LDV measurements (Jacquin, Blin, & Geffroy, 1991; Jacquin, Blin, & Geffroy, 1993; Jacquin & Geffroy, 1997). In the J-WT1 case, the amplification observed in Figure 7.12a is higher than the value expected based on an RDT analysis. In this case, the amplification based on the square of the experimentally measured density values Section 7.1.3 is lower than the measured amplification (Jacquin et al., 1991). It is assumed that the baroclinic effect has to be taken into account to explain the total amplification observed in the experiment. As expected for the heated jet case J-WT2, Figure 7.12b shows an amplification measured in the quasi-homogeneous region (at small values of k_0 in Figure 7.12b), and in agreement with the expected behavior described in the last section. However, in the sheared region, a damping effect can be observed. This region corresponds to the higher turbulence values in Figure 7.12b, where there is a large decrease (the damping can be almost a factor of 2). According to the authors, this behavior can be directly linked to the baroclinic effects described above.

7.2.1.2 Mixing-Layer/Shock Wave Interactions

There are fewer experimental studies on the interaction of shock waves with mixing-layers. One example, however, is a weak shock wave imposed at the trailing edge of a splitter plate used to generate a supersonic/subsonic mixing-layer at a convective Mach number of 1 (Barre, Braud, Chambres, & Bonnet, 1997). Figure 7.13 shows a schlieren visualization of the isobaric and shocked flows. With the deviation of the layer, no significant amplification of the turbulence was measured for several vorticity thicknesses, δ_ω downstream of the interaction. Within the uncertainty limits of the experiment, the decrease of longitudinal and shear fluctuations were the same as the natural configuration. The Mach number effect on the anisotropy appears to be unaffected by the shock interaction, consistent with the results presented in Figure 6.10.

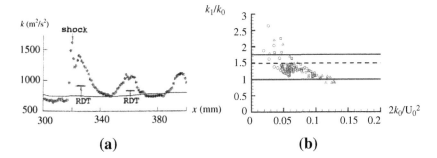

FIGURE 7.12 Evolution of the turbulent kinetic energy in jets interacting with a shock wave: (a) cold jet, case J-WT1. ——— Without shock and ∗ with shock (Jacquin, Blin, & Geffroy, 1991); (b) hot jet, case J-WT2, for different lateral positions corresponding to different values of the incoming turbulent kinetic energy k_0; symbols, results for different temperatures of the central jet; – – – amplification from LIA; ——— (upper line) amplification from RDT (L. Jacquin, private communication).

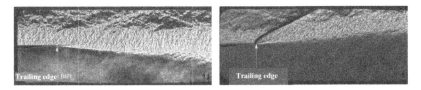

FIGURE 7.13 Schlieren visualization of a supersonic mixing layer at $M_c = 1$: (a) perfectly expanded case, (b) non-isobaric case. From Barre, Braud, Chambres, & Bonnet (1997), with permission.

7.2.1.3 Wake/Shock Wave Interactions

Most of the studies of shock wave interactions with wakes are devoted to transonic flows in turbine blades that include the additional complexity associated with unsteadiness, curvature and wall proximity. The studies of simpler (at least geometrically) wake/shock configurations were primarily devoted to the fundamental analysis of shock-wave/vortex interactions (Haas & Sturtevant, 1987). Turbulent wakes are often used for more realistic configurations such as a shock interacting with a vortex-like wake of an obstacle at Mach 3 (Kalkhoran & Sforza, 1994).

Two-dimensional flat plate wakes that have been described in Section 6.3 can also be investigated; although, only a few studies have been devoted to this wake interaction with a shock. Since the wake deficit that represents the driving velocity, rapidly decreases downstream from the trailing edge, such shock interactions have to be analyzed relative to the distance where the shock impacts the trailing edge. This distance is denoted as S in the schematic shown in Figure 7.14a from a theoretical investigation of the breakdown of the viscous

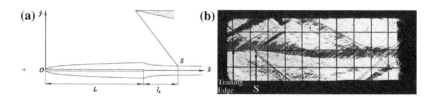

FIGURE 7.14 Supersonic wake/shock interaction: (a) schematic from Battam et al. (2004) with permission; (b) schlieren visualization of a $M = 2$ wake from Kumar (1975).

wake for either a shock impacting in the near wake region or farther downstream (Battam, Gorounov, Korolev, & Ruban, 2004). Closer to turbulent conditions, Kumar (1975) considered the interaction of a shock wave generated by a $8°$ wedge interacting with a Mach 2 turbulent wake. Figure 7.14b shows such a typical interaction that impacts the wake at $x/\theta = 20$ (θ is wake momentum thickness) with a deviation of the shock and a complex shock system driving an increase of the thickness of the wake immediately downstream of the interaction. When the interaction occurs farther from the trailing edge, the velocity is more uniform and the modification of the wake is qualitatively less pronounced in the thin flat plate wake experiment. When the interaction is coming from the shock waves emanating from the thick trailing edge (Nakagawa & Dahm, 2006), these shock waves reflect on the wind tunnel walls, and then interact at relatively large distances ($x/\theta = 90$). At that location (and periodically downstream for the other interactions), a strong increase of the spreading rate is observed. For both types of experiments (thin or thick flat plate wakes), one can anticipate that the turbulent field, although not measured, will be amplified. Clearly there is a need for more detailed investigations of these flows.

7.2.2 Wall-Bounded Flows

For wall-bounded flows, the effect of the shock can induce flow separation. This flow separation can be either open or followed by a reattachment region. Since the flow then assumes an elliptic character near the wall, and in the separated region, it opens the possibility for different modes of instability so that most of the wall-bounded shock wave interacting flows are subject to low frequency instabilities (Dupont, Haddad, & Debiève, 2006) with some complex behavior (see, for example, Robinet, Gressier, Casalis, & Moschetta, 2000). To illustrate these effects, two typical flow configurations of practical relevance within the aerospace field are of interest and shown in Figure 7.15.

The first deals with the (unsteady) buffeting phenomena already illustrated in Figure 1.6 using computational fluid dynamics (CFD) and schematically represented in Figure 7.15a. This phenomena occurs primarily in the transonic regime. A complete understanding of the instabilities of the buffeting phenomena is still an open question, primarily due to a flow configuration

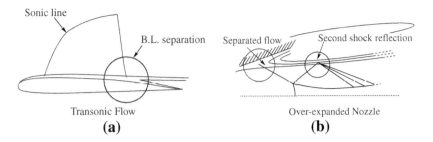

FIGURE 7.15 Shock wave/boundary layer interaction configurations: (a) airfoil transonic buffet; (b) rocket nozzles. From Barre, Bonnet, Gatski, and Sandham (2002) with permission.

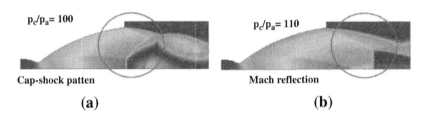

FIGURE 7.16 Mach number distribution from numerical simulation of an over-expanded supersonic nozzle (Vulcain engine). (a) Cap-shock pattern; (b) Mach reflection. From Hagemann and Frey (2008) with permission.

that requires a knowledge of the circulation around the airfoil, and which can affect any flow control that may be applied (see Section 8.3.3). The other flow configuration is the over-expanded nozzle flow shown in Figure 7.15b. This flow field occurs in the nozzles of rocket engines at sea level. Since the nozzles are designed to be fully expanded at the lower pressure in cruise conditions, the sea level behavior corresponds to an over-expanded condition. Depending on the ambient pressure, shock discs can be present inside the nozzle and separation can occur at the nozzle surface. Figure 7.16 shows the Mach number distribution from a numerical calculation of an over-expanded nozzle for two pressure ratios (chamber pressure/ambient pressure) that produce either a cap-shock or Mach reflection. It should also be noted that in the experiment (Nguyen, Deniau, Girard, & Alziary de Roquefort, 2003), the separation is never exactly axisymmetric. Thus, side loads are then present and low frequencies are experimentally observed (for a review see Reijasse, 2005) that cause severe problems during rocket take-off. In axisymmetric configurations, the solid walls of the nozzle make it difficult to probe the flow experimentally; however, simplified studies can be performed in two-dimensional nozzles in which schlieren visualizations can be done (see Figure 7.17).

Such experimental configurations are amenable to RANS solutions (Reijasse, 2005); however, it is often necessary to (re)calibrate closure

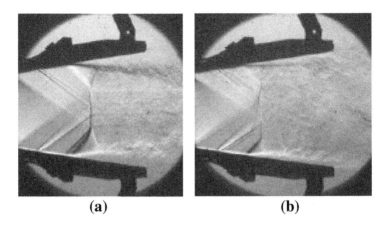

(a)   **(b)**

FIGURE 7.17 Schlieren visualization of a two-dimensional overexpanded nozzle: (a) free separation; (b) restricted separation. From Reijasse et al. (2005) with permission.

coefficients in order to optimize the comparisons between experiments and computations. This shows both the usefulness of such closures as well as an inherent problem for broader ranging flow field predictions. If sufficient information is available, it is usually possible to adjust coefficients in the various closure models to have a fairly reliable tool to augment experimental results; however, lacking this data, the closure models have not been developed to the level where general flow field prediction is possible.

In addition to these relatively simple building block flows, other complex flow fields have been explored. These include the two-dimensional expansion–compression corners, and single-fin plate configuration and the crossing shock wave/boundary layer interaction (double-fin/plate configurations). In the expansion–compression corner case, the added influence of favorable pressure gradient, convex streamline curvature and bulk dilatation are introduced into the dynamic balance. In the single- and double-fin configurations, the deflection of the flow by the fin(s) generates a swept shock system. Depending on the parameters involved, separation may occur which has a significant effect throughout the flow. Obviously, the double-fin geometry introduces added complexity and generates a wide range of shock–turbulence and shock–shock interactions. A thorough review of the experimental and numerical studies related to these configurations is given in Knight et al. (2003) (see also Panaras, 1996, for a thorough analysis of the physics of the single-fin geometry).

In general, the interaction of shocks with solid boundaries has been extensively studied for decades and reviews of the state-of-the-art at various times (Fernholz & Finley, 1980, 1981; Fernholz, Finley, Dussauge, & Smits, 1989; Green, 1971; Knight et al., 2003; Settles & Dolling, 1992), as well as compilation of (experimental) data (Fernholz & Finley, 1980, 1981;

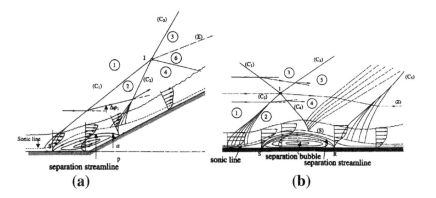

FIGURE 7.18 Shock wave/boundary-layer interaction: (a) two-dimensional compression ramp; (b) shock impingement. From Délery (2001) with permission.

Fernholz et al., 1989; Settles & Dodson, 1991) have provided a guide to what is known about such flow fields.

7.2.2.1 Shock Wave/Boundary-Layer Interaction: Compression Corner

In the buffeting and internal shock cases just highlighted, the basic building block that has to be considered as a guide to interpreting the physical phenomena is the shock wave/boundary-layer interaction (SWBLI). Such a flow field can be generated by a change in the wall geometry such as when there is a wall deflection (Figure 7.18a). As the figure shows, a complex shock structure is generated with separation (C_1), reattachment (C_2), and ramp induced (C_3) shocks present (Σ is the slip line). The various flow states (regions) are designated by the circled numbers (see Délery, 2001, for details).

Flows with surface alterations, such as compression corners, have been studied extensively experimentally and have been well documented (see Délery, 1985, for example). The current focus in studying such flows is directed towards numerical prediction using DNS, LES, hybrid methods and averaged methods. As such, the available literature is overwhelming and it is impossible to evaluate all the results. Nevertheless, there are some types of geometries that have been and continue to be the focus of study for probing the flow dynamics.

Two-dimensional mean flow fields are an important problem for studying boundary layer flows with shocks. For the two-dimensional compression corner, a wide range of mean and turbulent flow measurements are available for comparison (Settles & Dodson, 1991, 1994; Smits & Muck, 1987). Since parameters such as shape factor and sonic line location associated with the mean velocity ahead of the compression ramp have an impact on the separation zone, it is important in performing numerical comparisons to detail the incoming flow field carefully. In the vicinity of the separation zone, the shock structure naturally induces an amplification of the turbulence by the shock as well as a corresponding reduction of the turbulent fluctuations by the associated expansion fan. At the end of the reattachment region, the development of a new boundary layer begins although the relaxation of both the mean and turbulent fields will depend on several factors including the effect of any Taylor–Görtler vortices. As the ramp angle increases, the shock unsteadiness increases and causes enhanced velocity and thermal field oscillations (Dolling, 1993). Such unsteadiness implies that either direct or large eddy simulation methods may be more suitable solution methods in performing numerical computations. Averaged methods, if used, would need to be properly sensitized to such non-equilibrium phenomena to accurately replicate the physics and to be performed in a time accurate mode (basic unsteady Reynolds averaged Navier–Stokes (URANS) could be used but with uncertain accuracy). It is not surprising, that averaged methods, such as RANS, are incapable, in general, of predicting many aspects of such flows. As the discussion in earlier chapters has indicated, a variety of compressibility and variable density effects need to be considered and incorporated into models for them to be properly sensitized to the flow physics. Such an effort has yet to be undertaken for a complete compressible model.

Numerical simulations, such as DNS, have now reached a level of maturity that such flows can be undertaken with some confidence if the parameter range, for example Reynolds number, is limited to relatively modest values. If properly conducted, such numerical simulations can be used as a complementary tool to physical experiments. Filtered methods, while having important potential, have the added burden of accounting for the unresolved scales in some fashion— either through a MILES (monotone integrated LES) approach, ILES (implicit LES), or an explicit modeling approach. Previous simulations of compression ramp flows have been conducted for ramp angles from $8°$ to $25°$ and at inflow Mach numbers of 2.3 and 3 although the bulk of assessments have been conducted at $M = 3$ (Adams, 1998, 2000; Rizzetta & Visbal, 2002; Stolz, Adams, & Kleiser, 2001). Unfortunately, at ramp angles greater than $8°$, an assessment of the simulations results is disappointing due to both the inability to generally replicate the experimental data where available, and the diversity of inflow conditions that preclude relative comparison of results. An extensive assessment of such studies has been conducted (Knight et al., 2003), which while encouraging in identifying the interest in evaluating such flows, was disappointing in that accurate predictions were still not attainable. Nevertheless, simulations continue with an ever-increasing database to evaluate. Recently, both LES (Loginov, Adams, & Zheltovodov, 2006) and DNS (Wu & Martin, 2007, 2008 see also Wu & Martin, 2006) studies at $M_\infty \approx 2.9$ and ramp angle $\approx 24°$ have been performed with comparison to experimental results at similar conditions for validation. One of the interesting characteristics of any shock/boundary layer interaction is the relaxation length or "dynamic range" of the shock flow interaction. In compressible boundary layer flows, if the SRA holds upstream of the compression corner, then a relevant question is

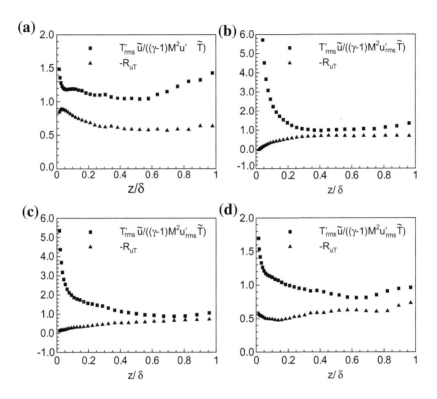

FIGURE 7.19 Variation of two SRA relations ahead of compression and behind compression corner along wall: (a) $x = -5.7\delta$; (b) $x = -2\delta$; (c) $x = 2\delta$; (d) $x = 6.1\delta$ (δ is inflow boundary layer thickness). From Wu and Martin (2007) with permission.

the length along the ramp before the relations hold in the post-shock flow. Figure 7.19 shows the behavior of the SRA relations given in Eqs. (3.99) and (3.106) in the regions ahead and behind the compression corner. The upstream distributions of both forms of the SRA are consistent with results from previous simulations of flat plate boundary layers (e.g. Guarini, Moser, Shariff, & Wray, 2000; Maeder, Adams, & Kleiser, 2001; Pirozzoli, Grasso, & Gatski, 2004). After the compression corner, a recovery zone many (incoming) boundary layer thicknesses downstream is suggested. For simulations, information about relaxation lengths is useful in establishing minimal sizes for computational domains, and for closure model development it provides an additional means of validation in assessing model performance. Some assessment of the effect on the turbulent velocity and thermal fields can be obtained from large eddy simulation (Loginov et al., 2006). In Figure 7.20, the streamwise variation of the distribution of *rms* density and streamwise velocity fluctuations gives at least some qualitative insight into the amplification downstream of the compression corner. It should be cautioned that Taylor–Görtler vortices can play a role in the

(a)

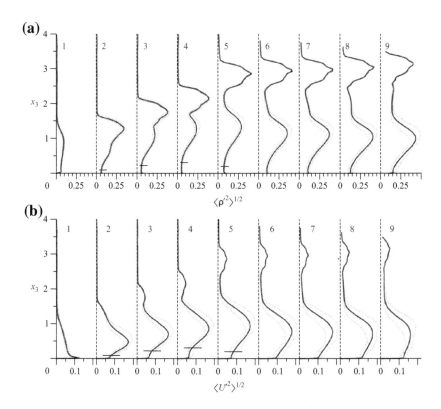

(b)

FIGURE 7.20 Downstream evolution of (a) density and (b) velocity fluctuations. Stations 1–4 are along the flat plate ahead of the corner, and stations 5–9 are downstream of the corner along the ramp. Thick solid line is time- and spanwise-averaged value, and the spanwise variation is indicated by the dotted lines. The horizontal line corresponds to the location of zero mean velocity. From Loginov et al. (2006) with permission.

energetic enhancement downstream of the corner. Other, more complex, flow fields of this type have been explored, and include two-dimensional expansion–compression corners. In the expansion–compression corner case, the added influence of favorable pressure gradient, convex streamline curvature and bulk dilatation are introduced into the dynamic balance (see Knight et al., 2003).

7.2.2.2 Shock Wave/Boundary-Layer Interaction: Impinging Shock

A flow with an imposed distortion, such as an impinging shock on a flat plate, is shown in Figure 7.18b. Once again, the figure shows a complex shock structure being generated resulting from the impinging oblique shock (C_1). In addition to the resulting reflected shock (C_3), a separation shock (C_2) is present which is followed downstream by a second reflected shock (C_4) and, further downstream by a reattachment shock (C_5). Σ is again the slip line, and as in Figure 7.18a, the various flow states (regions) are designated by the circled numbers (see Délery, 2001, for details).

In some ways, it is simpler than the compression corner case in that mean streamline curvature effects and the associated Taylor–Görtler vortices are not present. The impinging SWBLI problem has the unperturbed zero pressure gradient flat plate boundary layer flow as a basis so it is possible to accurately delimit the range of influence of the shock (in the same coordinate frame as the incoming flow). There have been several experimental studies of such flows dating from the experiments of Green (see Green, 1971 for a summary); however, more relevant are the recent studies of Bookey, Wyckham, Smits, and Martin (2005), Deleuze (1995), Humble, Scarano, and van Oudheusden (2007a), and Laurent (1996). The experiments are still very difficult, even with the rapid advances in particle image velocimetry (PIV) methods (Scarano, 2008); nevertheless, these experiments have provided sufficient information and data for quantitative replication and validation by numerical simulations (particularly the experiments of Bookey et al. (2005), that are performed at relatively low Reynolds numbers) and RANS computations.

Both direct (Li & Coleman, 2003; Pirozzoli, Grasso, & Gatski, 2005; Pirozzoli & Grasso, 2006; Shahab, Lehnasch, Gatski, & Comte, 2009; Wu, Bookey, Martin, & Smits, 2005), and large eddy simulation (Garnier et al., 2002) studies have been performed at conditions at or near the parameter ranges of these more recent experimental studies. As with the ramp flow case, the problem is truly three-dimensional with the issue of shock oscillation as important as in the ramp case. Other than mean pressure and skin friction comparisons with experiments, there does not appear to have been an extensive amount of quantitative comparisons with experiments or attempts at correlating amplification rates of the thermal and velocity correlations with the shock/turbulence simulations and theory discussed previously. Such analyses could be useful in understanding the dynamics of such flows. One of the fundamental questions to be answered in assessing and comparing data with experiments is whether the comparisons are with the same variables. This point has been raised in earlier chapters and it has been seen that at supersonic speeds the mass flux terms are not significant so that experimentally measured Reynolds variables and density-weighted numerical variables differed little. At hypersonic speeds, the situation may differ, but there have not been enough (direct) numerical simulations to adequately address this issue. However, in the presence of shocks, the mass flux can be significant and so quantitative comparisons with experiments may need to be performed more cautiously.

Recall the example in Section 5.3.1 of the effect of the impinging shock on the turbulence field using the anisotropy invariant map of the Reynolds stress tensor which provides a representation of the turbulence field based on the componentiality of the turbulent stress anisotropy defined in Eq. (5.29). Here, it is more informative to examine the invariant map in coordinates other than

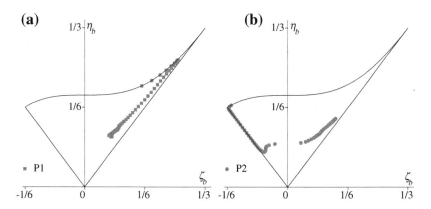

FIGURE 7.21 Anisotropy invariant map of Reynolds stress invariants upstream and downstream of impinging shock: (a) upstream location $P01$ at $-2.9\delta_\omega$ (δ_ω, vorticity thickness at this upstream station); (b) downstream location $P02$ at $0.4\delta_\omega$.

the $(III_b, -II_b)$. The reason is that instead of highlighting the wall-proximity behavior along the two-component boundary (cf. Figure 5.2), the dynamics in the shock impinging case dictate a focus along the axisymmetric boundaries of the invariant map. These invariant map boundaries are highlighted in (ζ_b, η_b) coordinates where $\zeta_b = (III_b/2)^{1/3}$ and $\eta_b = (-II_b/3)^{1/2}$. In order to quantify the difference between the (ζ_b, η_b) and $(III_b, -II_b)$ invariant map distributions, the anisotropy map upstream of the impinging shock in (ζ_b, η_b) coordinates (shown in Figure 7.21a at plane $P01$) should be compared with the corresponding upstream distribution in Figure 5.2.

In Figure 7.21b at plane $P02$, which is $0.4\delta_\omega$ downstream of the incident shock impingement location, a significant distortion of the invariant curve from that at plane $P01$ occurs. As the figure shows, very close to the wall the points still lie on the 2C boundary of the invariant map; however, the migration is no longer toward the 1C limit point and the axisymmetric boundary associated with the upstream invariant map, but now toward the axisymmetric boundary associated with a turbulent stress field dominated by one large (negative) normal stress anisotropy component and two smaller (positive) normal stress anisotropy components. (Strictly speaking the anisotropy components being referred to are those associated with the anisotropy tensor eigenvalues. The turbulent shear stress component is relatively small here so the normal components of the anisotropy tensor are closely related to the corresponding eigenvalues.) This means the turbulent stress field is representative of a (compressed) oblate spheroid in contrast to the undistorted stress field which is characteristic of an (elongated) prolate spheroid (Simonsen & Krogstad, 2005). In particular, this axisymmetric limit represents the action of two dominant normal stress components ($\overline{u_1' u_1'}$ and $\overline{u_3' u_3'}$) and a weaker third normal stress component

$(\overline{u_2' u_2'})$. As distance from the wall increases, the invariant curve in $P02$ migrates away from the (compressive) boundary toward the elongational boundary associated with the undistorted $P01$ upstream flow. As was shown in Figure 5.2, as distance downstream away from the impinging shock increases, there is a slow streamwise reconstruction of the turbulent stress anisotropy field. Examples such as this serve to emphasize the complementary role of both numerical simulations and experimental measurements. While experiments are an essential guide in providing data on engineering configurations, numerical simulations are also needed to provide the underlying base physics not readily accessible to experimental scrutiny.

While the anisotropy invariant map provides some insight into the componentiality of the turbulent stresses, further insight into the flow structure of such impinging shock flows is often desired. Higher-order moments of the velocity field, such as the skewness (third-moment, $S(u_i') = \overline{u_i'^3}/(\overline{u_i'^2})^{3/2}$) and flatness (fourth-moment, $F(u_i') = \overline{u_i'^4}/(\overline{u_i'^2})^2$) factors provide additional information about the fluctuating velocity fields. For example, for the skewness factor positive/negative values suggest (on average) positive/negative values dominate for the velocity component u_i'. In addition, regions where $S(u_1')$ changes sign have been found to be the locations where the Reynolds shear and normal stresses reach their maximum values (Simpson, Chew, & Shivaprasad, 1981; Skåre & Krogstad, 1994). Such features are in addition to the well-known relationship of the skewness to the asymmetry of the tails of the probability density function. The flatness (or kurtosis) factor is a relative measure of the weight in the tails of probability density function, and is inversely proportional to the intermittency of the turbulent velocity field. As such, in proximity to the wall, the flatness factor $F(u_1')$ increases due to the dominance of streamwise fluctuations (Simpson et al., 1981). As the wall is approached, the turbulence becomes more intermittent which corresponds to an increase in the flatness factor. Many of the features found in these higher-order moments that characterize the supersonic boundary layer flow with an impinging shock (Deleuze, 1995; Shahab, 2011) are similar to those obtained at lower speed flows with boundary layer separation (Simpson et al., 1981; Skåre & Krogstad, 1994). This is not surprising since such shock induced separations occur in near-wall regions dominated by low speed flow. Additional analyses of the flow structure can be performed using the quadrant analysis first developed by Lu and Willmarth (1973) (see also Pope, 2000), and has been applied in the case of supersonic boundary layer flows with and without shocks (e.g. Deleuze, Audiffren, & Elena, 1994; Shahab, 2011). This flow diagnostic provides information on the cross-correlation of the streamwise and wall-normal fluctuations and can, when coupled with the corresponding skewness and flatness factors, provide an averaged snapshot of the motion within the boundary layer at various distances from the wall. In comparison with the no shock boundary layer flow, the effect of the shock induced distortion on the flow structure can be quantified.

Another characterizing feature of the effect of the impinging shock is the extent of the upstream/downstream influence on the mean and turbulent fields. One approach originally proposed for incompressible flows (East & Sawyer, 1979) but later adapted to the compressible case (Délery, 1983; Délery & Marvin, 1986) is to examine a phase diagram of a stress coefficient, C_τ, and an equilibrium shape parameter, J, related to the mean flow field. Within the context of this discussion, equilibrium is simply assumed when the time scale of the mean shear, $(\partial \bar{u}/\partial y)^{-1}$, is long compared to the time scale of the turbulence (K/ε). This situation holds for fully developed boundary layers or ones evolving very slowly. For the stress coefficient, the maximum turbulent shear stress within the boundary layer is used, and for the shape parameter, the usual mean flow incompressible shape factor H_i (cf. Eq. (6.53)), is used. Both C_τ and J are then given by

$$C_\tau = \frac{-\left(\overline{\bar{\rho}u_1' u_2'}\right)_{\max}}{\frac{1}{2}\bar{\rho}_e \overline{U}_e^2}, \quad J = 1 - \frac{1}{H_i}. \tag{7.10}$$

Based on an analysis of zero pressure gradient, equilibrium turbulent (incompressible) boundary layers (East & Sawyer, 1979), it can be shown that the ratio $J/\sqrt{0.5C_\tau}$ is constant and equal to the value of 6.55. The extent to which the perturbed turbulent boundary layer deviates from its equilibrium state can be identified by plotting the evolution of the normalized shear stress against the equilibrium shape parameter J—specifically $\sqrt{C_\tau}$ versus J.

In Figure 7.22, the phase-plane results from experiments (Délery, 1983) and direct numerical simulations (Shahab, 2011) are shown. The experimental results represent the original adaptation of the phase-plane distortion history to compressible flows. The flow separation is generated by alteration of the tunnel geometry that produces a shock induced separation zone in a Mach number range of ≈ 1.5. The simulation results are representative of the current numerical state-of-the-art in SWBLI (the reader is also referred to Humble (2008) for a similar phase-plane plot from experiments), and are the result of a shock impinging on a boundary layer flow. In Figure 7.22b, the locations $P01$–$P06$ represent streamwise stations along flat plate upstream ($P01$ at $-2.9\delta_\omega$) to far downstream ($P06$ at $16.7\delta_\omega$) of the impingement point. While the overall qualitative trend is similar in the two plots, the detailed evolution for the impinging shock flow, Figure 7.22b, is quite different than the low supersonic Mach number tunnel study, and reflective of the intense distortion on the flow by the impinging shock. Nevertheless, Figure 7.22a is a useful and important basis for comparison.

In Figure 7.22b, the trajectory starts from a point "A" and parallels the equilibrium line until the separation point "S" is reached. This provides some measure of the upstream influence of the shock on the wall-proximity flow. After the separation a non-equilibrium region, characterized by a rapid increase in J, exists where the shear stress maximum lags below the equilibrium line. In contrast to the phase-plane portrait in Figure 7.22a, where a global maximum

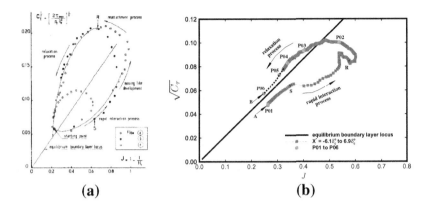

FIGURE 7.22 Phase-plane streamwise evolution of maximum shear stress in shock-induced separated boundary layer: (a) experiments, flow with incipient separation ○, separation ●, extended separation □, (from Délery, 1983, with permission); (b) DNS, with S designating separation point and R reattachment point of the separation zone (from Shahab, 2011).

in $\sqrt{C_\tau}$ was observed, the shock impingement case produces a local maximum slightly upstream of the reattachment point, and reflects the amplification of the turbulent shear stress. A second local maximum, which is similar to the global maximum in Figure 7.22a, occurs in Figure 7.22b and is due to amplification of shear stress by the convection of the large scale eddies downstream of the incident shock. Note that this maximum occurs after the mean flow (as represented by J) begins to recover from the shock distortion and coincides with the crossing of the equilibrium line. As distance downstream increases, the flow relaxes toward its undisturbed state (point "B"). Although this analysis was originally applied for the maximum turbulent shear stress, it can also be applied to other important components, such as the streamwise and wall-normal stresses (see Humble, 2008; Shahab, 2011).

Chapter 8

Elements of Compressible Flow Control

Compressible flow control is an important issue for the future development of aerodynamic vehicles as well as for other engineering applications where efficiency and optimal performance are required. An example of a flow control application specific to the supersonic regime is the *unstart* operational mode that occurs at inlets of ramjet and scramjet engines (Valdivia, Yuceil, Wagner, Clemens, & Dolling, 2009). In extreme cases, it can lead to aircraft stall due to an abrupt, severe degradation of engine performance. Such a degradation can be caused by the appearance of a bow shock located ahead of the engine inlet and in the intervening distance flow spillage occurs that significantly decreases the mass flow rate into the inlet. The appearance of the bow shock is a consequence of an upstream or downstream pressure disturbance that causes the shock train isolator to be ejected from the running engine. One of the causes of this ejection is the inhomogeneous burning in the combustor that causes the combustor pressure to rise and causes a destabilization of the shock system. Obviously, the unstart problem should be avoided and flow control approaches can be a novel solution to the problem.

Flow control is currently a rapidly evolving field of research yielding important technological advances. Although application to incompressible flows has evolved into a relatively mature topic (e.g. Gad-el-Hak, 2000), application in the compressible regime has not as yet attained the same level of maturity. Flow control strategies can be classified as either passive or active (see, for example, Gad-el-Hak, 2000; Gad-el-Hak, Pollard, & Bonnet, 1998). In the latter case energy input is needed, and the control can be either predetermined or reactive. For predetermined control, the loop is open and a fixed action is applied to the flow (variable), and no information is passed in the sensing system. For active control, the loop can be either open or close. In both the open- and close-loops, information is passed forward within the control system, but in a close-loop circuit there is a feedback mechanism allowing for a signal (flow) comparative action to occur.

Close-loop flow control, in particular, can be challenging due to the potential simultaneous influence of all the primitive flow variables (density, velocity, pressure, temperature) on the dynamics. Flow control based on the usual methods such as low-order modeling and linear approaches are then much more complex. For example the proper orthogonal decomposition (POD) approach, which is often used as a preferred tool for incompressible flow control strategies, was discussed in Section 4.3 for compressible flows and several issues associated with the correct implementation of the approach were identified. This suggests that for the foreseeable future close-loop control of high speed flows will be through "black box" devices. While such complexity may be a limiting factor in close-loop control some of these approaches have been applied to flat plate and jet flows (Kumar & Tewari, 2004; Lou, Alvi, Shih, Choi, & Annaswamy, 2002; Low, Berger, Lewalle, El-Hadidi, & Glauser, 2011; Samimy, Kim, Kastner, Adamovich, & Utkin, 2007b; Sinha, Kim, Kim, Serrani, & Samimy, 2010).

In this chapter, some of features that uniquely characterize the compressible flow control from its incompressible counterpart are discussed first. This includes issues associated with spatial and temporal scaling of control devices, and effects to be controlled unique to high speed flows. Then, the various types of control actuators used are discussed highlighting their effectiveness and relevant application areas. Finally, an overview of the control of different shear flows is presented further exemplifying both the diversity of application areas and the effectiveness of such control in high speed flows.

8.1 CHARACTERISTIC FEATURES

Flow control of high speed compressible flows inherently introduces features that are distinct from those found in the lower speed incompressible flows. Among these are scaling effects, sonic limit, positive, or negative aspects of shocks arising from the presence of (solid, temperature, pressure, or fluidic) actuators themselves, and pressure adaptations. Hopefully, this discussion will sensitize the reader to the various challenges and constraints in designing effective devices such as control jet actuators, energy deposition actuators and surface plasma actuators that will be presented in detail in Section 8.2.

8.1.1 Scaling Effects

Small spatial scales and correspondingly high frequencies are two limiting constraints imposed on wind tunnel testing of flow control strategies. As discussed in Chapter 4, sizes associated with most of the experimental setups can be an order of magnitude smaller than in the real configuration so that the scale of an actuator, for example, would then be an order of magnitude smaller than for a full size apparatus. As an illustration, consider a jet, which for a commercial plane, could have a diameter, D_r, of ≈ 2 m. Most experiments, however, are performed in academic wind tunnels with jets of diameter, d_e, of ≈ 5 cm, that results in a diameter ratio of 1/40. When flow control with pneumatic devices are performed in laboratory experiments with microjets, typical diameters of $d = 1$ mm are used, corresponding to 2% d_e or less. This size makes it difficult to operate due to high comparable viscous losses. A simple rescaling to operational size of such devices corresponds to engines with diameters $d \simeq 4$ cm, which are easily manageable without pressure losses. In the frequency domain, the same scaling ratio (1/40) applies as in the spatial estimate above. For a transonic flow study with the same jet exit velocities of about 400 m/s and (fixed) Strouhal number of ≈ 0.25, a frequency of 50 Hz for the full size jet (this frequency is easy to obtain) would correspond to a 2 kHz frequency in the wind tunnel experiment (this frequency is much more difficult to obtain with conventional means). (The reader is referred to Sections 4.1.1 and 8.2 for a further discussion on this and other constraints associated with high speed wind tunnel testing.)

For wind tunnel demonstrations, the spatial constraint on the microjet (actuator) control can be managed; however, the frequency constraint is much more difficult and suggests that demonstrations of high speed flow control would be much more difficult at laboratory scales than at real scales. Nevertheless, in contrast, control methods by plasmas, for example, can be better suited for the laboratory (small spatial scales), and where the attainable frequencies are sufficiently high, but in real conditions can be very difficult due to external conditions, size, and safety factors. As for sensors, high speed flows do not require specific sensors except for high frequency response. Most of the constraints have been described previously in Chapter 4; however, for actuators stringent constraints are required and are discussed next.

8.1.2 High Speed Effects of Actuators

A high speed flow that is to be controlled contains many of the same features of a subsonic flow. However, when compared to incompressible, low speed flows, compressible flows are susceptible to additional effects, such as shock waves and wave drag, as well as a heightened sensitivity to any perturbations. This enhanced sensitivity makes the use of actuators much more intrusive in compressible flows than in incompressible ones.

While the scaling effects just discussed can be constraining in compressible flows, there are features associated with compressibility that can be exploited. A typical illustration of the added complexity of supersonic or transonic flow control is the effect of control jets. When high velocity control jets are used either from pressure flowing gas or plasma discharges the jets can develop within three different regimes: under-, over- or fully-expanded (see Section 6.1).

In most cases, (parasitic) expansion or shock waves will be present and can lead to extra noise generation and pressure losses; although, microactuators can be used so that the size minimizes the intrinsic penalty (Hirt, Reich, & O'Connor, 2010). There are instances, however, where for some purposes the sensitivity of supersonic flows to perturbations can be used as an amplifier of the actuator effect. In such cases, flow vectoring or mixing enhancement can be obtained and will be discussed in Section 8.3.1. Fluidic control jets (either continuous, pulsed, or synthetic ones) are sometimes used as vortex generators (VGs), and are often preferred to mechanical obstacles since they can be turned off, avoiding the disadvantage of increasing drag when the effect is not required (as is the case in cruise conditions for airplanes).

Fluidic actuators can also be used as dynamical actuators for open- or close-loop operations, for better efficiency and reduced flow rates. Such actuators had been studied previously (Zukoski & Spaid, 1964), but there is renewed interest due to the development of new control strategies (Kumar & Tewari, 2004). As sketched in Figure 8.1 the flow from transverse jets promotes the usual flow perturbations observed in subsonic conditions (see Fric & Roshko, 1994), such as a jet shear layer vortex and the horseshoe vortex systems (see Figure 8.1a), but now several shock systems can appear (see Figure 8.1b). If the flow regime begins supersonic after the control jet exit, in addition to a separation shock upstream of the jet exit and a bow shock generated by the deviation of the primary flow, under some pressure and Mach conditions a barrel shock is observed downstream of the bow shock as well as a Mach disk (for pressure recovery at the end of the control jet region). Figure 8.1 shows the difference between these two regimes, and highlights the fact that such complexities clearly affect the entire flow to be controlled.

The trajectory of supersonic wall jets issuing normal to the wall is often a key issue for flow control. For subsonic flows, the penetration height of the jet

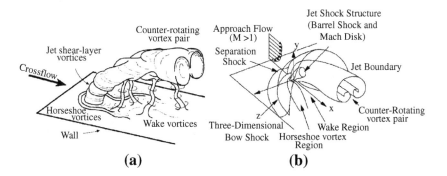

FIGURE 8.1 Illustrative sketches of transverse jet flow injection: (a) subsonic injection flow structure (from Fric & Roshko, 1994, with permission); (b) supersonic injection flow structure (from Gruber & Goss, 1999).

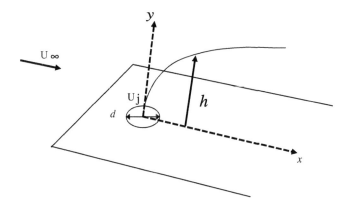

FIGURE 8.2 Sketch of wall jet trajectory in boundary layer flow.

can be estimated (Abramovich, 1963), as schematically depicted in Figure 8.2, from

$$\frac{h}{d} = J^m \left(\frac{x}{d}\right)^n, \tag{8.1}$$

with $m = 0.43$, $n = 1/3$ and J the dynamic pressure ratio given by

$$J = \frac{\rho_j U_j^2}{\rho_\infty U_\infty^2} \left(\frac{S_j}{S}\right) = \frac{\gamma_j p_j M_j^2}{\gamma_\infty p_\infty M_\infty^2} \left(\frac{S_j}{S}\right), \tag{8.2}$$

where the subscript j denotes the control jet characteristic (velocity, Mach number, and surface exit area), the subscript ∞ denotes the main flow, and S is a reference surface that should take into account the significant length scales of the flow to be controlled. For example, S can be the wetted surface in the case of a wing, the boundary layer thickness, the momentum thickness or even the same reference surface as S_j for unbounded flows. Note that the choice of S is somewhat arbitrary and, of course, can drastically change the value of J. When written in terms of the static pressure p, the dynamic pressure ratio is directly proportional to the Mach number (squared). However, for an isentropic process under adiabatic conditions, the total pressure is the same as the settling chamber pressure and the ratio of static to total pressure,

$$\frac{p}{p_t} = \left(1 + \frac{\gamma - 1}{2} M^2\right)^{-\gamma/(\gamma-1)}, \tag{8.3}$$

holds. This then yields the ratio of dynamic to total pressure for the jet,

$$\frac{\rho_j U_j^2 / 2}{p_t} = \frac{1}{2} \left(1 + \frac{\gamma - 1}{2} M_j^2\right)^{-\gamma/(\gamma-1)} M_j^2. \tag{8.4}$$

Interestingly, the Mach number dependent function on the right has a maximum, independent of the value of γ, and given by $M_j\big|_{max} = \sqrt{2}$ (Chauvet, 2007).

For supersonic (circular) control jets, the estimate for the penetration height is more difficult to determine. Even though there have been numerous proposals no real definitive consensus exists. Surprisingly, only one study has found a Mach number dependency (Povinelli, Povinelli, & Hersch, 1970) with the relationship, $h/d = 1.92 J^m (M_\infty/M_j)^{0.094} (x/d + 0.5)^n$ with $m = 0.35$ and $n = 0.227$. However, other studies have not found such a dependency; for example, at $M_\infty = 2$ and 3, the values of $m = 0.36$ and $n = 0.28$ were obtained (Papamoschou & Hubbard, 1993), and at $M_\infty = 5$ Schetz and Billig (1966) found $m = n = 0.43$. More recently, Bowersox, Fan, and Lee (2004) proposed for $M_\infty = 5$, a slightly different law $h/d = 2.1 J^m (x/d)^n$, with $m = 0.25$ and $n = 0.32$. It should be noticed that these results are obtained for given ranges of J. However, the main result derived is that there is no strong effect of the shape of the trajectory, a shift possibly being related to the flow regime. Similar relationships are available for the location of the shock disk.

Even surface plasma actuators can produce flow disturbances that are particularly strong in supersonic regimes, due to flow disturbances linked to the added velocity, pressure, and temperature introduced by the actuator. An illustration is shown in Figure 8.3 where a simple wall surface discharge (also "corona" discharge) into a $M = 2.85$ flow can promote a weak shock in the supersonic flow field (Shin, Narayanaswamy, Raja, & Clemens, 2007). In contrast, the sensitivity of supersonic flows to perturbations can be used to augment the effect of the actuator. As will be discussed later, such is the case when a shock system due to the actuation of control jets is used for flow vectoring or mixing enhancement.

When pneumatic actuators are used, another problem arises: the limitation of the sonic regime for converging jets is a severe limitation for some applications (such as separation control) when the velocity of the control jet should be several times higher (up to 3 times, Johari & Rixon, 2003) than the external velocity.

FIGURE 8.3 Flash-lamp schlieren imaging of flow interaction with the DC discharge. From Shin et al. (2007) with permission.

This velocity ratio is clearly not easy to obtain for controlling supersonic flows. On the other hand, whatever the velocity ratio, the flow rate is often an important parameter of flow control efficiency, and the presence of a sonic throat imposes a limitation on the (mass) flow rate, Q_m given by

$$Q_m = \left(\frac{2}{\gamma + 1}\right)^{\frac{1}{2}\left(\frac{\gamma+1}{\gamma-1}\right)} p_t \sqrt{\frac{\gamma}{RT_t}} A^*, \tag{8.5}$$

where A^* is the area of the sonic section, R the gas constant, and p_t, T_t, the total pressure and temperature, respectively, of the gas ahead of the throat. For a perfect gas under isentropic conditions, with the fluid taken at rest ahead of the throat (one-dimensional flow), $Q_m \approx 0.0404(p_t/\sqrt{T_t})A^*$ for air (with p_t in Pa, T_t in Kelvin and A^* in m^2).

8.2 ACTUATORS

As previously established, actuators for high speed flows are much more difficult to develop than for subsonic flows due to the higher velocities, higher frequencies, and higher flow control authority required. (For a thorough discussion of actuators that have been developed to date for both subsonic and high speed flows, the reader is referred to the recent review by Cattafesta and Sheplak, 2011.) The different actuators available can be classified as either issuing jets (continuous, pulsed, or synthetic) or actuators acting on thermodynamical properties of the fluid (temperature, pressure) fields. This is a rapidly evolving area of research due to the extensive technological progress in miniaturization and micro-electromechanical systems (MEMS) technology. The utilization of laser and plasma technologies has also opened new possibilities due to the multi-physics aspects introduced. In the following, some generic features and trends of each of these actuators are described, and will serve as a guide to control of a variety of shear flows including jets and mixing-layers, cavity and flows with shocks that will be discussed in Section 8.3.

8.2.1 Control Jets Issuing from Fluidic Actuators

Control jets can be created from pressure sources, with non-zero average flow rates, and can be steady or unsteady. They require mechanical energy to compress air or to bleed pressure from some part of the engine (aeronautical applications), tubing and vanes to operate—all of which then add complexity and mass to the system. Other possibilities include synthetic jets with zero net mass flux, less tubing, no need for pressure source, and thus less added mass. Nevertheless, irrespective of the actuator, the penetration properties of these control jets are the same as those described in Section 8.1.2. First, some characteristics of debiting jet actuators.

8.2.1.1 Steady Jet Actuators

In most laboratory experiments, the current size required for (sonic or supersonic) control jets is $\mathcal{O}(1$ mm) or less in diameter. There have been some attempts at machining converging–diverging microjets for supersonic control jets (Scroggs & Settles, 1996); however, this is technologically very complex, and since in most flow control strategies distributed actuators with a large number of microjets are required, simple fabrication processes are preferred. A less complex method, using simple converging microjets, has also been proposed (Phalnikar, Kumar, & Alvi, 2008). In this configuration, most of the jets will operate at sonic exit conditions with the flow rate limited by Eq. (8.5).

8.2.1.2 Unsteady Jet Actuators

In contrast to the steady blowing just discussed, unsteady blowing can be achieved through pulsed microjets that have enough authority for flow control provided the settling pressure can be adjusted accordingly. There are several ways to pulse microjets when trying to obtain the frequencies needed for flow control.

Rapid (mechanical) vanes are devices that have, in general, too small a bandwidth, the upper frequencies being less than 1 kHz. Several attempts to reach higher frequencies with rotating devices have been successful with frequencies up to 6.3 kHz (Ibrahim, Kunimura, & Nakamura, 2002). It should be pointed out that at the higher frequencies the modulations of the control jets of these devices can be incomplete with levels sometimes as low as 25% of the modulation. Nevertheless, these devices are under rapid development. Actuators based on micro-magnetomechanical systems (MMMS) for active flow control have been developed from MEMS technologies. For example, a magneto-dynamic microvalve of pulsed microjets provides up to 150 m/s jet velocities through sub-millimeter holes with frequencies up to 1 kHz (Pernod et al., 2010).

Several methods have been developed for synthetic jet actuators. All of them have the advantage that no tubing is needed and most have a fast response. Piezo-based resonant cavities have, in general, insufficient authority for flow control in the supersonic regime with the maximum velocity being of $\mathcal{O}(100$ m/s). Micro-combustion chambers have been developed with sufficient authority but generally with low frequency response (≈ 25 Hz); although more recent attempts have yielded frequencies ≈ 1.5 kHz (Cutler et al., 2005). For most high speed flow control, the velocity range is too low to exert significant authority over the flow; however, for transonic flows this type of synthetic jet can be used since the natural amplification characteristics of the flow can be exploited.

Such is the case for transonic jets in which frequencies up to 2.5 Hz can be achieved (Low, El-Hadidi, Andino, Berdanier, & Glauser, 2010).

Plasma discharges can be created inside closed cavities to provide a temperature spot that gives rise to a plasma-pulsed-jet. The increase in temperature and associated pressure increase can be used to create a synthetic jet outside a cavity. These devices are generally termed "spark jet" or "pulsed-plasma-jet" actuators, and act in three stages: energy deposition, plasma discharge (with fluid ejection), and refresh (cavity filling with air). Since being introduced (Cybyk, Wilkerson, Grossman, & Wie, 2003), this method has been further developed by several authors using different cavity geometries. Analytical models and numerical computations of such flows are quite difficult since there is a very complex interaction between rotational and vibrational modes involving several species (Haack, Taylor, Emhoff, & Cybyk, 2010). The exit velocities that can be achieved are expected to lie between 400 and 600 m/s, and in some cases up to 1000 m/s. For example, in Mach 3 boundary layers spark jet exit velocities of 250 m/s at 5 kHz have been observed (Narayanaswamy, Raja, & Clemens, 2010b). Typical gas expulsion times depend on the operating modes and are of the order of 200–800 µs (Belinger, Hardy, Barricau, Cambronne, & Caruana, 2011). These actuators can operate up to 10 kHz so that the duty cycle of such actuators can vary according to the actuator frequency.

Resonant tubes are adaptations of the Hartman whistle introduced over 90 years ago (for an historical perspective see Raman & Srinivasan, 2009) with bandwidths that are determined by their geometry. The powered resonant tube (PRT) actuator, for example, has been used with acoustic excitation up to 25 kHz (Chaudhari & Raman, 2010). A variant of this technique that creates a series of supersonic pulsed microjets operating with high mean momentum (up to 400 m/s), and high modulation (up to 30%) has been proposed and is shown in Figure 8.4a (Solomon, Kumar, & Alvi, 2010). This device has recently been improved and parametrized over a resonant frequency range of 1–60 kHz (Solomon, Foster, & Alvi, in press).

Lastly, fluidic oscillators (for example Gokoglu, Kuczmarski, Culley, & Raghu, 2010) are specially designed flow configurations (see Figure 8.4b) aimed at producing unsteady, periodic control jet velocities (sonic/supersonic) up to 10 kHz (with frequency fixed by device design). Once again, these actuators are not suited for closed-loop operation because they cannot be phase-locked.

8.2.2 Energy Deposition Type Actuators

The imposition of (high) energy into a flow is yet another approach toward controlling important flow dynamics. Here, three such actuators, that are representative of the extensive effort currently being directed in developing this methodology, are discussed. The laser based actuators impart high energy into the flow and thus often change the flow properties due to the increase

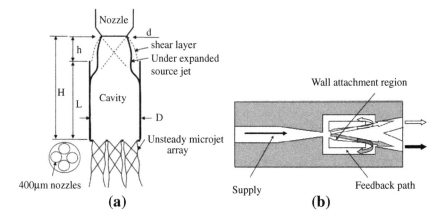

FIGURE 8.4 Examples of resonant tube and fluidic diverter actuators: (a) schematic of micro-actuator (from Solomon et al., 2010, with permission); (b) schematic of a fluidic oscillator (from Gokoglu et al., 2010).

in temperature. Such flow authority can strongly influence even the shock structure within the flow. Plasma based actuators, including surface plasma actuators, are also associated with the introduction of significant temperature increases in the flow that are then accompanied by increased pressure levels. Here also, such actuators can have a strong influence on the flow characteristics.

8.2.2.1 Laser Based Actuators

High energy can be deposited inside the flow in order to create (volume) flow perturbations by changing the temperature, or even physico-chemical properties of the fluid. As a practical illustration, control of hypersonic vehicles can be achieved by way of energy deposition, with the energy coming from either a plasma or laser (Bisek, Boyd, & Poggie, 2009). When high energy comes from laser pulses, Nd:YAG lasers are generally used (or other Nd based lasers) with energy on the order of 1 Joule. Such laser energy deposition (LED) can strongly modify shock systems. In one example, laser energy has been used (Lee, Jeung, Lee, & Kim, 2011) to deposit energy upstream of complex shock interactions occurring at Mach 6.5 when an oblique shock intersects a bow shock (Eldney-type interaction) to generate a blast wave and modify the shock structure. The pulse widths are fairly short $\mathcal{O}(\text{ns})$ with the repetition rate being less than 10 Hz (far from closed-loop control methods for supersonic flows, see Section 8.1).

8.2.2.2 Plasma Based Actuators

The physical properties of such actuators are quite complex and difficult to compute (Kandala & Candler, 2004). These methods are also limited to steady

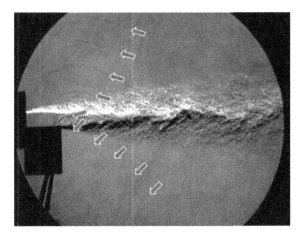

FIGURE 8.5 Phase-averaged schlieren image of a rectangular jet with LAFPAs firing and generating compression waves (marked with arrows). From Hahn, Kearney-Fischer, and Samimy (2011) with permission.

and point actuation, distributed flow control actuators are more difficult to achieve. Plasma generation can also be used in the filament discharge mode, depending on voltage and electrodes arrangement. Such localized arc filament plasma actuators (LAFPAs) have a discharge that creates a very intense increase of temperature, with a rapid temperature increase observed over the first 10–20 μs of arc operation, from below 1000 °C up to $\sim$ 2000 °C. The associated pressure field can then be used to act on the flow field, the authority of the actuator being quite high (Utkin et al., 2007). There are also strong pressure waves associated with the discharge as illustrated in Figure 8.5. Such a plasma actuator based method has been developed for noise abatement purposes (see Samimy, Kim, Kastner, Adamovich, & Utkin, 2007a) with the plasma distributed at the exit of a transonic jet (see Section 8.2.2).

The glow, or dielectric barrier discharge (DBD) plasma-type, actuators benefit from the advantages associated with electronic methods in that the design of the actuator (electrode) distribution is very flexible and can be directed toward various spatial modes (as is often required in flows, such as the azimuthal modes in jets (see Section 6.1)). In addition, the operational frequency domain is very compatible with the requirements of laboratory experiments, allowing for closed-loop operation.

8.2.3 Surface Plasma Actuators

These actuators are essentially plasma discharges on walls. Ionic wind is created by the surface discharge, and unless a "cold plasma" is used, temperature effects can be present (for a review see Moreau, 2007). If both electrodes lie on the same side of the wall, "glow" or "corona" discharges are observed that are often unstable. These are complex discharges that depend on the order in which the electrodes are distributed (that is, if the anode is placed upstream or downstream of the cathode). Numerous species are present in such discharges because of the local high temperature of the plasma (Mahadevan & Raja, 2012). Alternatively, if the two electrodes are placed on both sides of an insulated (thin) surface, the dielectric barrier discharge (DBD) is enhanced and is more stable. DBD can be actuated in either a continuous or unsteady (nanosecond scale) manner. At the surface, the wall jets have an induced velocity $\mathcal{O}(10 \text{ m/s})$ that minimizes the authority they exert on the flow, particularly in the supersonic regime (Pal, Sriram, Srisha Rao, & Jagadeesh, 2012); however, some efficiencies have been shown with these actuators for nearly transonic flows (Roupassov, Nikipelov, Nudnova, & Starikovskii, 2009).

The motivating interest in these types of actuators is that there is no tubing, added mass, or aerodynamic perturbation, although high frequencies can be reached including phase control useful for closed-loop control. Different arrangements on complex geometries are rapidly appearing in the literature, and some of these can create simple wall jets, or even normal jets, etc. but still at a low velocity (Kopiev, Ostrikov, Zaitsev, Kopiev, Belyaev, Bityurin, Klimov, Moralev, & Godin, 2011). Nevertheless, when the actuation takes place in a high receptivity region, such as the nozzle exit, the efficiency can be enhanced.

8.3 SHEAR FLOW CONTROL

As highlighted in Chapter 6, some of the most important features of free shear flows in the supersonic regime are the strong decrease of spreading rate, larger length of the potential core and consequently a general decrease of mixing properties (the latter being a very dramatic example for supersonic combustion). These features also have negative effects on the infrared (IR) signatures since the hot gases are less mixed with the ambient gas, and thus generally require hyper-mixing.

Mixing is not the only issue of flow control of jets. A fundamental issue regarding the performance of any tactical aircraft, for example, is its maneuverability. For this purpose, thrust vectoring as a fluidic control is a very efficient way to improve an aircraft's maneuverability. The first generation of thrust vectoring nozzles acted on the engine's exhaust through mechanical actuators in the compressible, transonic, and moderately supersonic regimes. Such techniques detrimentally increased the aircraft's weight. An alternative approach is through fluidic thrust vectoring: small secondary air streams are used to vectorize the primary jet. Such techniques can be expected to reduce nozzle weight up to 80% and maintenance cost up to 50% from the mechanical ones. Several kinds of fluidic thrust vectoring have been developed through control methods, such as counterflow thrust vectoring and secondary injection thrust vectoring. The first fluidic method uses a counterflow shear layer created between the primary jet and an additional exhaust collar to redirect the jet. The second one uses a secondary cross-flow jet into the supersonic part of the primary jet to create a new shock wave, causing asymmetry and unbalanced lateral forces on the nozzle.

Finally, aeroacoustic noise is strongly influenced by compressibility effects; particularly in the case when under-expanded jets enter and in which shock disks can be present. Although aeroacoustic properties of supersonic jets were not discussed previously in Chapter 6, for flow control it is necessary to make mention and to take into consideration the impact of aeroacoustic effects. Although noise reduction for compressible jets is an important and active area of research owing to its environmental impact for take-off and landing of civil aircraft, the impact of the space/time variations of the turbulence structure on noise emission is still an open question. This is particularly true when high speed, supersonic jets are involved; although, there is increasing evidence that modal solutions of the relevant conservation equations, that is wave packets, can be associated with the acoustics sources in such high speed flows (Jordan & Colonius, 2013, see also Tam, 1995). The Mach waves emitted from a supersonic jet are stronger when the wave packets have supersonic phase speeds which suggests a control strategy focused on these wave packet structures may be useful. It is obvious, however, that both spatial and temporal modes have to be controlled for noise abatement purposes. A unique challenge for noise control when contrasted with mixing enhancement, for example, is that the noise sources have to be suppressed without introducing extra parasitic noise. Thus, care must be taken in the selection of the control device. For passive control, for example, instead of tabs, chevrons are used, since they have less penetration and thus less device noise. For active flow control, microjets (steady, pulsed, or synthetic ones) distributed azimuthally are used with some noise suppression (Arakeri, Krothapalli, Siddavaram, Alkislar, & Lourenco, 2003; Callender, Gutmark, & Martens, 2005; Castelain, Sunyach, Juvé, & Béra, 2008; Laurendeau et al., 2008). Alternative devices using plasmas have also been explored (Kopiev, Ostrikov, Kopiev, Belyaev, & Faranosov, 2011; Utkin et al., 2007) since they offer more flexibility in terms of time response. For impinging jets, such as those associated with vertical take-off airplanes, the noise reduction problem is further complicated. The flow control in this case requires reducing the noise emitted, limiting the vibrations and unsteadiness, and reducing the erosion effects on the ground. The control strategies are then for both increasing the mixing and reducing the noise (e.g. Lou et al., 2002).

8.3.1 Jets and Mixing-Layers

The primary focus in the remainder of this chapter will be on flow control strategies of jets and mixing-layers. Such strategies necessarily involve actuators placed at the trailing edge of the nozzles, which then act preferentially on the mixing-layers that are present at the beginning of any jet.

In Section 6.1 it was pointed out that the initial stage of a developing jet contains high frequency instabilities (local, mixing-layer or shear layer modes) as well as relatively lower frequency instabilities corresponding to global or column modes. This dual behavior poses a difficult challenge for flow control since there is no well-defined strategy currently developed that addresses both (diverse) frequencies. Either high frequencies are addressed by the actuators in order to operate on the initial shear layer instability or lower frequencies (yet still elevated) are addressed to operate on the global modes. Consider again the example in Section 8.1.1, where a typical jet diameter in a wind tunnel experiment is ≈ 5 cm with an exit velocity of ≈ 400 m/s. The initial shear layer instability will depend on the initial shear layer thickness which typically depends on the thickness of the boundary layer immediately upstream of the jet exit. For a 1 mm initial vorticity thickness δ_ω, a 44 kHz shear layer instability frequency is found—which is very high for most actuators. Even the corresponding column mode has been evaluated and is $\mathcal{O}(2$ kHz), still a very high frequency.

8.3.1.1 Passive Control

In early experiments on flow control, passive devices were used. The principle was to modify the geometry of the exit in order to promote the generation of longitudinal vortices and enhance mixing. A limited number of experiments have also been performed on "pure" high speed mixing-layers and have included the utilization of lobes, tabs, indentations, and chevrons. While it is possible to enhance mixing to a large degree, the associated pressure losses and wave drag penalties, however, are very important (Reeder & Samimy, 1996).

The same methods can be used for mixing enhancement purposes in jets developing from mixing-layers. Several jet and tab geometries have been tested (for example Zaman, 1999), with the effects on the flow clearly visible with the creation of longitudinal vorticity enhanced by the presence of complex systems of three-dimensional shocks and expansion waves for supersonic regime. The modification of the spreading rates is important near the jet exit and rapidly recovers to undisturbed values downstream. These methods have been more recently used to control compressible transonic jets for noise reduction purposes, a subtle problem in terms of flow control as outlined earlier in this section. As an example, the "chevron" geometry allows for significant noise abatement by promoting longitudinal vortices at given azimuthal modes. It should be noted that passive control can also be performed by changing the global geometry of the jet exit; such as in the control of non-circular jets in the supersonic regime (Gutmark & Grinstein, 1999).

An alternative way to excite the column mode of the jet is to make use of a resonant cavity (Rossiter mode) placed underneath the jet exit (Yu & Schadow, 1994). This method is quite efficient in creating large-scale structures as can be observed in Figure 8.6a by comparing the controlled (lower) part with the uncontrolled (upper) part of the jet. This control method corresponds to a factor of three increase in terms of initial spreading rates.

8.3.1.2 Fluidic Control

The use of fluidic devices either in continuous or in unsteady mode can also be used. Such devices are considered as active control if some extra energy is added. The advantage of this type of control is that it can be adapted to the flow regime and canceled if not useful, thus avoiding aerodynamic penalties when not required. In particular, an interesting application would be to activate the control process during take-off or landing and stop it during cruise for noise reduction. Similarly in the case of buffet, control could be initiated only when the cruise conditions require it such as in emergency procedures. Another example in aeronautical applications, is the co-flow configuration in turbine engine exhausts. By designing a secondary flow containing a supersonic separation, the unstable mode of the main jet with the same flow rate as the original can be energized. The shear layer of a jet exhibits intense instability that grows into large scale eddies downstream of the nozzle which increase the spreading of the jet. In dual-stream jets, where the outer (secondary) stream flow is the most unstable, a mixing enhancement using secondary parallel injection (MESPI) method has been proposed (Debiasi & Tsai, 2007; Papamoschou, 2000) that takes advantage of such unstable mode excitation. The MESPI complements concurrent work with plugged nozzles on mixing enhancement using axial flow (MEAF) (Debiasi, Herberg, Tsai, & Papamoschou, 2008).

Another control strategy involves secondary jets. Such jets placed perpendicular to the main jet exit can be used to directly control the mixing-layer immediately at the (main) jet exit (Lardeau, Collin, Lamballais, & Bonnet, 2003). As with (passive) control with tabs, the spreading rates are enhanced for several diameters, but then recover to their unperturbed state after typically 2 diameters. Although the mixing areas can be increased in either subsonic or supersonic jets, compressibility effects, at least for moderate supersonic regimes, do not seem to be that important. The major effects arise from a resonant mechanism between the control jet instability and the instability of the mixing-layer of the main jet as shown in the schlieren visualization in Figure 8.6b where the natural (upper) and controlled (lower) parts of the jets are compared. As with the passive control by a cavity in Figure 8.6a, the fluidic flow control enhances the large-scale structure organization relative to the natural one. Fluidic control by microjets azimuthally distributed around the circumference of a jet in order to affect the different azimuthal modes has also been employed (Chauvet, Deck, & Jacquin, 2007). Localized arc filament plasma actuators (LAFPA) can be used to act in a manner similar to chevrons. This type of glow discharge has been used in supersonic jets (Samimy et al., 2004) with the advantage that it is possible to dynamically influence temporal modes as well as the

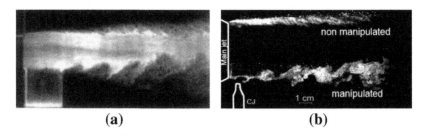

(a)   **(b)**

FIGURE 8.6 Effect of passive control on jet growth using planar Mie-scattering image (only lower part is controlled): (a) Mach 2 jet controlled through a cavity (from Yu & Schadow, 1994, with permission); (b) Mach 1.37 jet controlled through a secondary microjet (CJ) normal to main jet (adapted from Lardeau et al., 2003, with permission).

spatial (azimuthal) modes, thus acting on the different sources of instability in jets or mixing-layers (see Sections 6.1 and 6.2).

Jet mixing enhancement can also be achieved through a small divergence close to the exit plane. By controlling the boundary layer separation on the smallest side of the nozzle with an additional actuator, a supersonic annular jet can be controlled. Such an actuator consists of a short divergent located at the exhaust section of the nozzle and a secondary jet blowing out from a slot located in the middle of the divergent. The role of the divergent is to destabilize the boundary layer so it is easier for the secondary air stream to induce the boundary layer separation. A slight microjet acting in this diverging area then promotes localized separation that modifies the shock characteristics (depending on flow rate), and enhances the mixing of the jet. If only acting on one side of the jet, the pressure distribution on the two opposite sides of the nozzle becomes asymmetric and the main jet is vectorized. Initially introduced by Miller, Yagle, and Hamstra (1999), fluidic throat skewing for thrust vectoring in fixed geometry nozzles, boundary layer separation can be used as an actuator. Figure 8.7 shows a schematic of the resulting shock structure near the nozzle exit, with schlieren visualization from both experiments and numerical simulations shown in Figure 8.7b and c, respectively.

8.3.2 Cavity Flows

As just seen, cavities are often used as a means of control for jets, and as will be discussed in the next section, also in flows with shocks. Cavity flows are, nevertheless, of interest and technical importance in their own right and are encountered in many practical small-scale (probes, junctions, etc.) to large-scale (landing gear cavities, payload bays, etc.) applications. Such flows are represented by a sequential combination of boundary layer (upstream of cavity), mixing-layer (above cavity), and reattachment (downstream

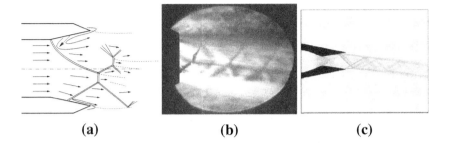

(a)   (b)   (c)

FIGURE 8.7 Fluidic thrust vectoring through localized separation: (a) schematic of shock structure at $M = 1.4$ with flow divergence $12°$ (from E. Collin, private communication, with permission); (b) schlieren from experiment—same conditions as (a) (from E. Collin, private communication, with permission). (c) CFD schlieren of supersonic jet with NPR $= 4.6$ (4% of flow rate) (from Waithe and Deere, 2003, with permission).

of cavity) flows. The underlying problems are noise, drag, and unsteady (cavity) interface dynamics. As an illustration, high fluctuating pressure effects are observed on supersonic tactical aircraft payload bays that can be detrimental to the structures, perturb the trajectory of the projectile, and generate high intensity noise (up to 180 dB). The noise problem arises from the acoustic coupling inside the cavity, known as "Rossiter modes." The mixing-layer that develops on top of the cavity plays a crucial role and compressibility effects are relevant (Forestier, Jacquin, & Geffroy, 2003) (see also Section 8.3.1). For subsonic cavity noise, flow control had originally been performed through aerodynamic modifications such as spoilers, tabs and flaps. A passive method for avoiding cavity resonance is to design the cavity itself in such a way that the inner (Rossiter) modes are modified and several shapes for that purpose have been proposed (Lada & Kontis, 2011). Another method consists of modifying the development of the mixing-layer by adding some artificial excitation. For example, by placing a rod that produces a wake with a given Strouhal number at the upstream edge of the cavity, it is possible to modify the mixing-layer's dynamic behavior and subsequent interaction with the trailing edge of the cavity (Dudley & Ukeiley, 2010; Illy, Jacquin, & Geffroy, 2008). This kind of flow control can be obtained from detached eddy simulation (DES) and large eddy simulation (LES), and as shown in Figure 8.8b, the important decrease in the overall pressure level (by 10 dB) can be computed.

Active control with microjets and slots through leading edge blowing can also be used (Arunajatesan et al., 2009) with results that are comparable to those obtained with rods (passive); however, the active methods are more flexible since no performance penalty is incurred when control is not needed (see Section 8.3.1, Fluidic control). The effect on the mixing-layer development is important for downstream distances up to a few mixing-layer thicknesses.

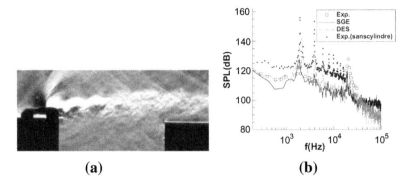

(a)   **(b)**

FIGURE 8.8   Control of transonic flow ($M = 0.78$) over a deep cavity ($L/D = 0.42$): (a) schlieren of cavity flow structure (from Illy et al., 2008, with permission); (b) pressure spectra at rear wall of cavity from DES and LES (labeled SGE) (from Comte, Daude, & Mary, 2008, with permission).

8.3.3 Flows with Shocks

In supersonic flows, the presence of shocks is hardly avoidable and flow control is often the only solution to minimize penalties associated with drag, vibrations, noise, etc. An obvious example is in supersonic flight of supersonic civil or tactical aircraft, re-entry vehicles, or missiles. The lead shock, whether attached or detached, causes a drag excess and heat exchange as well as parasitic effects on the ground. Lead shock control will be addressed in the next section.

In addition to the lead shock, shocks can emanate from specific geometries. For this category, two scenarios can arise: shocks due to the presence of walls, as in an air-intake, that then impinge on an opposite boundary layer; and shocks due to a wall deflection such as a compression ramp. Both of these shock wave/boundary layer interactions (SWBLI) have in common oblique shock wave patterns and an unsteadiness that is intrinsic to the system, and which often cause a separation region within the wall boundary layer. The origin of this unsteadiness is still not fully understood due to the complex coexistence of both high and low frequency components (Souverein et al., 2010). The major effects that are currently addressed by flow control strategies are spillage drag, inlet buzz and buffet. Spillage occurs when an inlet spills air around the outside instead of directing the air toward the compressor face (possibly leading to the unstart effect discussed at the beginning of this chapter). Inlet buzz (see for example Trapier, Deck, & Duveau, 2008) is the occurrence of shock oscillations at inlets that cause strong variations of inlet mass flow and possible losses in overall thrust. In such situations, flow control aims at reducing unsteadiness and the strength of the separated flow. Such unsteadiness and induced flow separations, with associated head losses, are also observed in normal shock configurations for transonic regimes. This is the case in transonic inlets and diffusers with dramatic effects on pressure recovery. Normal shocks can also be observed in more complex configurations such as the terminal shock of mixed-compression supersonic inlets. The buffet phenomenon is well known and also in this case a normal (lambda) shock is created due to the flow acceleration on the suction side of a wing profile evolving from a transonic to a sonic regime. The boundary layer separates downstream of the foot of the shock. This separation is unsteady, varying from a short to wide (up to the trailing edge) separation bubble, and then starting again with a small separation bubble. The major difference with the previous shock wave boundary layer interaction is from the flow geometry. In the case of a wing, the entire system is coupled between the subsonic regime (before the shock and on the pressure side) and the shock-induced separation. The circulation around the wing profile then imposes an added complexity that induces a low frequency behavior that is physically very different from the one observed in SWBLI. The buffet phenomenon induces structural vibrations called "buffeting" that appear at high angles of attack and certain cruise conditions (such as those occurring during emergency procedures) that have to be taken into account to define the cruise envelope of an aircraft. Flow control of all these effects is essential.

8.3.3.1 Control of Lead Shock

In supersonic cruise, the lead shock cannot be entirely avoided through design, but several methods for control have been developed over the decades (Bushnell, 2004). Initially, passive methods were investigated, such as shaping (minimizing the edge/nose diameter) that involved spikes and needles being placed upstream of the obstacle. Several active methods have been proposed; although, the principle is the same as the spikes and needles, with the non-material methods the suppression effect is on demand. One example of such an active method is a very thin water jet directed upstream of the obstacle (Bushnell & Huffman, 1968), and another is the injection of an inert gas from a sonic jet (on the obstacle axis) directed against a supersonic flow (Chen, Wang, & Lu, 2011) with the result in both cases being a "shock pointing" of the blunt nose (see Figure 8.9). The same effect can be obtained through other means such as energy deposition with a laser (Adelgren et al., 2005), or localized combustion. With all these methods, drag reduction of up to 50% can be achieved.

8.3.3.2 Control of Normal (Transonic) SWBLI

The generic configuration is characterized by the presence of a lambda shock, and occurring in transonic flows such as inlets, divergent channels, or transonic wings. The methods mostly consist of energizing the boundary layer, and making it more resistant to separation. As in subsonic flows (Gad-el-Hak, 2000), this is achieved by introducing longitudinal vortices, for example through VGs that can be mechanical or fluidic devices. In Figure 8.10 the

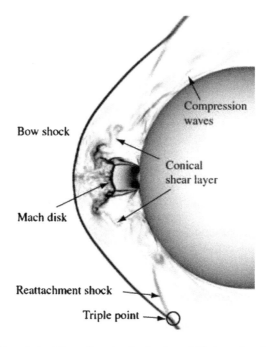

FIGURE 8.9 Numerical schlieren-like visualization from LES data of sonic jet opposing a supersonic flow impinging on a hemispherical nose. From Chen et al. (2011) with permission.

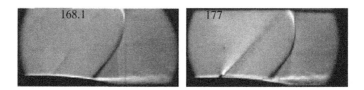

FIGURE 8.10 Schlieren visualization for steady shock wave configuration without control (left) and with VGs control (right). From Bur et al. (2009) with permission.

influence of mechanical VGs on the shock structure is shown (Bur, Coponet, & Carpels, 2009, see also Lee, Loth, & Babinsky, 2011). The behavior of the main shock is altered through the presence of the upstream control shock and the overall extent of the shock structure is reduced. Another approach is to control the normal shock through laser energy deposition. An important application is the mixed-compression inlet where the ability to control the (upstream) movement of the normal shock is desired (Yan, Knight, Kandala, & Candler, 2007).

While these methods are not that different from the ones successfully developed in subsonic flow control of separation, an alternative method can be used that takes into account the strong pressure gradient associated with the normal shock. It involves the introduction of a cavity beneath the shock which is open to the mean flow through a porous surface and ranges in the streamwise direction from upstream to downstream of the lambda shock. The cavity placement allows the pressure to be better balanced between its value before and after the interaction.

As described earlier in this section, the buffet phenomenon is specific to transonic wings. Its control can be achieved using the same principles and methods used in the channel flows. Examples of buffet control by mechanical or fluidic VGs have been assessed, from both experiments and computational fluid dynamics (CFD) (Dandois, Brunet, Molton, Abart, & Lepage, 2010). However, due to the circulation properties of the wing configuration, the trailing edge conditions are known to play an important role in the aerodynamic behavior. These properties allow for flow control strategies that are based on trailing edge actuators (Caruana, Mignosi, Robitaillié, & Corrége, 2003). One example is a "trailing edge deflector" (TED), that is, a small mechanical deflector (small flap) located at the trailing edge. With these deflectors, the lift coefficient for the same pressure fluctuation level (corresponding to buffet onset) can be increased by 35% with this device. The same concept that mimics the TED, now using fluidic injection, has also been developed (Browaeys, Deniau, Collin, & Bonnet, 2008).

8.3.3.3 Control of Oblique (Supersonic) SWBLI

As was discussed in Section 7.2.2, oblique shock interactions often originate from a wall modification such as a compression corner, or from the generation of a shock outside of the boundary layer with a shock generator. Passive methods such as (mechanical) VGs with different arrangements, such as Wheeler doublets (see McCormick, 1993), and also porous surfaces can be used for control (Délery, 2011a, 2011b). (Cavities were discussed in the previous subsection in connection with their applications to normal (transonic) shock configurations.) Active methods are used to impose artificial shock

FIGURE 8.11 Investigation of a pulsed-plasma-jet for a $M = 3$ shock/boundary layer control: (1) unperturbed shock location and (2) position of the shock under plasma actuation. From Narayanaswamy, Clemens, and Raja (2010) with permission.

wave patterns upstream of the natural SWBLI, but are mainly applicable to compression corner interactions. Unsteady mechanical "pulsed" VGs (moving up and down through slots) can be used, but better results are obtained with fluidic devices such as rapid vanes (with nominal response up to 1 ms, see Section 8.2) as VGs (Bueno et al., 2006). Different arrangements of pulsed vortex generator jets (VGJs) are used to produce a virtual shock shaping (VSS) (Kumar, Ali, Alvi, & Venkatakrishnan, 2011). The use of spark jets (plasma-based synthetic jets, see Section 8.2) with higher frequency response up to several kHz in a $M = 3$ flow (Narayanaswamy, Raja, & Clemens, 2010) can be used for controlling the position of the main shock in SWBLI (see Figure 8.11).

Bibliography

Abe, K., Kondoh, T., & Nagano, Y. (1995). A new turbulence model for predicting fluid flow and heat transfer in separating and reattaching flows—ii. Thermal field calculations. *International Journal of Heat and Mass Transfer, 38*, 1467–1481.

Abe, K., Kondoh, T., & Nagano, Y. (1996). A two-equation heat transfer model reflecting second-moment closures for wall and free turbulent flows. *International Journal of Heat and Fluid Flow, 17*, 228–237.

Abramovich, G. N. (1963). *The theory of turbulent jets*. Boston: MIT Press.

Adams, N. A. (1998). Direct simulation of turbulent compression ramp flow. *Theoretical and Computational Fluid Dynamics, 12*, 109–129.

Adams, N. A. (2000). Direct simulation of the turbulent boundary layer along a compression ramp at $M = 3$ and $Re_\theta = 1685$. *Journal of Fluid Mechanics, 420*, 47–83.

Adelgren, R., Yan, H., Elliott, G. S., Knight, D. D., Beutner, T. J., & Zheltovodov, A. A. (2005). Control of energy IV interaction by pulsed laser energy deposition. *AIAA Journal, 43*, 256–269.

Adrian, R. J. (2007). Conditional averages and stochastic estimation. In C. Tropea, A. L. Yarin, & J. F. Foss (Eds.), *Springer Handbook of Experimental Fluid Mechanics* (pp. 1370–1378). Berlin: Springer.

Adumitroaie, V., Ristorcelli, J. R., & Taulbee, D. B. (1999). Progress in Favre-Reynolds stress closures for compressible flows. *Physics of Fluids, 11*, 2696–2719.

Agui, J. H., Briassulis, G., & Andreopoulos, Y. (2005). Studies of interactions of a propagating shock wave with a decaying grid turbulence: Velocity and vorticity fields. *Journal of Fluid Mechanics, 524*, 143–195.

Aguilar, C., Azpeitia, C., Alvarado, J. M., & Stern, C. (2005). Study of Mach lines and acoustic waves in a jet using Rayleigh scattering. Paper No. 2005-1351, AIAA.

Albrecht, H., Borys, M., Damaschke, N., & Tropea, C. (2003). *Laser Doppler and Phase Doppler Measurement Techniques*. Berlin: Springer.

Allemand, B., Bruchet, P., Champinot, C., Melen, S., & Porzucek, F. (2001). Theoretical and experimental study of supersonic oxygen jets. Industrial application in EAF. *Revue De Metallurgie-Cahiers D Informations Techniques*, 571–587.

Alvi, F. S., Shih, C., Elavarasan, R., Garg, G., & Krothapalli, A. (2003). Control of supersonic impinging jet flows using supersonic microjets. *AIAA Journal, 41*, 1347–1355.

Amatucci, V. A., Dutton, J. C., Kuntz, D. W., & Addy, A. L. (2003). Two-stream, supersonic, wake flow field behind a thick base, Part I: General features. *AIAA Journal, 30*, 2039–2046.

Andreopoulos, Y., Agui, J., & Briassulis, G. (2000). Shock wave turbulence interactions. *Annual Review of Fluid Mechanics* (pp. 309–345). New York: Annual Reviews.

Andreopoulos, Y., Agui, J. H., Wang, Z., & Hermening, K. (2000). Interactions of turbulent jets with traveling shock waves. Paper No. 2000-0929, AIAA.

293

Arakeri, V. H., Krothapalli, A., Siddavaram, V., Alkislar, M. B., & Lourenco, L. M. (2003). On the use of microjets to suppress turbulence in a mach 0.9 axisymmetric jet. *Journal of Fluid Mechanics, 490*, 75–98.

Aris, R. (1962). *Vectors, tensors and the basic equations of fluid mechanics.* New York: Dover Publications Inc.

Arnette, S., Samimy, M., & Elliott, G. (1998). The effect of expansion regions on the turbulence structure of a supersonic boundary layer. *Journal of Fluid Mechanics, 367*, 67–105.

Arunajatesan, S., Kannepalli, C., Sinha, N., Sheehan, M., Alvi, F., Shumway, G., Ukeiley, L. (2009). Suppression of cavity loads using leading-edge blowing. *AIAA Journal, 47*, 1132–1144.

Aupoix, B. (2000). Introduction to turbulence modelling for compressible flows. *Lecture Series*, Vol. 2000–04. von Karman Institute for Fluid Mechanics.

Aupoix, B. (2004). Modeling of compressibility effects in mixing layers. *Journal of Turbulence, 5*. doi:10.1088/1468-5248/5/1/007.

Aupoix, B., & Bézard, H. (2006). Compressible mixing layers: Data analysis and modelling. Bulletin70, ERCOFTAC.

Barone, M. F., Oberkampf, W. L., & Blottner, F. G. (2006). Validation case study: Prediction of compressible turbulent mixing layer growth rate. *AIAA Journal, 44*, 1488–1497.

Barre, S. (1993). *Action de la compressibilité sur la structure des couches de mélange turbulentes supersoniques.* Ph.D. thesis, Université de Poitiers.

Barre, S., Alem, D., & Bonnet, J.-P. (1996). Experimental study of a normal shock/homogeneous turbulence interaction. *AIAA Journal, 34*, 968–974.

Barre, S., Alem, D., & Bonnet, J.-P. (1998). Reply by the authors to H.S. Ribner. *American Institute of Aeronautics and Astronautics Journal, 36*, 495.

Barre, S., Bonnet, J.-P., Gatski, T. B., & Sandham, N. D. (2002). Compressible, high speed flows. In B. E. Launder, & N. D. Sandham (Eds.), *Closure Strategies for Modelling Turbulent and Transitional Flows* (pp. 522–581). Cambridge: Cambridge University Press.

Barre, S., Braud, P., Chambres, O., & Bonnet, J.-P. (1997). Influence of inlet pressure conditions on supersonic turbulent mixing layers. *Experimental Thermal and Fluid Science, 14*, 68–74.

Barre, S., Dupont, P., & Dussauge, J.-P. (1992). Hot wire measurements in transonic turbulent flows. *European Journal of Mechanics B – Fluid, 11*, 439–454.

Barre, S., Dupont, P., & Dussauge, J.-P. (1997). Estimates of convection velocity of large turbulent structures in supersonic mixing layers. *Aerospace Science and Technology, 1*, 355–366.

Barre, S., Quine, C., & Dussauge, J.-P. (1994). Compressibility effects on the structure of supersonic mixing layers: Experimental results. *Journal of Fluid Mechanics, 259*, 47–78.

Battam, N. W., Gorounov, D., Korolev, G., & Ruban, A. (2004). Shock wave interaction with a viscous wake in supersonic flow. *Journal of Fluid Mechanics, 504*, 301–341.

Bauer, P. T., Zumwalt, G. W., & Fila, L. J. (1968). A numerical method and an extension of the Korst jet mixing theory for multispecies turbulent jet mixing. Paper No. 68-112. AIAA.

Béguier, C., Dekeyser, I., & Launder, B. E. (1978). Ratio of scalar and velocity dissipation time scales in shear flow turbulence. *Physics of Fluids, 21*, 307–315.

Belinger, A., Hardy, P., Barricau, P., Cambronne, J.-P., & Caruana, D. (2011). Influence of the energy dissipation rate in the discharge of a plasma synthetic jet actuator. *Journal of Physics D – Applied Physics, 44*. doi:10.1088/0022-3727/44/36/365201.

Bellaud, S. (1999). *Mesures et analyses détaillées des champs turbulents en couches de mélange annulaires supersoniques.* Ph.D. thesis, Université de Poitiers.

Bendat, J. S., & Piersol, A. G. (2000). *Random Data: Analysis and Measurement Procedures.* New York: John Wiley & Sons.

Bergmann, M., Cordier, L., & Brancher, J.-P. (2005). Optimal rotary control of the cylinder wake using proper orthogonal decomposition reduced-order model. *Physics of Fluids, 17*, 1–21.

Berkooz, G., Holmes, P., & Lumley, J. L. (1993). The proper orthogonal decomposition in the analysis of turbulent flows. *Annual Review of Fluid Mechanics*, (pp.539–575). New York: Annual Reviews.

Bestion, D., Gaviglio, J., & Bonnet, J.-P. (1983). Comparison between constant-current and constant-temperature hot wire anemometers in high speed flows. *Review of Scientific Instruments, 54*, 1513–1524.

Bilger, R. W. (1975). A note on Favre averaging in variable density flows. *Combustion Science Technology, 11*, 215–217.

Birch, S. F., & Eggers, J. M. (1972). A critical review of the experimental data for developed free turbulent shear flows. In *Free turbulent shear flows* (pp. 11–40). No. 321 in NASA SP.

Bisek, N. J., Boyd, I. D., & Poggie, J. (2009). Numerical study of plasma-assisted aerodynamic control for hypersonic vehicles. *Journal of Spacecraft Rockets, 46*, 568–576.

Blaisdell, G. A., Mansour, N. N., & Reynolds, W. C. (1991). *Numerical simulations of a compressible homogeneous turbulence.* Report No. TF-50, Department of Mechanical Engineering, Stanford University.

Blaisdell, G. A., Mansour, N. N., & Reynolds, W. C. (1993). Compressibility effects on the growth and structure of homogeneous turbulent shear flow. *Journal of Fluid Mechanics, 256*, 443–485.

Blaisdell, G. A., & Sarkar, S. (1993). Investigation of the pressure-strain correlation in compressible homogeneous turbulent shear flow. In L. D. Kral, & T. A. Zang (Eds.), *Transitional and Turbulent Compressible Flows* (Vol. 151, pp. 133–138). New York: ASME.

Blin, E. (1993). *Étude expérimentale de l'interaction entre une turbulence libre et une onde de choc.* Ph.D. thesis, Université de Paris VI.

Bogdanoff, D. W. (1983). Compressibility effects in turbulent shear layers. *AIAA Journal, 21*, 926–927.

Bogey, C., & Bailly, C. (2006). Investigation of downstream and sideline subsonic jet noise using Large Eddy Simulation. *Theoretical and Computational Fluid Dynamics, 20*, 23–40.

Bogulawski, L., & Popiel, C. O. (1979). Flow structure of the free round turbulent jet in the initial region. *Journal of Fluid Mechanics, 90*, 531–539.

Boguszko, M., & Elliott, G. S. (2005). On the use of filtered Rayleigh scattering for measurements in compressible flows and thermal fields. *Experiments in Fluids, 28*, 33–49.

Bonnet, J.-P. (1982). *Étude théoretique et expérimentale de la turbulence dans un sillage supersonique.* Ph.D. thesis, Université de Poitiers.

Bonnet, J.-P., & Alziary de Roquefort, T. (1980). Determination and optimization of the frequency response of constant-temperature anemometer in supersonic flow. *Review of Scientific Instruments, 51*, 234–249.

Bonnet, J.-P., & Chaput, E. (1986). Large-scale structures visualization in a high reynolds number, turbulent flat-plate wake at supersonic speed. *Experiments in Fluids, 4*, 350–356.

Bonnet, J.-P., & Debisschop, J.-R. (1992). Analyse expérimentale de la turbulence dans les couches de mélange supersoniques. *Revue Scientifique et Technique Défense 3e trimestre*, 65–84.

Bonnet, J.-P., Debisschop, J.-R., & Chambres, O. (1993). Experimental studies of the turbulent structure of supersonic mixing layers. Paper No. 93-0217, AIAA.

Bonnet, J.-P., & Delville, J. (1996). General concepts on structure identification. In J.-P. Bonnet (Ed.), *Eddy Structure Identification* (pp. 1–60). Wien: Springer Verlag.

Bonnet, J.-P., Delville, J., Sapin, S., Sullivan, P., & Yeru, R. (1991). Compressibility effects in turbulent far wakes. In *Eighth Symposium on Turbulent Shear flows* (pp. 23.1.1–23.1.6).

Bonnet, J.-P., Grésillon, D., Cabrit, B., & Frolov, V. (1995). Collective light scattering as non-particle laser velocimetry. *Measurement Science and Technology, 6*, 620–636.

Bonnet, J.-P., Grésillon, D., & Taran, J.-P. (1998). Nonintrusive measurements for high-speed supersonic and hypersonic flows. *Annual Review of Fluid Mechanics, 20*, 231–273 New York: Annual Reviews.

Bonnet, J.-P., & Knani, M. A. (1988). Calibration and use of inclined hot wire in a supersonic turbulent wake. *Experiments in Fluids, 6*, 179–188.

Bonnet, J.-P., Moser, R., & Rodi, W. (1998). Free shear flows. In *A Selection of Test Cases for the Validation of Large-Eddy Simulations of Turbulent Flows* (pp. 29–36), AR-345, AGARD.

Bookey, P., Wyckham, C., Smits, A., & Martin, M. P. (2005). New experimental data of STBLI at DNS/LES accessible Reynolds numbers. Paper No. 2005-0309, AIAA.

Bourguet, R., Braza, M., & Dervieux, A. (2007). Reduced-order modeling for unsteady transonic flows around an airfoil. *Physics of Fluids, 11*, CID: 111701

Bowersox, R. D. (1996). Combined laser doppler velocimetry and cross-wire anemometry analysis for supersonic turbulent flow. *AIAA Journal, 34*, 2269–2275.

Bowersox, R. D., Fan, H., & Lee, D. (2004). Sonic injection into a Mach 5.0 freestream through diamond orifices. *Journal of Propulsion and Power, 20*, 280–287.

Bradshaw, P. (1974). The effect of mean compression or dilatation on the turbulence structure of supersonic boundary layers. *Journal of Fluid Mechanics, 63*, 449–464.

Bradshaw, P. (1977). Compressible turbulent shear layers. *Annual Review of Fluid Mechanics*, (pp.33–54). New York: Annual Reviews.

Bradshaw, P. (1981). Compressibility effects on free shear layers. In *Conference on Complex Turbulent Flows* (Vol. 1, pp. 364–368). Stanford University.

Bradshaw, P., Ferriss, D. H., & Johnson, R. F. (1964). Turbulence in the noise-producing region of a circular jet. *Journal of Fluid Mechanics, 19*, 591–624.

Bray, K. N. C. & Libby, P. A. (1994). Recent developments in the BML model of premixed turbulent combustion. In *Turbulent Reacting Flows* (pp. 115–151). New York: Academic Press.

Breidenthal, R. (1990). The sonic eddy: A model of compressible turbulence. Paper No. 90-0495, AIAA.

Breidenthal, R. E. (1992). Sonic eddy—a model for compressible turbulence. *AIAA Journal, 30*, 101–104.

Briassulis, G., Agui, J., & Andreopoulos, Y. (2001). The structure of weakly compressible grid-generated turbulence. *Journal of Fluid Mechanics, 432*, 219–283.

Briassulis, G. K., & Andreopoulos, J. (1996). High resolution measurements of isotropic turbulence interacting with shock waves. Paper No. 96-0042, AIAA.

Browaeys, G., Deniau, H., Collin, E., & Bonnet, J.-P. (2008). Pneumatic control strategies for transonic buffet—A numerical approach. In *Proceedings of the Seventh International ERCOFTAC Symposium on Engineering Turbulence Modelling and Measurements* (pp. 805–810).

Browand, F. K., & Latigo, B. O. (1979). Growth of the two-dimensional mixing layer from a turbulent and nonturbulent boundary layer. *Physics of Fluids, 22*, 1011–1019.

Brown, G. L., & Roshko, A. (1974). On density effects and large structure in turbulent mixing layers. *Journal of Fluid Mechanics, 64*, 775–816.

Brown, J. D., Bogdanoff, D. W., Yates, L. A., & Chapman, G. T. (2008). Free-flight dynamic aero data for a lifting CEV capsule. Paper No. 2008-1232, AIAA.

Brun, C., Boiarciuc, M. P., Haberkorn, M., & Comte, P. (2008). Large eddy simulation of compressible channel flow: Arguments in favour of universality of compressible turbulent wall bounded flows. *Theoretical and Computational Fluid Dynamics, 22*, 189–212.

Bueno, P. C., Wagner, J. L., Searcy, J. A., Ganapathisubramani, B., Clemens, N. T., & Dolling, D. S. (2006). Experiments in unsteady forcing of Mach 2 shock wave/boundary layer interactions. Paper No. 2006-0878, AIAA.

Bur, R., Coponet, D., & Carpels, Y. (2009). Separation control by vortex generator devices in a transonic channel flow. *Shock Waves, 19*, 521–530.

Bushnell, D. M. (2004). Shock wave drag reduction. *Annual Review of Fluid Mechanics,* (pp.81–96.). New York: Annual Reviews.

Bushnell, D. M., & Huffman, J. K. (1968). *Forward penetration of liquid water and liquid nitrogen injected from an orifice at the stagnation point of a hemispherically blunted body in hypersonic flow.* Technical Memorandum X-1493, NASA.

Cabrit, B. (1992). *Diffusion collective de la lumière par un gaz turbulent: dispersion moléculaire et turbulente.* Ph.D. thesis, Université de Paris VI.

Callender, B., Gutmark, E., & Martens, S. (2005). Far-field acoustic investigation into chevron nozzle mechanisms and trends. *AIAA Journal, 43*, 87–95.

Cambon, C., Coleman, G. N., & Mansour, N. N. (1993). Rapid distortion analysis and direct simulation of compressible homogeneous turbulence at finite Mach number. *Journal of Fluid Mechanics, 257*, 641–665.

Cambon, C., & Jacquin, L. (2006). Is compressibility always stabilizing and why? Open issues. Bulletin70, ERCOFTAC.

Caruana, D., Mignosi, A., Robitaillié, C., & Corrège, M. (2003). Separated flow and buffeting control. *Flow Turbulence and Combustion, 71*, 221–245.

Carvin, C., Debière, J.-F., & Smits, A. J. (1988). The near-wall temperature profile of turbulent boundary layers. Paper No. 88-0136, AIAA.

Castelain, T., Sunyach, M., Juvé, D., & Béra, J.-C. (2008). Jet-noise reduction by impinging microjets: An acoustic investigation testing microjet parameters. *AIAA Journal, 46*, 1081–1087.

Catris, S., & Aupoix, B. (2000a). Density corrections for turbulence model. *Aerospace Science and Technology, 4*, 1–11.

Catris, S., & Aupoix, B. (2000b). Towards a calibration of the length-scale equation. *International Journal of Heat and Fluid Flow, 21*, 606–613.

Cattafesta, L. N, & Sheplak, M. (2011). Actuators for active flow control. *Annual Review of Fluid Mechanics, 71*, 247–272. New York: Annual Reviews.

Cebeci, T., & Bradshaw, P. (1977). *Momentum transfer in boundary layers.* Washington: Hemisphere Publishing.

Cebeci, T., & Bradshaw, P. (1984). *Physical and computational aspects of convective heat transfer*. New York: Springer-Verlag.

Cebeci, T., & Smith, A. M. O. (1974). *Analysis of Turbulent Boundary Layers*. New York: Academic Press.

Chambres, O. (1997). *Analyse expérimentale de la modélisation de la turbulence en couche de mélange*. Ph.D. thesis, Université de Poitiers.

Chambres, O., Barre, S., & Bonnet, J.-P. (1997). Balance of kinetic energy in supersonic mixing layer compared to subsonic mixing layer and subsonic jets with variable density. In *IUTAM Symposium on Variable Density Low Speed Turbulent Flows* (pp. 303–308).

Chang, C. T. (1957). Interaction of a plane shock and oblique plane disturbances with special reference to entropy waves. *Journal of the Aeronautical Sciences, 24*, 675–682.

Chapman, S., & Cowling, T. G. (1970). *The Mathematical Theory of Non-Uniform Gases*. Cambridge: Cambridge University Press.

Chassaing, P. (1985). Une alternative à la formulation des équations du mouvement turbulent d'un fluide à masse volumique variable. *Journal de Mécanique Théorique et Appliquée, 4*, 375–389.

Chassaing, P. (2001). The modeling of variable density turbulent flows. *Flow Turbulence and Combustion, 18*, 293–332.

Chaudhari, K., & Raman, G. (2010). Ultrasonic powered resonance tube actuators for flow control applications. Paper No. 2010-4413, AIAA.

Chauvet, N. (2007). *Simulation numérique et analyse physique dun jet propulsif contrôlé par des injections radiales*. Ph.D. thesis, Université de Poitiers.

Chauvet, N., Deck, S., & Jacquin, L. (2007). Numerical study of mixing enhancement in a supersonic round jet. *AIAA Journal, 45*, 1675–1687.

Chen, L.-W., Wang, G.-L., & Lu, X.-Y. (2011). Numerical investigation of a jet from a blunt body opposing a supersonic flow. *Journal of Fluid Mechanics, 684*, 85–110.

Chinzei, N., Nasuya, G., Komuro, T., Murakami, A., & Kudou, K. (1986). Spreading of two-stream supersonic turbulent mixing layers. *Physics of Fluids, 29*, 1345–1347.

Chu, B.-T., & Kovasznay, L. S. G. (1958). Non-linear interactions in a viscous heat-conducting compressible gas. *Journal of Fluid Mechanics, 3*, 494–514.

Clemens, N. T., & Mungal, M. G. (1995). Large scale structure and entrainment in the supersonic mixing layer. *Journal of Fluid Mechanics, 284*, 171–216.

Coleman, G. N., Kim, J., & Moser, R. D. (1995). A numerical study of turbulent supersonic isothermal-wall channel flow. *Journal of Fluid Mechanics, 305*, 159–183.

Coleman, G. N., & Mansour, N. N. (1991a). Modeling the rapid spherical compression of isotropic turbulence. *Physics of Fluids A, 3*, 2255–2259.

Coleman, G. N., & Mansour, N. N. (1991b). Simulation and modeling of homogeneous compressible turbulence under isotropic mean compression. In *Eighth Symposium on Turbulent Shear Flows* (pp. 21.3.1–21.3.6).

Coles, D. (1954). Measurement of turbulent friction on a smooth flat plate in supersonic flow. *Journal of the Aeronautical Sciences, 7*, 433–448.

Coles, D. (1956). The law of the wake in the turbulent boundary layer. *Journal of Fluid Mechanics, 1*, 191–226.

Coles, D. E. (1964). *The turbulent boundary layer in a compressible fluid*. *Physics of Fluids, 7* 1403–1423. See also Report R-403-PR, The Rand Corporation, Santa Monica, CA.

Coles, D. (1981). Prospects for useful research on coherent structure in turbulent shear flow. *Proceedings of the Indian Academy of Sciences-Engineering Sciences, 4*, 111–127.

Coles, D. E. (1962). *The turbulent boundary layer in a compressible fluid*. Report R-403-PR, The Rand Corporation.

Comte, P., Daude, F., & Mary, I. (2008). Simulation of the reduction of unsteadiness in a passively controlled transonic cavity flow. *Journal of Fluid and Structures, 24*, 1252–1261.

Comte, P., Fouillet, Y., & Lesieur, M. (1992). Simulation numérique des zones de mélange compressibles. *Revue Scientifique et Technique Défense 3e trimestre*, 43–63.

Comte-Bellot, G., & Sarma, G. R. (2001). Constant voltage anemometer practice in supersonic flow. *AIAA Journal, 39*, 261–270.

Cordier, L. (2008). Proper orthogonal decomposition: An overview. Lecture Series 2008-5. von Karman Institute for Fluid Mechanics.

Cordier, L., Delville, J., & Bonnet, J.-P. (2007). Proper orthogonal decomposition. In C. Tropea, A. L. Yarin, & J. F. Foss (Eds.), *Springer Handbook of Experimental Fluid Mechanics* (pp. 1346–1370). Berlin: Springer.

Corrsin, S., & Uberoi, M. S. (1949). *Further experiments on the flow and heat transfer in a heated turbulent air jet*. Technical Note 1865, NACA.

Cousteix, J. (1989). *Turbulence et Couche Limite*. Toulouse: Cepadues-Editions.

Cousteix, J., & Aupoix, B. (1990). *Turbulence models for compressible flows*. Report No. 764, AGARD.

Cutler, A. D., Beck, B. T., Wilkes, J. A., Drummond, J.-P., Alderfer, D. W., & Danehy, P. M. (2005). Development of a pulsed combustion actuator for high-speed flow control. Paper No. 2005-1084, AIAA.

Cybyk, B. Z., Wilkerson, J. T., Grossman, K. R., & Wie, D. M. V. (2003). Computational assessment of the sparkjet flow control actuator. Paper No. 2003-3711, AIAA.

Dahlburg, J.-P., Dahlburg, R. B., Gardner, J. H., & Picone, J. M. (1990). Inverse cascade in two-dimensional compressible turbulence: I. Incompressible forcing at low Mach number. *Physics of Fluids A, 2*, 1481–1486

Dandois, J., Brunet, V., Molton, P., Abart, J.-C., & Lepage, A. (2010). Buffet control by means of mechanical and fluidic vortex generators. Paper No. 2010-4975, AIAA.

De Chant, L. J. (1998). An analytical skin friction and heat transfer model for compressible, turbulent, internal flows. *International Journal of Heat and Fluid Flow, 19*, 623–628.

Deardorff, J. W. (1970). Turbulence measurements in supersonic two-dimensional wake. *Journal of Fluid Mechanics, 41*, 453–480.

Debiasi, M., Herberg, R., Tsai, H. M., & Papamoschou, D. (2008). Mixing characteristics of elliptical jets with plug. Paper No. 2008-0787, AIAA.

Debiasi, M., & Tsai, H. M. (2007). Effects of tangential blowing on high-speed jets. Paper No. 2007-3861, AIAA.

Debiève, J.-F., Dupont, P., Smith, D. R., & Smits, A. J. (1997). Supersonic turbulent boundary layer subjected to step changes in wall temperature. *AIAA Journal, 35*, 51–57.

Debiève, J.-F., & Lacharme, J. P. (1986). A shock-wave/free turbulence interaction. In *IUTAM Symposium on Turbulent Shear-Layer/Shock-Wave Interactions* (pp. 393–403).

Debisschop, J.-R. (1993). *Comportement de la turbulence en couches de mélange supersoniques*. Ph.D. thesis, Université de Poitiers.

Debisschop, J. R., & Bonnet, J.-P. (1993). Mean and fluctuating velocity measurements in supesonic mixing layers. In W. Rodi, & F. Martelli (Eds.), *Engineering Turbulence Modelling and Experiments 2* (pp. 467–478). Amsterdam: Elsevier Science Publishers.

Debisschop, J.-R., Chambres, O., & Bonnet, J.-P. (1994). Velocity field characteristics in supersonic mixing layers. *Experimental Thermal and Fluid Science, 9*, 147–155.

Debisschop, J.-R., Sapin, S., Delville, J., & Bonnet, J.-P. (1993). Supersonic mixing layer analysis by laser planogram and hot-wire based POD. In J.-P. Bonnet & M. N. Glauser (Eds.), *Eddy Structure Identification in Free Turbulent Shear Flows* (pp. 453–462). Dordrecht and Boston: Kluwer.

Deck, S. (2005a). Numerical simulation of transonic buffet over a supercritical airfoil. *AIAA Journal, 43*, 1556–1566.

Deck, S. (2005b). Zonal detached-eddy simulation of a flow around a high-lift configuration. *AIAA Journal, 43*, 2372–2384.

Délery, J. (1983). Experimental investigation of turbulence properties in transonic shock/boundary-layer interaction. *American Institute of Aeronautics and Asronautics Journal, 21*(2), 180–185.

Délery, J. (1985). Shock wave/turbulent boundary layer interaction and its control. *Progress in Aerospace Sciences, 22*, 209–280.

Délery, J., & Marvin, J. G. (1986). Shock-wave boundary layer interactions. AGARDograph No. AG-280, AGARD.

Délery, J. M. (2001). Shock wave/boundary layer interactions. In G. Ben-Dor, O. Igra & T. Elperin (Eds.),. *Handbook of Shock Waves* (Vol. 2, pp. 205–264) Academic Press.

Délery, J. M. (2011a). Control of transonic sblis. In H. Babinsky, & J. K. Harvey (Eds.), *Shock Wave-Boundary-Layer Interactions* (pp. 251–265). Cambridge: Cambridge University Press.

Délery, J. M. (2011b). SBLI control. In H. Babinsky, & J. K. Harvey (Eds.), *Shock Wave-Boundary-Layer Interactions* (pp. 202–211). Cambridge: Cambridge University Press.

Deleuze, J. (1995). *Structure d'une couche limite turbulente soumise à une onde de choc incidente*. Ph.D. thesis, Université de la Mediterranee Aix—Marseille.

Deleuze, J., Audiffren, N., & Elena, M. (1994). Quadrant analysis in a heated-wall supersonic boundary layer. *Physics of Fluids, 6*, 4031–4041.

Delville, J., & Bonnet, J.-P. (2008). Two point correlation in fluid dynamics: POD, LSE and related methods. Lecture series 2008-5. von Karman Institute for Fluid Mechanics.

Delville, J., Ukeiley, L., Cordier, L., Bonnet, J.-P., & Glauser, M. N. (1999). Examination of the large scale structure in a turbulent mixing layer. Part 1: Proper orthogonal decomposition. *Journal of Fluid Mechanics, 391*, 91–122.

Demetriades, A. (1970). Turbulence measurements in supersonic two-dimensional wake. *Physics of Fluids, 13*, 1672–1678.

Demetriades, A. (1976). Turbulence correlations in a compressible wake. *Journal of Fluid Mechanics, 74*, 251–267.

Denis, S., Delville, J., Garem, J. H., Barre, S., & Bonnet, J.-P. (Décembre, 1998). étude du contrôle des couches de mélange planes et axisymétriques-(partie subsonique), rapport final—contrat CNRS 780-441.

Deville, M. O., & Gatski, T. B. (2012). *Mathematical Modeling for Complex Fluids and Flows.* Heidelberg: Springer.

Dimotakis, P. E. (1984). Entrainment into a fully developed, two-dimensional shear layer. Paper No. 84-0368, AIAA.

Dimotakis, P. E. (1986). Two-dimensional shear-layer entrainment. *AIAA Journal, 24*, 1791–1796.

Dimotakis, P. E. (1991). Turbulent free shear layer mixing and combustion. In *High Speed Flight Propulsion Systems* (pp. 265–340). AIAA.

Dolling, D. (1993). Fluctuation loads in shock wave turbulent boundary layer interactions: Tutorial and update. Paper No. 93-0284, AIAA.

Doris, L., Tenaud, C., & Ta-Phuoc, L. (2000). LES of spatially developing 3D compressible mixing layer. *Comptes Rendus De L Academie Des Sciences Serie II Fascicule B-Mecanique, 328*, 567–573.

Dubois, T., Domaradzki, J. A., & Honein, A. (2002). The subgrid-scale estimation model applied to large eddy simulation of compressible turbulence. *Physics of Fluids, 14*. doi: 10.1063/1.1466465.

Ducros, F., Ferrand, V., Nicoud, F., Weber, C., Darracq, D., Gacherieu, C., et al. (1999). Large-eddy simulation of the shock/turbulence interaction. *Journal of Computational Physics, 152*, 517–549.

Dudley, J. G., & Ukeiley, L. (2010). Suppression of fluctuating surface pressures in a supersonic cavity flow. Paper No. 2010-4974, AIAA.

Dumas, R. (1976). La turbulence dans les écoulements incompressibles. In A. Favre, L. Kovasznay, R. Dumas, J. Gaviglio & M. Coantic (Eds.), *La Turbulence en Mécanique des Fluides* (pp. 174–213). Paris: Gauthier-Villars.

Dupont, P., & Debiève, J.-F. (1992). A hot wire method for measuring turbulence in transonic or supersonic heated flows. *Experiments in Fluids, 13*, 84–90.

Dupont, P., Haddad, C., & Debiève, J.-F. (2006). Space and time organization in a shock-induced separated boundary layer. *Journal of Fluid Mechanics, 559*, 255–277.

Dupont, P., Muscat, P., & Dussauge, J.-P. (1999). Localisation of large scale structures in a supersonic mixing layer: A new method and first analysis. *Flow Turbulence and Combustion, 62*, 335–358.

Durbin, P. A., & Pettersson-Reif, B. A. (2001). *Statistical Theory and Modeling for Turbulent Flows.* Chichester: John Wiley & Sons, Ltd.

Durbin, P. A., & Zeman, O. (1992). Rapid distortion theory for homogeneous compressed turbulence with application to modeling. *Journal of Fluid Mechanics, 242*, 349.

Dussauge, J.-P., & Gaviglio, J. (1987). The rapid expansion of a supersonic turbulent flow: role of bulk dilatation. *Journal of Fluid Mechanics, 174*, 81–112.

Dutton, J., Burr, R., Goebel, S., & Messersmith, N. (1990). Compressibility and mixing in turbulent free shear layers. In *Proceedings of the 12th Symposium on Turbulence* (pp. A22.1–A22.12). University of Missouri-Rolla.

East, L. F., & Sawyer, W. G. (1979). An investigation of the structure of equilibrium turbulent boundary layers. In *Conference Proceedings No. CP-271*, AGARD.

El Baz, A. M, & Launder, B. E (1993). Second-order modeling of compressible mixing layers. In W. Rodi, & F. Martelli (Eds.), *Engineering Turbulence Modelling and Experiments* (Vol. 2, pp. 63–72). Amsterdam: Elsevier Science.

Elliott, G., Samimy, M., & Arnette, S. (1992). Study of compressible mixing layers using filtered rayleigh scattering based visualizations. *AIAA Journal, 30*, 2567–2569.

Elliott, G. S., & Samimy, M. (1990). Compressibility effects in free shear layers. *Physics of Fluids A, 2*, 1231–1240.

Ennix, K. A. (1993). *Engine exhaust characteristics evaluation in support of aircraft acoustic testing.* Technical Memorandum 104263, NASA.

Erlebacher, G., Hussaini, M. Y., Kreiss, H. O., & Sarkar, S. (1990). The analysis and simulation of compressible turbulence. *Theoretical and Computational Fluid Dynamics, 2,* 73–95.

Erlebacher, G., Hussaini, M. Y., Speziale, C. G., & Zang, T. A. (1987). *Toward the large eddy simulations of compressible turbulent flows.* Contractor Report 178273, NASA.

Erlebacher, G., Hussaini, M. Y., Speziale, C. G., & Zang, T. A. (1992). Toward the large eddy simulation of compressible turbulent flows. *Journal of Fluid Mechanics, 238,* 155–185.

Fabre, D., Jacquin, L., & Sesterhenn, J. (2001). Linear interaction of a cylindrical entropy spot with a shock. *Physics of Fluids, 13,* 2403–2422.

Fasel, H. F., von Terzi, D. A., & Sandberg, R. D. (2006). A methodology for simulating compressible turbulent flows. *Journal of Applied Mechanics – Transactions of the ASME, 73,* 405–412.

Favre, A. (1965a). Equations des gaz turbulents compressibles I. Formes générales. *Journal de Mecanique, 4,* 361–390.

Favre, A. (1965b). Equations des gaz turbulents compressibles II. Méthode des vitesses moyennes; méthode des vitesses macroscopiques pondérées par la masse volumique. *Journal de Mecanique, 4,* 391–421.

Favre, A. (1969). Statistical equations of turbulent gases. *Problems of Hydrodynamics and Continuum Mechanics* (pp. 231–266). Philadelphia: SIAM.

Favre, A. (1976). équations fondamentales des fluides a masse volumique variable en écoulements turbulents. In A. Favre, L. Kovasznay, R. Dumas, J. Gaviglio & M. Coantic (Eds.), *La Turbulence en Mécanique des Fluides* (pp. 23–78). Paris: Gauthier-Villars.

Feiereisen, W. J., Reynolds, W. C., & Ferziger, J. H. (1981). *Numerical simulation of a compressible homogeneous shear flow.* Report No. TF-13, Department of Mechanical Engineering, Stanford University.

Fernholz, H. H., & Finley, P. J. (1977). *A critical compilatation of compressible turbulent boundary layer data.* Report No. AG-223, AGARD.

Fernholz, H. H., & Finley, P. J. (1980). *A critical commentary on mean flow data for two-dimensional compressible turbulent boundary layers.* Report No. AG-253, AGARD.

Fernholz, H. H., & Finley, P. J. (1981). *A further compilatation of compressible turbulent boundary layer data with a survey of turbulence data.* Report No. AG-263, AGARD.

Fernholz, H. H., & Finley, P. J. (1996). The incompressible zero-pressure-gradient turbulent boundary layer: An assessment of the data. *Progress in Aerospace Sciences, 32,* 245–311.

Fernholz, H. H., Finley, P. J., Dussauge, J.-P., & Smits, A. J. (1989). *A survey of measurements and measuring techniques in rapidly distorted compressible turbulent boundary layers.* Report No. AG-315, AGARD.

Forestier, N., Jacquin, L., & Geffroy, P. (2003). The mixing layer over a deep cavity at high-subsonic speed. *Journal of Fluid Mechanics, 475,* 101–145.

Fourguette, D. C., Mungal, M. G., & Dibble, R. W. (1991). Time evolution of the shear layer of a supersonic axisymmetric jet. *AIAA Journal, 29,* 1123–1130.

Foysi, H., Sarkar, S., & Friedrich, R. (2004). Comprssibility effects and scaling in supersonic flows. *Journal of Fluid Mechanics, 509,* 207–216.

Frankl, F., & Voishel, V. (1943). *Turbulent friction in the boundary layer of a flat plate in a two-dimensional compressible flow at high speeds.* Technical Memorandum 1053, NACA.

Freund, J. B., Lele, S. K., & Moin, P. (2000a). Compressibility effects in a turbulent annular mixing layer Part 1. Turbulence and growth rate. *Journal of Fluid Mechanics, 421,* 229–267.

Freund, J. B., Lele, S. K., & Moin, P. (2000b). Compressibility effects in a turbulent annular mixing layer Part 2. Mixing of a passive scalar. *Journal of Fluid Mechanics, 421,* 269–292.

Fric, T. F., & Roshko, A. (1994). Vortical structure in the wake of a transverse jet. *Journal of Fluid Mechanics, 279,* 1–47.

Friedrich, R. (1998). Modelling of turbulence in compressible flows. In A. Hanifi, P. H. Alfredsson, A. V. Johanssoa & D. S. Henningson (Eds.), *Transition Turbulence and Combustion Modeling* (pp. 243–348). Dordrecht: Kluwer Academic Publishers.

Friedrich, R. (2007). Compressible turbulent flows: Aspects of prediction and analysis. *Zeitschift fur Angewandte Mathematik und Mechanik, 87,* 189–211.

Frohnapfel, B., Lammers, P., Jovanović, J., & Durst, F. (2007). Interpretation of the mechanism associated with turbulent drag reduction in terms of anisotropy invariants. *Journal of Fluid Mechanics, 577,* 457–466.

Frössel, W. (1938). *Flow in Smooth Straight Pipes at Velocities above and below sound velocity.* Technical Memorandum 844, NACA.

Gad-el-Hak, M. (2000). *Flow Control: Passive, Active, and Reactive Flow Management.* Cambridge: Cambridge University Press. Reprinted in papaerback 2007.

Gad-el-Hak, M., Pollard, A., & Bonnet, J. (Eds.) (1998). *Flow Control: Fundamentals and Practices Lecture Notes in Physics* (Vol. m53). Berlin: Springer-Verlag.

Gai, S. L., Hughes, D. P., & Perry, M. S. (2002). Large-scale structures and growth of a flat plate compressible wake. *AIAA Journal, 40,* 1164–1169.

Garnier, E., Sagaut, P., & Deville, M. (2002). Large eddy simulation of shock/homogeneous turbulence interaction. *Computers and Fluids, 31,* 245–268.

Gatski, T. B. (1996). Turbulent flows: Model equations and solution methodology. In R. Peyret (Ed.), *Handbook of Computational Fluid Dynamics* (pp. 339–415). London: Academic Press.

Gatski, T. B. (2004). Constitutive equations for turbulent flows. *Theoretical and Computational Fluid Dynamics, 18,* 345–369.

Gatski, T. B. (2009). Second-moment and scalar flux representations in engineering and geophysical flows. *Fluid Dynamics Research, 41,* 012202.

Gatski, T. B., & Rumsey, C. L. (2002). Linear and nonlinear eddy viscosity models. In B. E. Launder, & N. D. Sandham (Eds.), *Closure Strategies for Modelling Turbulent and Transitional Flows* (pp. 9–46). Cambridge: Cambridge University Press.

Gatski, T. B., Rumsey, C. L., & Manceau, R. (2007). Current trends in modelling research for turbulent aerodynamic flows. *Philosophical Transactions of the Royal Society A: Mathematical, Physical and Engineering Sciences, 365,* 2389–2418.

Gaviglio, J. (1976). La turbulence dans les écoulements compressibles des gaz. In A. Favre, L. Kovasznay, R. Dumas, J. Gaviglio & M. Coantic (Eds.), *La Turbulence en Mécanique des Fluides* (pp. 232–262). Paris: Gauthier-Villars.

Gaviglio, J. (1987). Reynolds analogies and experimental study of heat transfer in the supersonic boundary layer. *International Journal of Heat and Mass Transfer, 30,* 911–926.

Gaviglio, J., Dussauge, J.-P., Debiève, J.-F., & Favre, A. (1977). Behavior of a turbulent flow, strongly out of equilibrium, at supersonic speeds. *Physics of Fluids, 20,* S179–S192.

George, W. K., Beuther, P. D., & Lumley, J. L. (1978). Processing of random signals. In *Proceedings Dynamic Flow Conference* (pp. 20–63).

George, W. K., & Davidson, L. (2004). Role of initial conditions in establishing asymptotic flow behavior. *AIAA Journal, 42,* 438–446.

Gerolymos, G. A., Sénéchal, D., & Vallet, I. (2007). Pressure, density and temperature fluctuations in compressible turbulent flow—i. Paper No. 2007-3408, AIAA/CEAS.

Gerolymos, G. A., Sénéchal, D., & Vallet, I. (2008). Pressure, density and temperature fluctuations in compressible turbulent flow—ii. Paper No. 2008-647, AIAA.

Geurts, B. J. (2004). *Elements of Direct and Large-Eddy Simulation.* Philadelphia: R.T. Edwards, Inc.

Ghosh, S., Foysi, H., & Friedrich, R. (2010). Compressible turbulent channel and pipe flow: Similarities and differences. *Journal of Fluid Mechanics, 648,* 155–181.

Glauser, M. N., & George, W. K. (1992). Compressible turbulent channel and pipe flow: Similarities and differences. *Experimental Thermal and Fluid Science, 5,* 617–632.

Goebel, S. G., & Dutton, J. C. (1991). Experimental study of compressible turbulent mixing layers. *AIAA Journal, 29,* 538–546.

Gokoglu, S. A., Kuczmarski, M. A., Culley, D. E., & Raghu, S. (2010). Numerical studies of a supersonic fluidic diverter actuator for flow control. Paper No. 2010-4415, AIAA.

Goldstein, M. (1976). *Aeroacoustics.* Boston: McGraw-Hill.

Goodyer, M. J. (1997). *Introduction to cryogenic wind tunnels.* Report R-812, AGARD.

Green, J. E. (1971). Intereactions between shock waves and turbulent boundary layers. *Progress in Aerospace Sciences, 11,* 235–341.

Grégoire, O., Souffland, D., & Gauthier, S. (2005). A second-order turbulence model for gaseous mixtures induced by Richtmyer-Meshkov instability. *Journal of Turbulence, 6.* doi:10.1088/14685240500307413.

Grésillon, D. (1994). Collective scattering in gases and plasmas. In *Waves and Instabilities in Plasmas.* pp.171–226. Berlin: Springer.

Grésillon, D., Gémaux, G., Cabrit, B., & Bonnet, J.-P. (1990). Observation of supersonic turbulent wakes by Laser fourier Densitometry (LFD). *European Journal of Mechanics B – Fluid, 9,* 415–436.

Grube, N. E., Taylor, E. M., & Martin, M. P. (2007). Assessment of WENO methods with shock-confining filtering for LES of compressible turbulence. Paper No. 2007-4198, AIAA.

Gruber, M. R., & Goss, L. P. (1999). Surface pressure measurements in supersonic transverse injection flowfields. *Journal of Propulsion and Power, 15,* 633–641.

Gruber, M. R., Messersmith, N. L., & Dutton, J. C. (1993). Three-dimensionnal velocity field in a compressible mixing layers. *AIAA Journal, 31,* 2061–2067.

Guarini, S. E., Moser, R. D., Shariff, K., & Wray, A. (2000). Direct numerical simulation of a supersonic turbulent boundary layer at Mach 2.5. *Journal of Fluid Mechanics, 414,* 1–33.

Gutmark, E. J., & Grinstein, F. F. (1999). Flow control with noncircular jets. *Annual Review of Fluid Mechanics,* (pp. 239–272). New York: Annual Reviews.

Ha Minh, H., Launder, B. E., & MacInnes, J. (1981). The turbulence modeling of variable density flows—A mixed-weighted decomposition. In *Selected Papers from 3rd Symposium on Turbulent Shear Flows* (pp. 291–308). Berlin: Springer-Verlag.

Haack, S. J., Taylor, T., Emhoff, J., & Cybyk, B. (2010). Development of an analytical sparkjet model. Paper No. 2010-4979, AIAA.

Haas, J., & Sturtevant, B. (1987). Interaction of weak shock waves with cylindrical and spherical gas inhomogeneities. *Journal of Fluid Mechanics, 181,* 41–76.

Hagemann, G., & Frey, M. (2008). Shock pattern in the plume of rocket nozzles: Needs for design consideration. *Shock Waves, 17,* 387–395.

Hahn, C., Kearney-Fischer, M., & Samimy, M. (2011). On factors influencing arc filament plasma actuator performance in control of high speed jets. *Experiments in Fluids, 51,* 1591–1603.

Hall, J. L., Dimotakis, P. E., & Rosemann, H. (1993). Experiments in non-reacting compressible shear layers. *AIAA Journal, 31,* 2247–2254.

Hallbäck, M., Groth, J., & Johansson, A. V. (1990). An algebraic model for nonisotropic turbulent dissipation rate in reynolds stress closures. *Physics of Fluids, 2,* 1859–1866.

Hanjalič, K., & Launder, B. E. (1976). Contribution towards a Reynolds-stress closure for low-Reynolds-number turbulence. *Journal of Fluid Mechanics, 74,* 593–610.

Hanjalič, K., & Launder, B. E. (2011). *Modelling Turbulence in Engineering and the Environment: Second-Moment Routers to Closure.* Cambridge: Cambridge University Press.

Hannappel, R., & Friedrich, R. (1992). Interacting of isotropic turbulence with a normal shock wave. *Applied Scientific Research, 51,* 507–512.

Hannappel, R., & Friedrich, R. (1995). Direct numerical simulation of a Mach 2 shock interacting with isotropic turbulence. *Applied Scientific Research, 54,* 205–221.

Hartung, L. C., & Duffy, R. E. (1986). Effects of pressure on turbulence in shock-induced flows. Paper No. 86-0127, AIAA.

Heinz, S. (2003). *Statistical Mechanics of Turbulent Flows.* Berlin: Springer.

Hermanson, J., & Cetegen, B. (2000). Shock-induced mixing of nonhomogeneous density turbulent jets. *Physics of Fluids, 12,* 1210–1225.

Herrin, J. L., & Dutton, J. C. (1994). Supersonic base flow experiments in the near wake of a cylindrical afterbody. *AIAA Journal,* 77–83.

Hirt, S., Reich, D. B., & O'Connor, M. B. (2010). Micro-ramp flow control for oblique shock interactions: Comparisons of computational and experimental data. Paper No. 2010-4973, AIAA.

Holmes, P., Lumley, J. L., & Berkooz, G. (1998). *Turbulence, Coherent Structures, Dynamical Systems and Symmetry.* Cambridge: Cambridge University Press.

Honkan, A., & Andreopoulos, J. (1992). Rapid compression of grid generated turbulence by a moving shock wave. *Physics of Fluids A,* 2562–2572.

Huang, P. G., & Bradshaw, P. (1995). The law of the wall for turbulent flows in pressure gradients. *AIAA Journal, 33,* 624–632.

Huang, P. G., Bradshaw, P., & Coakley, T. J. (1992). *Assessment of closure coefficients for compressible flow turbulence models.* Technical Memorandum 103882, NASA.

Huang, P. G., Bradshaw, P., & Coakley, T. J. (1994). Turbulence models for compressible flows. *AIAA Journal, 32,* 735–740.

Huang, P. G., & Coakley, T. J. (1993). Calculations of supersonic and hypersonic flows using compressible wall functions. In W. Rodi, & F. Martelli (Eds.), *Engineering Turbulence Modelling and Experiments* (Vol. 2, pp. 731–739). Amsterdam: Elsevier Science.

Huang, P. G., & Coleman, G. N. (1994). Van Driest transformation and compressible wall-bounded flows. *AIAA Journal, 32*, 2110–2113. Errata *AIAA Journal, 33*, 1756.

Huang, P. G., Coleman, G. N., & Bradshaw, P. (1995). Compressible turbulent channel flows: DNS results and modeling. *Journal of Fluid Mechanics, 305*, 185–218.

Huilgol, R. R., & Phan-Thien, N. (1997). *Fluid mechanics of viscoelasticity*. Amsterdam: Elsevier.

Humble, R. A. (2008). *Unsteady flow organization of a shock wave/boundary layer interaction*. Ph.D. thesis, Delft University of Technology.

Humble, R. A., Scarano, F., & van Oudheusden, B. W. (2007a). Particle image velocimetry measurements of a shock-wave/turbulent boundary layer interaction. *Experiments in Fluids, 43*, 173–183.

Humble, R. A., Scarano, F., & van Oudheusden, B. W. (2007b). Unsteady flow organization of compressible planar base flows. *Physics of Fluids, 19*, 076101.

Hussain, A. K. M. F., & Reynolds, W. C. (1970). The mechanics of an organized wave in turbulent shear flow. *Journal of Fluid Mechanics, 41*, 241–258.

Hussain, A. K. M. F., & Zedan, M. F. (1978a). Effects of initial conditions on the axisymmetric free shear layer: Effect of the initial fluctuation level. *Physics of Fluids, 21*, 1475–1481.

Hussain, A. K. M. F., & Zedan, M. F. (1978b). Effects of initial conditions on the axisymmetric free shear layer: Effects of the initial momentum thickness. *Physics of Fluids, 21*, 1100–1112.

Ibrahim, M. K., Kunimura, R., & Nakamura, Y. (2002). Mixing enhancement of compressible jets by using unsteady microjets as actuators. *AIAA Journal, 40*, 681–688.

Ikawa, H., & Kubota, T. (1975). Investigation of supersonic turbulent mixing layers with zero pressure gradient. *AIAA Journal, 13*, 566–572.

Illy, H., Jacquin, L., & Geffroy, P. (2008). Observations on the passive control of flow oscillations over a cavity in a transonic regime by means of a spanwise cylinder. Paper No. 2008-3774, AIAA.

Iollo, A., Lanteri, S., & Désidéri, J. (2000). Stability properties of POD-Galerkin approximations for the compressible Navier-Stokes equations. *Theoretical and Computational Fluid Dynamics, 13*, 377–396.

Jacquin, L., Blin, E., & Geffroy, P. (1991). Experiments on free turbulence/shock wave interactions. In *Eighth Symposium on Turbulent Shear Flows* (pp. 1.2.1–1.2.6).

Jacquin, L., Blin, E., & Geffroy, P. (1993). Experiments on free turbulence/shock wave interactions. In F. Durst, R. Friedrich, B. Launder, F. Schmidt, U. Schumann & J. Whitelaw (Eds.), *Proceedings of the Eighth Symposium on Turbulent Shear Flows* (pp. 229–248). Heidelberg: Springer-Verlag.

Jacquin, L., Cambon, C., & Blin, E. (1993). Turbulence amplification by a shock wave and rapid distortion theory. *Physics of Fluids A, 5*, 2539–2550.

Jacquin, L., & Geffroy, P. (1997). Amplification and reduction of turbulence in a heated jet/shock interaction experimental study of free shear turbulence/shock wave interaction. In *Proceedings of the 11th Symposium on Turbulent Shear Flows* (pp. L12–L17).

Jameel, M. I., Cormack, D. E., Tran, H., & Moskal, T. E. (1994). Sootblower optimization Part 1: Fundamental hydrodynamics of a sootlower nozzle and jets. *Tappi Journal, 77*, 135–142.

Jamme, S. (1998). *Étude de l'interaction entre une turbulence homogène isotrope et une onde de choc*. Ph.D. thesis, Université de Poitiers.

Jamme, S., Cazalbou, J.-B., Torres, F., & Chassaing, P. (2002). Direct numerical simulation of the interaction between a shock wave and various types of isotropic turbulence. *Flow Turbulence and Combustion, 68*, 227–268.

Jin, L. H., So, R. M. C., & Gatski, T. B. (2003). Equilibrium states of turbulent homogeneous buoyant flows. *Journal of Fluid Mechanics, 482*, 207–233.

Johansson, P. B. V., George, W. K., & Gourlay, M. J. (2003). Equilibrium similarity, effects of initial conditions and local reynolds number on the axisymmetric wake. *Physics of Fluids, 15*, 603–617.

Johari, H., & Rixon, G. S. (2003). Effects of pulsing on a vortex generator jet. *AIAA Journal, 41*, 2309–2315.

Johnson, D. A. (1989). Laser-doppler anemometry. In *A survey of measurements and measuring techniques in rapidly distorted compressible turbulent boundary layers*. AGARDograph No. 315. AGARD.

Johnson, J. A., Zhang, Y., & Johnson, L. E. (1988). Evidence of Reynolds number sensitivity in supersonic turbulent shocklets. *AIAA Journal*, 502–504.

Jordan, P., & Colonius, T. (2013). Wavepackets and turbulent jet noise. *Annual Review of Fluid Mechanics*, 231–273.

Jordan, P., Schlegel, M., Stalnov, M., Noack, B. R., & Tinney, C. E. (2007). Identifying noisy and quiet modes in a jet. Paper No. 2007-3602, AIAA/CEAS.

Jovanović, J. (2003). *Statistical dynamics of turbulence.* Berlin: Springer.

Jung, D., Gamard, S., & George, W. K. (2004). Downstream evolution of the most energetic modes in a turbulent axisymmetric jet at high Reynolds number. Part 1. The near-field region. *Journal of Fluid Mechanics, 514,* 173–204.

Kalkhoran, I., & Sforza, P. (1994). Airfoil pressure measurements during oblique shock-wave/vortex interaction in a Mach 3 stream. *AIAA Journal, 32,* 783–796.

Kampé de Fériet, J., & Betchov, R. (1951). Theoretical and experimental averages of turbulent functions. *Proceedings of the Koninklijke Nederlandse Akademie Van Wetenschappen, 53,* 389–398.

Kandala, R., & Candler, G. V. (2004). Numerical studies of laser-induced energy deposition for supersonic flow control. *AIAA Journal, 42,* 2266–2275.

Kastengren, A. L., & Dutton, J. C. (2004). *Wake topology in a three-dimensional base flow.* Paper No. 2004-2340, AIAA.

Kawai, S., & Fujii, K. (2005). Computational study of supersonic base flow using hybrid turbulence methodology. *AIAA Journal, 43,* 1265–1275.

Kegerise, M. A., & Spina, E. F. (2000). A comparative study of constant-voltage and constant temperature anemometers. *Experiments in Fluids, 29,* 165–177.

Keller, J., & Merzkirch, W. (1990). Interaction of a normal shock wave with a compressible turbulent flow. *Experiments in Fluids, 8,* 241–248.

Kida, S., & Orszag, S. A. (1990a). Energy and spectral dynamics in forced compressible turbulence. *Journal of Scientific Computing, 5,* 85–125.

Kida, S., & Orszag, S. A. (1990b). Enstrophy budget in decaying compressible turbulence. *Journal of Scientific Computing, 5,* 1–34.

Kida, S., & Orszag, S. A. (1992). Energy and spectral dynamics in decaying compressible turbulence. *Journal of Scientific Computing, 7,* 1–34.

Kim, J. (1989). On the structure of pressure fluctuations in simulated turbulent channel flow. *Journal of Fluid Mechanics, 205,* 421–451.

Kim, S. C. (1990). New mixing-length model for supersonic shear layers. *AIAA Journal, 28,* 1999–2000.

Kirby, M., Boris, J.-P., & Sirovich, L. (1990). Energy and spectral dynamics in decaying compressible turbulence. *International Journal for Numerical Methods in Fluids, 10,* 411–428.

Kistler, A. L. (1959). Fluctuation measurements in a supersonic turbulent boundary layer. *Physics of Fluids, 2,* 290–296.

Knight, D., Yan, H., Panaras, A. G., & Zheltovodov, A. (2003). Advances in CFD prediction of shock wave turbulent boundary layer interactions. *Progress in Aerospace Sciences, 39,* 121–184.

Knight, D., Zhou, G., Okong'o, N., Shukla, V. (1998). Compressible large eddy simulation using unstructured grids. Paper No. 1998-0535, AIAA.

Kopiev, V., Ostrikov, N., Zaitsev, M., Kopiev, V., Belyaev, I., Bityurin, V., Klimov, A., Moralev, I., & Godin, S. (2011). Jet noise control by nozzle surface HF DBD actuators. Paper No. 2011-0911, AIAA.

Kopiev, V. F., Ostrikov, N. N., Kopiev, V. A., Belyaev, I. V., & Faranosov, G. A. (2011). Instability wave control by plasma actuators: Problems and prospects. Paper No. 2011-0973, AIAA.

Kosović, B., Pullin, D. I., & Samtaney, R. (2002). Subgrid-scale modeling for large-eddy simulations of turbulence. *Physics of Fluids, 14,* 1511–1522.

Kovasznay, L. S. G. (1950). The hot-wire anemometer in supersonic flow. *Journal of the Aeronautical Sciences, 17,* 565–584.

Kovasznay, L. S. G. (1953). Turbulence in supersonic flow. *Journal of the Aeronautical Sciences, 20,* 657–674.

Kreuzinger, J., Friedrich, R., & Gatski, T. B. (2006). Compressibility effects in the solenoidal dissipation rate equation: A priori assessment and modeling. *International Journal of Heat and Fluid Flow, 27,* 696–706.

Krishnamurty, V. S., & Shyy, W. (1997). Study of compressibility modifications to the $k - \epsilon$ turbulence model. *Physics of Fluids, 9,* 2769–2788.

Kruse, R. L. (1968). *Transition and flow reattachment behind an Apollo-like body at Mach numbers to 9.* Technical Note D-4645, NASA.

Kumar, G. V. R., & Tewari, A. (2004). Active closed-loop control of supersonic flow with transverse injection. Paper No. 2004-2699, AIAA.

Kumar, R., Ali, M. Y., Alvi, F. S., & Venkatakrishnan, L. (2011). Generationa dn control of oblique shocks using microjets. *AIAA Journal, 49,* 2751–2759.

Kumar, S. (1975). *Contribution á l'étude de l'interaction onde de choc-sillage turbulent supersonique*. Ph.D. thesis, Université Poitiers.

Lada, C., & Kontis, K. (2011). Experimental studies of open cavity configurations at transonic speeds with flow control. *Journal of Fluid Mechanics, 48*, 719–724.

Lai, Y. G., & So, R. M. C. (1990). On near-wall turbulent flow modeling. *Journal of Fluid Mechanics, 221*, 641–673.

Lamarri, M. (1996). *Mesures par vélocimétrie Laser-Doppler dans une couche de mélange turbulente supersonique: quelques aspects du processus de mesures*. Ph.D. thesis, Université de Poitiers.

Lardeau, S., Collin, E., Lamballais, E., & Bonnet, J. (2003). Analysis of a jet-mixing layer interaction. *International Journal of Heat and Fluid Flow, 24*, 520–528.

Lau, J. C. (1981). Effects of exit Mach number and temperature on mean-flow and turbulence characteristics in round jets. *Journal of Fluid Mechanics, 93*, 193–218.

Lau, J. C., Morris, P. J., & Fisher, M. J. (1979). Measurements in subsonic and supersonic free jets using a laser velocimeter. *Journal of Fluid Mechanics, 93*, 1–27.

Laufer, J. A. (1961). Aerodynamic noise in supersonic wind tunnels. *Journal of Aerospace Science, 28*, 685–692.

Launder, B. E., & Sandham, N. D. (Eds.) (2000). *Closure Strategies for Turbulent and Transitional Flows*. Cambridge: Cambridge University Press

Laurendeau, E., Jordan, P., Bonnet, J.-P., Delville, J., Parnaudeau, P., & Lamballais, E. (2008). Subsonic jet noise reduction by fluidic control: The interaction region and the global effect. *Physics of Fluids* 20 10.1063/1.300624.

Laurent, H. (1996). *Turbulence d'une interaction onde de choc/couche limite sur une paroi plane adiabatique ou chauffée*. Ph.D. thesis, Université de la Méditerranée Aix, Marseille.

Lee, H., Jeung, I., Lee, S., & Kim, S. (2011). Experiment of flow control using laser energy deposition around high speed propulsion system. In *Proceedings of the Seventh International Symposium on Beamed Energy Propulsion. AIP Conference Proceedings* (Vol. 1402, pp. 416–423).

Lee, S., Lele, S., & Moin, P. (1997). Interaction of isotropic turbulence with shock wave: Effects of shock strength. *Journal of Fluid Mechanics, 340*, 225–247.

Lee, S., Lele, S. K., & Moin, P. (1991). Eddy shocklets in decaying compressible turbulence. *Physics of Fluids A, 3*, 657–684.

Lee, S., Lele, S. K., & Moin, P. (1993). Direct numerical simulation of isotropic turbulence interacting with a weak shock wave. *Journal of Fluid Mechanics, 251*, 533.

Lee, S., Loth, E., & Babinsky, H. (2011). Normal shock boundary layer control with various vortex generator geometries. *Computers and Fluids, 49*, 233–246.

Lele, S. K. (1994). Compressibility effects on turbulence. *Annual Review of Fluid Mechanics*, 211–254.

Lenormand, E., Sagaut, P., Ta Phuoc, L., & Comte, P. (2000). Subgrid-scale models for large-eddy simulations of compressible wall bounded flows. *AIAA Journal, 25*, 1340–1350.

Lenonard, A. (1974). Energy cascade in large-eddy simulations of turbulent fluid flows. In *Advances in Geophysics* (pp. 237–248). New York: Academic Press.

Lepicovsky, J., Ahuja, K. K., Brown, W. H., & Burrin, R. H. (1987). Coherent large-scale structures in high reynolds number supersonic jets. *AIAA Journal, 25*, 1419–1425.

Lesieur, M., Comte, P., & Normand, X. (1991). Direct and large-eddy simulations of transitioning and turbulent shear flows. Paper No. 91-0335, AIAA.

Lesieur, M., Métais, O., & Comte, P. (2005). *Large-Eddy Simulations of Turbulence*. Cambridge: Cambridge Universitry Press.

Leuchter, O. (1998). Interaction of shock waves with grid turbulence. In *A selection of test cases for the validation of large-eddy simulations of turbulent flows* (pp. 19–22). AR-345. AGARD.

Lewkowicz, A. K. (1982). An improved universal wake equation for turbulent boundary layers and some of its consequences. *Zeitschrift Fur Flugwissenschaften Und Weltraumforschung, 6*, 261–266.

Li, Q., Chen, H., & Fu, S. (2003). Large-scale vortices in high-speed mixing layers. *Physics of Fluids, 15*, 3240–3243.

Li, Q., & Coleman, G. N. (2003). DNS of an oblique shock impinging upon a turbulent boundary layer. In R. Friedrich, B. J. Geurts, & O. Metais (Eds.), *Direct and Large Eddy Simulation V* (pp. 387–396). Dordrecht: Kluwer Academic.

Liepmann, D., & Gharib, M. (1992). The role of streamwise vorticity in the near-field entrainment of round jets. *Journal of Fluid Mechanics, 245*, 643–668.

Liepmann, H. W., & Laufer, J. (1947). *Turbulence models for compressible flows*. TN 1257, NACA.

Liepmann, H. W., & Roshko, A. (1957). *Elements of Gas Dynamics*. New York: John Wiley & Sons.

Lilly, D. K. (1966). On the application of the eddy-viscosity concept in the inertial sub-range of turbulence. Manuscript 123, NCAR.

Lilly, D. K. (1992). A proposed modification of the Germano subgrid-scale closure method. *Physics of Fluids A, 4*, 633–635.

Loginov, M., Adams, N. A., & Zheltovodov, A. A. (2006). Large-eddy simulation of shock-wave/turbulent-boundary-layer interaction. *Journal of Fluid Mechanics, 565*, 135–169.

Lou, H., Alvi, F. S., Shih, C., Choi, J., & Annaswamy, A. (2002). Active control of supersonic impinging jets: Flowfield properties and closed-loop strategies. Paper No. 2002-2728, AIAA.

Low, K. R., Berger, Z. P., Lewalle, J., El-Hadidi, B., & Glauser, M. N. (2011). Correlations and wavelet based analysis of near-field and far-field pressure of a controlled high speed jet. Paper No. 2011-4020, AIAA.

Low, K. R., El-Hadidi, B., Andino, M. Y., Berdanier, R., & Glauser, M. N. (2010). Investigation of different active control strategies for high speed jets using synthetic jet actuators. Paper No. 2010-4267, AIAA.

Lu, S. S., & Willmarth, W. W. (1973). Measurements of the structure of the Reynolds stress in a turbulent boundary layer. *Journal of Fluid Mechanics, 60*, 481–511.

Lu, Y. (2001). Including non-Boussinesq ocean circulation models. *Journal of Physical Oceanography, 31*, 1616–1622.

Lumley, J. L. (1967). The structure of inhomogeneous turbulent flows. In A. M. Yaglom, & V. I. Tatarsky (Eds.), *Atmospheric Turbulence and Radio Wave Propagation* (pp. 166–176). Moscow: Nauka.

Lumley, J. L. (1970). *Stochastic Tools in Turbulence*. New York: Academic Press.

Lumley, J. L. (1978). Computational modeling of turbulent flows. *Advances in Applied Mechanics, 18*, 123–176.

Lumley, J. L., & Panofsky, H. A. (1964). *The Structure of Atmospheric Turbulence*. New York: Interscience Publishers.

Lumley, J. L., & Poje, A. (1997). Low-dimensional models for flows with density fluctuations. *Physics of Fluids, 9*, 2023–2031.

Lund, T. S., Wu, X., & Squires, K. D. (1998). Generation of turbulent inflow data for spatially-developing boundary layer simulations. *Journal of Comparative Physiology, 140*, 233–258.

Maeder, T., Adams, N. A., & Kleiser, L. (2001). Direct simulation of turbulent supersonic boundary layers by an extended temporal approach. *Journal of Fluid Mechanics, 429*, 187–216.

Mahadevan, S., & Raja, L. L. (2012). Simulation of direct-current surface plasma discharges in air for supersonic flow control. *AIAA Journal, 50*, 325–337.

Mahesh, K., Lee, S., Lele, S., & Moin, P. (1995). The interaction of an isotropic field of acoustic waves with a shock wave. *Journal of Fluid Mechanics, 300*, 383–407.

Mahesh, K., Lele, S., & Moin, P. (1997). The influence of entropy fluctuations on the interaction of turbulence with a shock wave. *Journal of Fluid Mechanics, 334*, 353–379.

Mahesh, K., Lele, S. K., & Moin, P. (1994). The response of anisotropic turbulence to rapid homogeneous one-dimensional compression. *Physics of Fluids, 6*, 1052–1062.

Mahesh, K., Moin, P., & Lele, S. K. (1996). *The interaction of a shock wave with a turbulent shear flow*. Report No. TF-69, Department of Mechanical Engineering, Stanford University.

Mansour, N. N., Kim, J., & Moin, P. (1988). Reynolds-stress and dissipation-rate budgets in turbulent channel flow. *Journal of Fluid Mechanics, 194*, 15–44.

Martin, M. P. (2005). *Preliminary study of the SGS time scales for compressible boundary layers using DNS data*. Paper No. 2005-0665, AIAA.

Martin, M. P. (2007). Direct numerical simulation of hypersonic turbulent boundary layers Part 1. Initialization and comparison with experiments. *Journal of Fluid Mechanics, 570*, 347–364.

Martin, M. P., Piomelli, U., & Candler, G. V. (2000). Subgrid-scale models for compressible large-eddy simulations. *Theoretical and Computational Fluid Dynamics, 13*, 361–376.

Marusic, I., & Kunkel, G. J. (2003). Streamwise turbulence intensity formulation for flat-plate boundary layers. *Physics of Fluids, 15*, 2461–2464.

Marusic, I., Uddin, A. K. M., & Perry, A. E. (1997). Similarity law for the streamwise turbulence intensity in zero-pressure-gradient turbulent boundary layers. *Physics of Fluids, 9*, 3718–3726.

Mathew, J., Foysi, H., & Friedrich, R. (2006). A new approach to LES based on explicit filtering. *International Journal of Heat and Fluid Flow, 27,* 594–602.

McCormick, D. C. (1993). Shock/boundary-layer interaction control with vortex generators and passive cavity. *AIAA Journal, 31,* 91–96.

Melling, A. (1997). Tracer particles and seeding for particle image velocimetry. *Measurement Science and Technology, 8,* 1406–1416.

Miles, R., & Lempert, W. (1990). Two-dimensional measurement of density, velocity and temperature in turbulent high-speed air flows by UV Rayleigh scattering. *Applied Physics B, 51,* 1–7.

Miles, R., & Lempert, W. (1997). Quantitative flow visualization in unseeded flows. *Annual Review of Fluid Mechanics, 29,* 285–326.

Miles, R., Lempert, W., & Forkey, J. (2001). Laser Rayleigh scattering. *Measurement Science and Technology, 12*(5), R33–R51.

Miller, D. N., Yagle, P. J., & Hamstra, J. W. (1999). Fluidic throat skewing for thrust vectoring in fixed geometry nozzles. Paper No. 99-0365, AIAA.

Mistral, S. (1993). *Étude expérimentale et simulation numérique des transferts de quantité de mouvement et ther miques dans les jets supersoniques coaxiaux.* Ph.D. thesis, I.N.P.T. Toulouse.

Moin, P., & Mahesh, K. (1998). Direct numerical simulation: A tool in turbulence research. *Annual Review of Fluid Mechanics, 30,* 539–578. New York: Annual Reviews.

Moin, P., Squires, K., Cabot, W., & Lee, S. (1991). A dynamic subgrid-scale model for compressible turbulence and scalar transport. *Physics of Fluids A, 3,* 2746.

Moore, F. K. (1954). *Unsteady oblique interaction of a shock wave with a plane disturbance.* Report 1165, NACA (supersedes NACA TN 2879, 1953).

Moreau, E. (2007). Airflow control by non-thermal plasma actuators. *Journal of Physics D-Applied Physics, 40,* CID: 026304.

Moreno, D., Krothapalli, A., Alkislar, M. B., & Lourenco, L. M. (2004). Low-dimensional model of a supersonic rectangular jet. *Physical Review E, 69,* CID: 026304.

Morinishi, Y., Tamano, S., & Nakabayashi, K. (2004). Direct numerical simulation of compressible turbulent channel flow between adiabatic and isothermal walls. *Journal of Fluid Mechanics, 502,* 273s–308s.

Morinishi, Y., Tamano, S., & Nakamura, E. (2007). New scalig of turbulent statistics fro incompressible channel flow with different total heat flux gradients. *International Journal of Heat and Mass Transfer, 50,* 1781–1789.

Morkovin, M. V. (1956). Fluctuations and hot-wire anemometry in compressible flows. AGAR-Dograph24, AGARD.

Morkovin, M. V. (1962). Effects of compressibility on turbulent flows. In A. Favre (Ed.), *Mecanique de la Turbulence* (pp. 367–380). CNRS.

Morkovin, M. V. (1964). Effects of compressibility on turbulent flows. In A. Favre (Ed.), *The Mechanics of Turbulence* (pp. 367–380). New York: Gordon and Breach.

Morrison, J. H., Gatski, T. B., Sommer, T. P., Zhang, H. S., & So, R. M. C. (1993). Evaluation of a near-wall turbulent closure model in predicting compressible ramp flows. In R. M. C. So, C. G. Speziale, & B. E. Launder (Eds.), *Near-Wall Turbulent Flows* (pp. 239–250). Amsterdam: Elsevier Science Publishers.

Nagano, Y. (2002). Modeling heat transfer in near-wall flows. In B. E. Launder, & N. D. Sandham (Eds.), *Closure Strategies for Modelling Turbulent and Transitional Flows* (pp. 189–247). Cambridge: Cambridge University Press.

Nagano, Y., & Shimada, M. (1995). Rigorous modeling of dissipation-rate equation using direct simulations. *JSME International Journal Series B – Fluids and Thermal Engineering, 38,* 51–59.

Nagano, Y., & Shimada, M. (1996). Development of a two-equation heat transfer model based on direct simulations of turbulent flows with different prandtl numbers. *Physics of Fluids, 8,* 3379–3402.

Nakagawa, M., & Dahm, W. J. A. (2006). Scaling properties and wave interactions in confined supersonic bluff body wakes. *AIAA Journal, 44,* 1299–1309.

Narayanaswamy, V., Clemens, N. T., & Raja, L. L. (2010a). Investigation of a pulsed-plasma jet for shock/boundary layer control. Paper No. 2010-1089, AIAA.

Narayanaswamy, V., Raja, L. L., & Clemens, N. T. (2010b). Characterization of a high-frequency pulsed-plasma jet actuator for supersonic flow control. *AIAA Journal, 48,* 297–305.

Nguyen, A. T., Deniau, H., Girard, S., & Alziary de Roquefort, T. (2003). Unsteadiness of flow separation and end-effects regime in a thrust-optimized contour rocket nozzle. *Flow Turbulence and Combustion, 71,* 161–181.

Normand, X., & Lesieur, M. (1992). Direct and large-eddy simulations of transition in the compressible boundary layer. *Theoretical and Computational Fluid Dynamics, 3*, 231–252.

Oberlack, M. (1997). Non-isotropic dissipation in non-homogeneous turbulence. *Journal of Fluid Mechanics, 350*, 351–374.

Oster, D., & Wygnanski, I. (1982). The forced mixing layer between parallel streams. *Journal of Fluid Mechanics, 123*, 91–130.

Owen, F. K. (1990). Turbulence and shear stress measurements in hypersonic flow. Paper No. 90-1394, AIAA.

Pal, S., Sriram, R., Srisha Rao, M. V., Jagadeesh, G. (2012). Effect of dielectric barrier discharge plasma in supersonic flow. In K. Kontis (Ed.), *Proceedings of the 28th International Symposium on Sock Waves* (pp. 867–872). Heidelberg: Springer.

Panaras, A. G. (1996). Review of the physics of swept-shock/boundary layer interactions. *Progress in Aerospace Sciences, 32*, 173–244.

Panda, J., & Seasholtz, R. G. (2006). Experimental investigation of Reynolds and Favre averaging in high-speed jets. *AIAA Journal, 44*, 1952–1959.

Pantano, C., & Sarkar, S. (2002). A study of compressibility effects in the high-speed turbulent shear layer using direct simulation. *Journal of Fluid Mechanics, 451*, 329–371.

Papamoschou, D. (1989). Structure of the compressible turbulent shear layer. Paper No. 89-0126, AIAA

Papamoschou, D. (1993). Zones of influence in the compressible shear layer. *Fluid Dynamics Research, 11*, 217–228.

Papamoschou, D. (1995). Evidence of shocklets in a counterflow supersonic shear layer. *Physics of Fluids, 7*, 233–235.

Papamoschou, D. (2000). Mixing enhancement using axial flow. Paper No. 2000-0093, AIAA.

Papamoschou, D., & Hubbard, D. G. (1993). Visual observations of supersonic transverse jets. *Experiments in Fluids, 14*, 468–476.

Papamoschou, D., & Lele, S. K. (1993). Vortex-induced disturbance field in a compressible shear layer. *Physics of Fluids A, 5*, 1412–1419.

Papamoschou, D., & Roshko, A. (1986). Observations of supersonic free shear layers. Paper No. 86-0162, AIAA.

Papamoschou, D., & Roshko, A. (1988a). The compressible turbulent shear layer: An experimental study. *Journal of Fluid Mechanics, 197*, 453–477.

Papamoschou, D., & Roshko, A. (1988b). Observations of supersonic free shear layers. *Sandhana-Academy Proceedings in Engineering Sciences, 12*, 1–14.

Passot, T., & Pouquet, A. (1987). Numerical simulation of compressible homogeneous flows in the turbulent regime. *Journal of Fluid Mechanics, 181*, 441–466.

Passot, T., & Pouquet, A. (1991). Numerical simulations of three-dimensional supersonic flows. *European Journal of Mechanics B – Fluids, 10*, 377–394.

Pernod, P., Preobrazhensky, V., Merlen, A., Ducloux, O., Talbi, A., Gimeno, L., et al. (2010). MEMS magneto-mechanical microvalves (MMMS) for aerodynamic active flow control. *Journal of Magnetism and Magnetic Materials, 322*, 1642–1646.

Perret, L., Delville, J., Manceau, R., & Bonnet, J.-P. (2006). Generation of turbulent inflow conditions for large eddy simulation from stereoscopic PIV measurements. *International Journal of Heat and Fluid Flow, 27*, 576–584.

Perry, A. E., Henbest, S., & Chong, M. S. (1986). A theoretical and experimental study of wall turbulence. *Journal of Fluid Mechanics, 181*, 441–466.

Perry, A. E., & Li, J. D. (1990). Experimental support for the attached-eddy hypothesis in zero-pressure-gradient turbulent boundary layers. *Journal of Fluid Mechanics, 181*, 441–466.

Petrie, H. L., Samimy, M., & Addy, A. L. (1986). Compressible separated flows. *AIAA Journal, 24*, 1971–1978.

Phalnikar, K. A., Kumar, R., & Alvi, F. S. (2008). Experiments on free and impinging supersonic microjets. *Experiments in Fluids, 44*, 819–830.

Piomelli, U., Cabot, W. H., Moin, P., & Lee, S. (1991). Subgrid-scale backscatter in turbulent and transitional flows. *Physics of Fluids A, 3*, 1766–1771.

Pirozzoli, S., & Grasso, F. (2004). Direct numerical simulations of isotropic compressible turbulences: Influence of compressibility on dynamics and structures. *Physics of Fluids, 16*, 4386–4407.

Pirozzoli, S., & Grasso, F. (2006). Direct numerical simulations of impinging shock wave/turbulent boundary layer interaction at $M = 2.25$. *Physics of Fluids, 18*. doi: 10.1063/1.2216989.

Pirozzoli, S., Grasso, F., & Gatski, T. B. (2004). Direct numerical simulation and analysis of a spatially evolving supersonic turbulent boundary layer at $M = 2.25$. *Physics of Fluids, 16*, 530–545.

Pirozzoli, S., Grasso, F., & Gatski, T. B. (2005). DNS analysis of shock wave/turbulent boundary layer interaction at $M = 2.25$. In *Proceedings of the fourth international symposium on turbulent shear flow phenomena* (pp. 1207–1211).

Poggi, F., Thorembey, M.-H., & Rodriguez, G. (1998). Velocity measurements in turbulent gaseous mixtures induced by Richtmyer-Meshkov instability. *Physics of Fluids, 10*, 2698–2700.

Pope, S. B. (2000). *Turbulent Flows*. Cambridge: Cambridge University Press.

Pot, P. J. (1979). *Measurements in a 2-D wake merging into a boundary layer*. TR 79063 L, NRL.

Povinelli, F. P., Povinelli, L. A., & Hersch, M. (1970). Supersonic jet penetration (up to Mach 4) into a Mach 2 airstream. *Journal of Spacecraft, 7*, 988–992.

Prandtl, L., & Tietjens, O. G. (1934). *Fundamentals of Hydro- and Aeromechanics*. New york: Dover Publications, Inc.

Pruett, C. D. (2000). On Eulerian times-domain filtering for spatial large-eddy simulation. *American Institute of Aeronautics and Astronautics Journal, 38*, 1634–1642.

Pruett, C. D., Gatski, T. B., Grosch, C. E., & Thacker, W. D. (2003). The temporally filtered Navier-Stokes equations: Properties of the residual stress. *Physics of Fluids, 15*, 2127–2140.

Pui, N. K., & Gartshore, I. S. (1979). Measurements of the growth rate and structure in plane mixing layers. *Journal of Fluid Mechanics, 91*, 111–130.

Raman, G., & Srinivasan, K. (2009). The powered resonance tube: From HartmannGs discovery to current active flow control applications. *Progress in Aerospace Sciences, 45*, 97–123.

Reeder, M. F., & Samimy, M. (1996). The evolution of a jet with vortex generating tabs: Real time visualizations and qualitative measurements. *Journal of Fluid Mechanics, 311*, 73–118.

Reichardt, H. (1951). Complete representation of the turbulent velocity distribution in smooth pipes. *Zeitschrift Fur Angewandte Mathematik Und Mechanik, 31*, 208–219.

Reijasse, P. (2005). *Aérodynamique des tuyères propulsives en sur-détente: décollement et charges latérales en régime stabilisé*. Ph.D. thesis, Université Paris VI.

Reijasse, P., Kachler, T., Boccaletto, L., & Lambare, H. (2005). Afterbody and nozzle aerodynamics for launchers through the CNES-ONERA ATAC programme. In *Sixth International Symposium on Launcher Technologies* (pp. 1–18).

Reynolds, O. (1895). On the dynamical theory of incompressible viscous fluids and the determination of the criterion. *Philosophical Transactions of the Royal Society A: Mathematical, Physical and Engineering Sciences, 186*, 123–164.

Reynolds, W. C. (1976). Computation of turbulent flows, *Annual Review of Fluid Mechanics*, 183–208. New York: Annual reviews.

Reynolds, W. C., & Hussain, A. K. M. F. (1972). The mechanics of an organized wave in turbulent shear flow. Part 3. Theoretical models and comparisons with experiments. *Journal of Fluid Mechanics, 54*, 263–288.

Ribner, H. S. (1954a). *Convection of a pattern of vorticity through a shock wave*. Report 1164, NACA, Supersedes NACA TN 2864, 1953.

Ribner, H. S. (1954b). *Shock-turbulence interaction and the generation of noise*. Technical Note 3255, NACA.

Riethmuller, M. L., & Scarano, F. (Eds.) (2005). *Advanced Measurement Techniques for Supersonic Flows*. Rhode-St-Genèse: VKI.

Ringuette, M. J., Martin, M. P., Smits, A. J., & Wu, M. (2006). Characterization of the turbulent structure in supersonic boundary layers using DNS data. Paper No. 2006-3359, AIAA.

Ringuette, M. J., Wu, M., & Martin, M. P. (2007). Coherent structures in DNS of turbulent boundary layers at Mach 3. Paper No. 2007-1138, AIAA.

Ristorcelli, J. R. (1993). A representation for the turbulent mass flux contribution to reynolds-stress and two-equation closures for compressible turbulence. Contractor Report 191569, NASA.

Rizzetta, D. P., & Visbal, M. R. (2002). Application of large-eddy simulation to supersonic compression ramps. *American Institute of Aeronautics and Astronautics Journal, 40*, 1574–1581.

Robinet, J. C., Gressier, J., Casalis, G., & Moschetta, J. (2000). Shock wave instability and the carbuncle phenomenon: Same intrinsic origin? *Journal of Fluid Mechanics, 417*, 237–263.

Robinson, D. F., Harris, J. E., & Hassan, H. A. (1995). Unified turbulence closure model for axisymmetric and planar free shear flows. *AIAA Journal, 33*, 2325–2331.

Robinson, D. F., & Hassan, H. A. (1998). Further development of the $k - \zeta$ (enstrophy) turbulence closure model. *AIAA Journal, 36*, 1825–1833.

Rodi, W. (1975). A review of experimental data of uniform density free turbulent boundary layers. In B. Launder (Ed.), *Studies in convection* (Vol. 1, pp. 79–165). New York: Academic Press.

Rodi, W., & Mansour, N. N. (1993). Low Reynolds number $k - \epsilon$ modelling with the aid of direct simulation data. *Journal of Fluid Mechanics, 250*, 509–529.

Rotman, D. (1991). Shock wave effects on a turbulent flow. *Physics of Fluids A, 3*, 1792–1806.

Roupassov, D. V., Nikipelov, A. A., Nudnova, M. M., & Starikovskii, A. Y. (2009). Flow separation control by plasma actuator with nanosecond pulsed-periodic discharge. *AIAA Journal, 47*, 168–185.

Rowley, C. W., Colonius, T., & Murray, R. M. (2004). Model reduction for compressible flows using POD and Galerkin projection. *Physica D, 189*, 115–120.

Roy, C. J., & Blottner, F. G. (2006). Review and assessment of turbulence models for hypersonic flows. *Progress in Aerospace Sciences, 42*, 469–530.

Rubesin, M. W. (1976). A one-equation model of turbulence for use with the compressible Navier-Stokes equations. Technical Memorandum X-73,128, NASA.

Rubesin, M. W. (1990). *Extra compressibility terms for Favre-averaged two-equation models of inhomogeneous turbulent flows*. Contractor Report 177556, NASA.

Rubesin, M. W., & Rose, W. C. (1973). *The turbulent mean-flow, Reynolds-stress, and heat-flux equations in mass-averaged dependent variables*. Technical Memorandum X-62248, NASA.

Sabin, C. M. (1965). An analytical and experimental study of the plane, incompressible, turbulent free shear layer with arbitrary velocity ratio and pressure gradient. *Transactions of the American Society of Mechanical Engineers Series D, 87*, 421–428.

Sagaut, P. (2006). *Large Eddy Simulation for Incompressible Flows*. Berlin: Springer.

Sahu, J., Nietubicz, C. J., & Steger, J. L. (1985). Navier-stokes computations of projectile base flow with and without mass injection. *AIAA Journal, 23*, 1348–1355.

Samimy, M., Adamovich, I., Webb, B., Kastner, J., Hileman, J., Keshav, S., et al. (2004). Development and characterization of plasma actuators for high-speed jet control. *Experiments in Fluids, 37*, 577–588.

Samimy, M., Debiasi, M., Caraballo, E., Serrani, A., Yuan, X., Little, J. Myatt, J. H. (2007). Feedback control of subsonic cavity flows using reduced-order models. *Journal of Fluid Mechanics, 579*, 315–346.

Samimy, M., & Elliott, G. (1990). Effects of compressibility in free shear layers. *Physics of Fluids A, 2*, 1231–1240.

Samimy, M., Kim, J.-H., Kastner, J., Adamovich, L., & Utkin, Y. (2007a). Active control of a Mach 0.9 jet for noise mitigation using plasma actuators. *AIAA Journal, 45*, 890–901.

Samimy, M., Kim, J.-H., Kastner, J., Adamovich, L., & Utkin, Y. (2007b). Active control of high-speed and high-Reynolds-number jets using plasma actuators. *Journal of Fluid Mechanics, 578*, 305–330.

Samimy, M., & Wernet, M. (2000). Review of planar multiple-component velocimetry in high-speed flows. *AIAA Journal, 38*(4), 553–574.

Samtaney, R., Pullin, D. I., & Kosović, B. (2001). Direct numerical simulation of decaying compressible turbulence and shocklet statistics. *Physics of Fluids, 13*, 1415–1430.

Sandberg, R. D., & Fasel, H. F. (2006). Direct numerical simulations of transitional supersonic base flows. *AIAA Journal, 44*, 848–858.

Sandham, N. D., & Reynolds, W. C. (1990). Compressible mixing layer: Linear theory and direct simulation. *AIAA Journal, 28*, 618–624.

Sandham, N. D., & Reynolds, W. C. (1991). Three-dimensional simulations of large eddies in the compressible mixing layer. *Journal of Fluid Mechanics, 224*, 133–158.

Saric, W. S., Dussauge, J.-P., Smith, R. W., Smits, A. J., Fernholz, H., Finley, P. J., et al. (1996). Turbulent boundary layers in subsonic and supersonic flow. AG 335, AGARD.

Sarkar, S. (1992). The pressure-dilatation correlation in compressible flows. *Physics of Fluids A, 4*, 2674–2682.

Sarkar, S. (1995). The stabilizing effect of compressibility in turbulent shear flow. *Journal of Fluid Mechanics, 282*, 163.

Sarkar, S., Erlebacher, G., Hussaini, M. Y., & Kreiss, H. O. (1991). The analysis and modelling of dilatational terms in compressible turbulence. *Journal of Fluid Mechanics, 227*, 473–493.

Sarkar, S., & Lakshmanan, B. (1991). Application of a Reynolds stress turbulence model to the compressible shear layer. *AIAA Journal, 29*, 743–749.

Sarma, G. R. (1998). Transfer function analysis of the constant voltage anemometer. *Review of Scientific Instruments, 69*, 2385–2391.

Scarano, F. (2008). Overview of PIV in supersonic flows. *Particle Image Velocimetry: New Developments and Recent Applications. Topics in Applied Physics* (Vol. 112, pp. 445–463). Heidelberg: Springer.

Scarano, F., & van Oudheusden, B. W. (2003). Planar velocity measurements of a two-dimensional compressible wake. *Experiments in Fluids, 34*, 430–441.

Schetz, J. A., & Billig, F. S. (1966). Penetration of gaseous jets injected into a supersonic stream. *Journal of Spacecraft Rockets, 8*, 1658–1665.

Schneider, S. P. (2006). Laminar-turbulent transition on reentry capsules and planetary probes. *Journal of Spacecraft Rockets, 43*, 1153–1173.

Schrijer, F. F. J., Scarano, F., & van Oudheusden, B. W. (2006). Application of PIV in a Mach 7 double-ramp flow. *Experiments in Fluids, 41*, 353–363.

Scroggs, S. D., & Settles, G. S. (1996). An experimental study of supersonic microjets. *Experiments in Fluids, 21*, 401–409.

Settles, G., & Dodson, L. (1991). *Hypersonic shock boundary layer interaction database*. Contractor Report 117557, NASA.

Settles, G. S., & Dodson, L. J. (1994). Supersonic and hypersonic shock/boundary layer interaction database. *American Institute of Aeronautics and Astronautics Journal, 32*, 1377–1383.

Settles, G. S., & Dolling, D. S. (1992). Swept shock-wave/boundary-layer interactions. *Tactical missile aerodynamics: General topis of progress in astronautics and aeronautics* pp. 505–574 (Vol. 141) American Institute of Aeronautics and Astronautics.

Shahab, M. F. (2011). *Numerical investigation of the influence of an impinging shock wave and heat transfer on a developing turbulent boundary layer*. Ph.D. thesis, L'École Nationale Supérieure de Mécanique et d'Aérotechnique.

Shahab, M. F., Lehnasch, G., Gatski, T. B., & Comte, P. (2009). Statistical characteristics of the interaction of an impinging shock wave and turbulent boundary layer. In *Proceedings of the Sixth International Symposium on Turbulent Shear Flow Phenomena* (pp. 1351–1356).

Shahab, M. F., Lehnasch, G., Gatski, T. B., & Comte, P. (2011). Statistical characteristics of an isothermal, supersonic developing boundary layer flow from DNS data. *Flow, Turbulence and Combustion, 86*, 369–397.

Shih, T.-H. (1996). Constitutive relations and realizability of single-point turbulence closures. *Turbulence and Transition Modelling* (pp. 155–192). Amsterdam: Kluwer-Springer.

Shih, T.-H., & Lumley, J. L. (1993). Remarks on turbulent constitutive relations. *Mathematical Computer Modeling, 18*, 9–16.

Shih, T.-H., Povinelli, L. A., & Liu, N.-S., (2003). Application of generalized wall function for complex turbulent flows. *Journal of Turbulence* 4. doi:10.1088/1468-5248/4/1/015.

Shih, T.-H., Povinelli, L. A., Liu, N.-S., Potapczuk, M. G., & Lumley, J. L. (1999). A generalized wall function. Technical Memorandum 209398, NASA.

Shih, T.-H., Reynolds, W. C., & Mansour, N. N. (1990). A spectrum model for weakly anisotropic turbulence. *Physics of Fluids A, 8*, 1500–1502.

Shin, J., Narayanaswamy, V., Raja, L. L., & Clemens, N. T. (2007). Characterization of a direct-current glow discharge plasma actuator in low-pressure supersonic flow. *AIAA Journal, 45*, 1596–1605.

Simonsen, A. J., & Krogstad, P.-A. (2005). Turbulent stress invariant analysis: Clarification of existing terminology. *Physics of Fluids, 17*, 088103-1–088103-4.

Simpson, R. L., Chew, Y. T., & Shivaprasad, B. G. (1981). The structure of a separating turbulent boundary layer: Part 2. Higher-order turbulence results. *Journal of Fluid Mechanics, 113*, 53–73.

Sinha, A., Kim, K., Kim, J., Serrani, A., & Samimy, M. (2010). Extremizing feedback control of a high-speed and high Reynolds number jet. *AIAA Journal, 48*, 387–399.

Sinha, K., & Candler, G. V. (2003). Turbulent dissipation-rate equation for compressible flows. *AIAA Journal, 41*, 1017–1021.

Siriex, M., & Solignac, J. L. (1966). Contribution à l'étude expérimentale de la couche de mélange turbulente isobare d'un ecoulement supersonique. *Symposium on separated flows*, CP 4(1), AGARD.

Sirovich, L. (1987). Turbulence and the dynamics of coherent structures. *Quarterly of Applied Mathematics, 45*, 561–590.

Sjögren, T., & Johansson, A. V. (2000). Development and calibration of algebraic nonlinear models for terms in the reynolds stress transport equations. *Physics of Fluids, 12*, 1554–1572.

Skåre, P. E., & Krogstad, P.Å. (1994). A turbulent equilibrium boundary layer near separation. *Journal of Fluid Mechanics, 272*, 319–348.

Slessor, M. D., Zhuang, M., & Dimotakis, P. E. (2000). Turbulent shear-layer mixing; growth-rate compressibility scaling. *Journal of Fluid Mechanics, 414*, 35–45.

Smagorinsky, I. (1963). General circulation experiments with the primitive equations. I. The basic experiment. *Monthly Weather Review, 91*, 99–164.

Smith, B. R. (1995). Prediction of hypersonic shock wave turbulent boundary layer interactions with the $k - \ell$ two equation turbulence model. Paper No. 95-0232, AIAA.

Smith, D. R., & Smits, A. J. (1993). Simultaneous measurement of velocity and temperature fluctuations in the boundary layer of a supersonic flow. *Experimental Thermal and Fluid Science, 7*, 221–229.

Smits, A. J., & Dussauge, J.-P. (1989). Hot-wire anemometry in supersonic flow. In *A survey of measurements and measuring techniques in rapidly distorted compressible turbulent boundary layers*. AGARDograph No. 315. AGARD (pp. 5.1–5.14).

Smits, A. J., & Dussauge, J.-P. (2006). *Turbulent Shear Layers in Supersonic Flow* (2nd ed.). New York: Springer.

Smits, A. J., & Muck, K.-C. (1987). Experimental study of three shock wave/turbulent boundary layer interactions. *Journal of Fluid Mechanics, 182*, 291–314.

So, R. M. C., Zhang, H. S., Gatski, T. B., & Speziale, C. G. (1994). Logarithmic laws for compressible turbulent boundary layers. *AIAA Journal, 32*, 2162–2168.

Solomon, J. T., Foster, C., & Alvi, F. S. (2012). Design and characterization of high-bandwidth, resonance enhanced pulsed microactuators: A parametric study. *AIAA Journal*, doi: 10.2514/1.J051806.

Solomon, J. T., Kumar, R., & Alvi, F. S. (2010). High-bandwidth pulsed microactuators for high-speed flow control. *AIAA Journal, 48*, 2386–2396.

Souverein, L. J., Dupont, P., Debiève, J.-F., Dussauge, J.-P., vanOudheusden, W., & Scarano, F. (2010). Effect of interaction strength on unsteadiness in turbulent shock-wave-induced separations. *AIAA Journal, 48*, 1480–1493.

Spalart, P. R. (1988). Direct simulation of a turbulent boundary layer up to $R_\theta = 1410$. *Journal of Fluid Mechanics, 187*, 61–98.

Spalding, D. B., & Chi, S. W. (1964). The drag of a compressible turbulent boundary layer on a smooth flat plate with heat transfer. *Journal of Fluid Mechanics, 18*, 117–143.

Speziale, C. G. (1998). A combined large-eddy simulation and time-dependent RANS capability for high-speed compressible flows. *Journal of Science and Computer, 13*, 253–274.

Speziale, C. G., Erlebacher, G., Zang, T. A., & Hussaini, M. Y. (1988). The subgrid-scale modeling of compressible turbulence. *Physics of Fluids, 31*, 940–942.

Speziale, C. G., & Gatski, T. B. (1997). Analysis and modelling of anisotropies in the dissipation rate of turbulence. *Journal of Fluid Mechanics, 344*, 155–180.

Spina, E. F., & McGinley, C. B. (1994). Constant-temperature anemometry in hypersonic flows: Critical issues and sample results. *Experiments in Fluids, 17*, 365–374.

Spina, E. F., Smits, A. J., & Robinson, S. K. (1994). The physics of supersonic turbulent boundary layers. *Annual Review of Fluid Mechanics*, 287–319.

Spyropoulos, E. T., & Blaisdell, G. A. (1996). Evaluation of the dynamic model for simulations of compressible decaying isotropic turbulence. *AIAA Journal, 34*, 990–998.

Spyropoulos, E. T., & Blaisdell, G. A. (1998). Large-eddy simulation of a spatially evolving supersonic turbulent boundary-layer flow. *AIAA Journal, 36*, 1983–1990.

Staroselsky, I., Yakhot, V., Kida, S., & Orszag, S. A. (1990). Log-time, large-scale properties of a randomly stirred compressible fluid. *Physical Review Letters, 65*, 171–174.

Stern, C. E., Aguilar, C., & Alvarado, J. M. (2006). Study of Mach lines and acoustic waves in a jet using Rayleigh scattering. Paper No. 2006-2532, AIAA/CEAS.

Stolz, S. (2005). High-pass filtered eddy-viscosity models for large-eddy cimulations of compressible wall-bounded flows. *Journal of Fluids Engineering – Transactions of the ASME, 127*, 666–673.

Stolz, S., & Adams, N. A. (2003). Large-eddy simulation of high-Reynolds number supersonic boundary layers using the approximate deconvolution model and a rescaling and recycling technique. *Physics of Fluids, 15*, 2398–2412.

Stolz, S., Adams, N. A., & Kleiser, L. (2001). The approximate deconvolution model for large-eddy simulations of compressible flows and its application to shock-turbulent-boundary-layer interaction. *Physics of Fluids, 13*, 2985–3001.

Strasser, W. (2011). Toward the optimization of a pulsatile three-stream coaxial airblast injector. *International Journal of Multiphase Flow, 37*, 831–844.

Sun, X.-B., & Lu, X.-Y. (2006). A large eddy simulation approach of compressible turbulent flow without density weighting. *Physics of Fluids, 18*. doi:10.1063/1.2391839.

Tam, C. K. W. (1991). Jet noise generated by large-scale coherent motions. In H. H. Hubbard (Ed.), *Aeroacoustics of Flight Vehicles, Theory and Practice. Noise Sources* (Vol. 1, pp. 311–390). NASA.

Tam, C. K. M. (1995). Supersonic jet noise. *Annual Review of Fluid Mechanics, 17*–43.

Tamano, S., & Morinishi, Y. (2006). Effect of different thermal wall boundary conditions on compressible turbulent channel flow at $m = 15$. *Journal of Fluid Mechanics, 548*, 361–373.

Taulbee, D., & VanOsdol, J. (1991). Modeling turbulent compressible flows: The mass fluctuating velocity and squared density. Paper No. 91-0524, AIAA.

Tedeschi, G., Gouin, H., & Elena, M. (1999). Motion of tracer particles in supersonic flows. *Experiments in Fluids, 26*, 288–296.

Tennekes, H., & Lumley, J. L. (1972). *A First Course in Turbulence*. Boston: The MIT Press.

Thacker, W. D., Sarkar, S., & Gatski, T. B. (2007). Analyzing the influence of compressibility on the rapid pressure-strain rate correlation in turbulent shear flows. *Theoretical and Computational Fluid Dynamics, 21*, 171–199.

Thompson, P. A. (1988). *Compressible-Fluid Dynamics*. Troy, NY: Philip A. Thompson.

Thurow, B. S., Jiang, N., Kim, J.-W., Lempert, W., & Samimy, M. (2008). Issues with measurements of the convective velocity of large-scale structures in the compressible shear layer of a free jet. *Physics of Fluids, 20*. doi:10.1063/1.2926757.

Townsend, A. (1976). *The Structure of Turbulent Shear Flows*. Cambridge: Cambridge University Press.

Tran, H., Tandra, D., & Jones, A. K. (2007). Development of low-pressure sootblowing technology. In *International Chemical Recovery Conference* (pp. 1–7). Pulp and Paper Association of Canada & Tappi.

Trapier, S., Deck, S., & Duveau, P. (2008). Delayed detached-eddy simulation and analysis of supersonic inlet buzz. *AIAA Journal, 46*, 118–131.

Troiler, J. W., & Duffy, R. E. (1985). Turbulent measurements in shock induced flows. *American Institute of Aeronautics and Astronautics Journal, 23*, 1172–1178.

Tropea, C., Yarin, A. L., & Foss, J. F. (Eds.) (2007a). *Springer Handbook of Experimental Fluid Mechanics*. pp. (229–287). Springer. (Thermal anemometry)

Tropea, C., Yarin, A. L., & Foss, J. F. (Eds.) (2007b). *Springer Handbook of Experimental Fluid Mechanics*. Berlin: Springer.

Tropea, C., Yarin, A. L., & Foss, J. F. (Eds.) (2007c). Working principle of shock tubes/tunnels and shock exapnsion tubes/tunnels. In *Springer Handbook of Experimental Fluid Mechanics* (pp. 1082–1095). Berlin: Springer.

Troutt, T. R., & McLaughlin, D. K. (1982). Experiments on the flow and acoustic properties of a moderate reynolds number supersonic jet. *Journal of Fluid Mechanics, 116*, 123–156.

Truesdell, C. (1991). *A first course in rational continuum mechanics*. New York: Academic Press.

Truesdell, C., & Rajagopal, K. R. (2000). *An Introduction to the Mechanics of Fluids*. Boston: Birkäuser.

Urbin, G., & Knight, D. (2001). Large-eddy simulation of a supersonic boundary layer using an unstructured grid. *AIAA Journal, 39*, 1288–1295.

Utkin, Y. G., Keshav, S., Kim, J.-H., Kastner, J., Adamovich, I. V., & Samimy, M. (2007). Development and use of localized arc filament plasma actuators for high-speed flow control. *Journal of Physics D-Applied Physics, 40*, 685–694.

Valdivia, A., Yuceil, K. B., Wagner, J. L., Clemens, N. T., & Dolling, D. S. (2009). Active control of supersonic inlet unstart using vortex generator jets. Paper No. 2009-4022, AIAA.

Vallet, I. (2008). Reynolds-stress modelling of m = 225 shock-wave/turbulent boundary-layer interaction. *International Journal for Numerical Methods in Fluids, 56*, 525–555.

van der Bos, F. & Geurts, B. J. (2006). Computational turbulent stress closure for large-eddy simulation of compressible flow. *Journal of Turbulence, 7*. doi:10.1080/14685240600573120.

van Driest, E. R. (1951). Turbulent boundary layer in compressible fluids. *Journal of the Aeronautical Sciences, 18*, 145–160.

van Driest, E. R. (1952). *Investigation of laminar boundary layer in compressible fluids using the Crocco method*. Technical Note 2597, NACA.

van Driest, E. R. (1954). On the boundary layer with variable Prandtl number. In *1954 Jahrbuch wissenschaftlichen gesellschaft fur luftfahrt e.v. (wgl). friedr* (pp. 66–75). Vieweg & Sohn.

van Driest, E. R. (1955). The turbulent boundary layer with variable Prandtl number. In H. Görtler, & W. Tollmien (Eds.), *50 Jahre Grenzschichtforschung. Friedr* (pp. 257–271). Vieweg & Sohn.

van Driest, E. R. (1956a). On turbulent flow near a wall. *Journal of the Aeronautical Sciences, 15*, 1007–1011,1036.

van Driest, E. R. (1956b). The problem of aerodynamic heating. *Aeronautical Engineering Review, 15*, 26–41.

Viegas, J. R., & Rubesin, M. W. (1983). Wall-function boundary conditions in the solution of the Navier-Stokes equations for complex compressible flows. Paper No. 83-1694, AIAA.

Viegas, J. R., Rubesin, M. W., & Horstman, C. C. (1985). On the use of wall functions as boundary conditions for two-dimensional separated compressible flows. Paper No. 85-0180, AIAA.

Vigo, G., Dervieux, A., Mallet, M., Ravachol, M., & Stoufflet, B. (1998). Extension of methods based on the proper orthogonal decomposition to the simulation of unsteady compressible Navier-Stokes flows. In K. Papailiou, D. Tsahalis, J. Périaux & D. Knoerzer (Eds.), *Computational Fluid Dynamics'98, Proceedings of the Fourth ECCOMAS Conference* (pp. 648–653). New York: John Wiley and Sons.

von Kaenel, R., Adams, N. A., Kleiser, L., & Vos, J. B. (2002). The approximate deconvolution model for large-eddy simulation of compressible flows with finite volume schemes. *Journal of Fluids Engineering – Transactions of the ASME, 124*, 829–834.

Vreman, A. W., Sandham, N. D., & Luo, K. H. (1996). Compressible mixing layer growth rate and turbulence characteristics. *Journal of Fluid Mechanics, 320*, 235–258.

Vreman, B., Geurts, B., & Kuerton, H. (1994). On the formulation of the dynamic mixed subgrid-scale model. *Physics of Fluids, 6*, 4057–4059.

Vreman, B., Geurts, B., & Kuerton, H. (1995a). A priori tests of large eddy simulation of the compressible plane mixing layer. *Journal of Engineering Mathematics, 29*, 299–327.

Vreman, B., Geurts, B., & Kuerton, H. (1995b). Subgrid-modelling in LES of compressible flow. *Applied Scientific Research, 54*, 191–203.

Vreman, B., Geurts, B., & Kuerton, H. (1997). Large-eddy simulation of the turbulent mixing layer. *Journal of Fluid Mechanics, 339*, 357–390.

Waithe, K. A., & Deere, K. A. (2003). Experimental and computational investigation of multiple injection ports in a convergent-divergent nozzle for fluidic thrust vectoring. Paper No. 2003-3802, AIAA.

Weiss, J., Chokani, N., & Comte-Bellot, G. (2005). Constant-temperature and constant-voltage anemometer use in a Mach 2.5 flow. *AIAA Journal, 43*, 1140–1143.

Weiss, J., Knauss, H., & Wagner, S. (2003). Experimental determination of the free-stream disturbance field in a short-duration supersonic wind tunnel. *Experiments in Fluids, 35*, 291–302.

White, F. M. (1991). *Viscous Fluid Flow* (2nd ed.). Boston: McGraw-Hill.

Wikström, P. M., Wallin, S., & Johansson, A. V. (2000). Derivation and investigation of a new explicit algebraic model for the passive scalar flux. *Physics of Fluids, 12*, 688–702.

Wilcox, D. C. (1992). Dilatation-dissipation corrections for advanced turbulence models. *AIAA Journal, 30*, 2639–2646.

Wu, M., Bookey, P., Martin, M. P., & Smits, A. J. (2005). Assessment of numerical methods for DNS of shockwave/turbulent boundary layer interaction. Paper No. 2005-2145, AIAA.

Wu, M., & Martin, M. P. (2006). Assessment of numerical methods for DNS of shockwave/turbulent boundary layer interaction. Paper No. 2006-0717, AIAA.

Wu, M., & Martin, M. P. (2007). Direct numerical simulation of supersonic turbulent boundary layer flow over a compression ramp. *American Institute of Aeronautics and Astronautics Journal, 45*, 879–889.

Wu, M., & Martin, M. P. (2008). Analysis of shock motion in shockwave and turbulent boundary layer interaction using direct numerical simulation data. *Journal of Fluid Mechanics, 594*, 71–83.

Wu, P., Lempert, W. R., & Miles, R. B. (2000). MHz-rate pulse-burst laser system and visualization of shock/wave boundary layer interaction in a Mach 2.5 wind tunnel. *AIAA Journal, 38*, 672–679.

Yan, H., Knight, D., Kandala, R., & Candler, G. (2007). Effect of a laser pulse on a normal shock. *AIAA Journal, 45*, 1270–1280.

Yang, Q., & Fu, S. (2008). Analysis of flow structures in supersonic plane mixing layers using the POD method. *Science in China Series G, 51*, 541–558.

Yoshizawa, A. (1986). Statistical theory for compressible turbulent shear flows, with the application to subgrid modeling. *Physics of Fluids, 29*, 2152–2164.

Yoshizawa, A. (2003). Statistical theory of compressible turbulence based on mass-weighted averaging, with an emphasis on a cause of countergradient diffusion. *Physics of Fluids, 15*, 585–596.

Young, A. D. (1953). Boundary layers. In L. Howarth (Ed.), *Modern Developments in Fluid Mechanics – High Speed Flows* (pp. 453–455). London: Oxford University Press.

Yu, K. H., & Schadow, K. C. (1994). Cavity-actuated supersonic mixing and combustion control. *Combustion and Flame, 383*, 295–301.

Zaman, K. B. M. Q. (1999). Spreading characteristics of compressible jets from nozzles of various geometries. *Journal of Fluid Mechanics, 383*, 197–228.

Zang, T. A., Dahlburg, R. B., & Dahlburg, J.-P. (1992). Direct and large-eddy simulations of three-dimensional compressible Navier-Stokes turbulence. *Physics of Fluids A, 4*, 127–140.

Zang, Y., Street, R. L., & Koseff, J. R. (1993). A dynamic mixed subgrid-scale model and its application to turbulent recirculating flows. *Physics of Fluids A, 5*, 3186–3196.

Zeman, O. (1990). Dilatation dissipation: The concept and application in modeling compressible mixing layers. *Physics of Fluids A, 2*, 178–188.

Zeman, O. (1991a). On the decay of compressible isotropic turbulence. *Physics of Fluids A, 3*, 951–955.

Zeman, O. (1991b). The role of pressure-dilatation correlation in rapidly compressed turbulence and in boundary layers. In *Annual research briefs—1991 Center for Turbulence Research* (pp. 105–117). NASA and Stanford University.

Zeman, O., & Blaisdell G. A. (1991). New physics and models for compressible turbulent flows. In A. V. Johansson, & P. H. Alfredsson, *Advances in Turbulence* (Vol. 3, pp. 445–454). Heidelberg: Springer-Verlag.

Zeman, O., & Coleman, G. N. (1993). Compressible turbulence subjected to shear and rapid compression. In F. Durst, R. Friedrich, B. E. Launder, F. W. Schmidt, U. Schumann, & J. H. Whitelaw (Eds.), *Proceedings of the Eighth Symposium on Turbulent Shear Flows* (pp. 283–296). Heidelberg: Springer-Verlag.

Zhang, H. S., So, R.M. C., Gatski, T. B., & Speziale, C. G. (1993). A near-wall second-order closure for compressible turbulent flows. In R. M. C. So, C. G. Speziale & B. E. Launder (Eds.), *Near-Wall Turbulent Flows* (pp. 209–218). Amsterdam: Elsevier Science Publishers.

Zhang, H. S., So, R. M. C., Speziale, C. G., & Lai, Y. G. (1993). A near-wall two-equation model for compressible turbulent flows. *AIAA Journal, 31*, 196–199.

Zhao, C. Y., So, R. M. C., & Gatski, T. B. (2001). Turbulence modeling effects on the prediction of equilibrium states of buoyant shear flows. *Theoretical and Computational Fluid Dynamics, 14*, 399–422.

Zukoski, E. E., & Spaid, F. W. (1964). Secondary injection of gases into a supersonic flow. *AIAA Journal, 2*, 1689–1696.

Index

Printed in the United States
By Bookmasters